MATH MASTERS

Everyday Mathematics®

The University of Chicago School Mathematics Project

The University of Chicago School Mathematics Project

Max Bell, Director, *Everyday Mathematics* First Edition; James McBride, Director, *Everyday Mathematics* Second Edition; Andy Isaacs, Director, *Everyday Mathematics* Third, CCSS, and Fourth Editions; Amy Dillard, Associate Director, *Everyday Mathematics* Third Edition; Rachel Malpass McCall, Associate Director, *Everyday Mathematics* CCSS and Fourth Editions; Mary Ellen Dairyko, Associate Director, *Everyday Mathematics* Fourth Edition

Authors
Jean Bell, Max Bell, John Bretzlauf, Mary Ellen Dairyko, Amy Dillard, Robert Hartfield, Andy Isaacs, Kathleen Pitvorec, James McBride, Peter Saecker

Fourth Edition Grade 3 Team Leader
Mary Ellen Dairyko

Writers
Lisa J. Bernstein, Camille Bourisaw, Julie Jacobi, Gina Garza-Kling, Cheryl G. Moran, Amanda Louise Ruch, Dolores Strom

Open Response Team
Catherine R. Kelso, Leader; Amanda Louise Ruch, Andy Carter

Differentiation Team
Ava Belisle-Chatterjee, Leader; Martin Gartzman, Barbara Molina, Anne Sommers

Digital Development Team
Carla Agard-Strickland, Leader; John Benson, Gregory Berns-Leone, Juan Camilo Acevedo

Virtual Learning Community
Meg Schleppenbach Bates, Cheryl G. Moran, Margaret Sharkey

Technical Art
Diana Barrie, Senior Artist; Cherry Inthalangsy

UCSMP Editorial
Lila K. S. Goldstein, Senior Editor; Kristen Pasmore, Molly Potnick, Rachel Jacobs

Field Test Coordination
Denise A. Porter

Field Test Teachers
Eric Bachmann, Lisa Bernstein, Rosemary Brockman, Nina Fontana, Erin Gilmore, Monica Geurin, Meaghan Gorzenski, Deena Heller, Lori Howell, Amy Jacobs, Beth Langlois, Sarah Nowak, Lisa Ringgold, Andrea Simari, Renee Simon, Lisa Winters, Kristi Zondervan

Digital Field Test Teachers
Colleen Girard, Michelle Kutanovski, Gina Cipriani, Retonyar Ringold, Catherine Rollings, Julia Schacht, Christine Molina-Rebecca, Monica Diaz de Leon, Tiffany Barnes, Andrea Bonanno-Lersch, Debra Fields, Kellie Johnson, Elyse D'Andrea, Katie Fielden, Jamie Henry, Jill Parisi, Lauren Wolkhamer, Kenecia Moore, Julie Spaite, Sue White, Damaris Miles, Kelly Fitzgerald

Contributors
John Benson, Jeanne Mills DiDomenico, James Flanders, Lila K. S. Goldstein, Funda Gonulates, Allison M. Greer, Catherine R. Kelso, Lorraine Males, Carole Skalinder, John P. Smith III, Stephanie Whitney, Penny Williams, Judith S. Zawojewski

Center for Elementary Mathematics and Science Education Administration
Martin Gartzman, Executive Director; Meri B. Fohran, Jose J. Fragoso, Jr., Regina Littleton, Laurie K. Thrasher

External Reviewers
The *Everyday Mathematics* authors gratefully acknowledge the work of the many scholars and teachers who reviewed plans for this edition. All decisions regarding the content and pedagogy of *Everyday Mathematics* were made by the authors and do not necessarily reflect the views of those listed below.

Elizabeth Babcock, California Academy of Sciences; Arthur J. Baroody, University of Illinois at Urbana-Champaign and University of Denver; Dawn Berk, University of Delaware; Diane J. Briars, Pittsburgh, Pennsylvania; Kathryn B. Chval, University of Missouri-Columbia; Kathleen Cramer, University of Minnesota; Ethan Danahy, Tufts University; Tom de Boor, Grunwald Associates; Louis V. DiBello, University of Illinois at Chicago; Corey Drake, Michigan State University; David Foster, Silicon Valley Mathematics Initiative; Funda Gönülateş, Michigan State University; M. Kathleen Heid, Pennsylvania State University; Natalie Jakucyn, Glenbrook South High School, Glenview, IL; Richard G. Kron, University of Chicago; Richard Lehrer, Vanderbilt University; Susan C. Levine, University of Chicago; Lorraine M. Males, University of Nebraska-Lincoln; Dr. George Mehler, Temple University and Central Bucks School District, Pennsylvania; Kenny Huy Nguyen, North Carolina State University; Mark Oreglia, University of Chicago; Sandra Overcash, Virginia Beach City Public Schools, Virginia; Raedy M. Ping, University of Chicago; Kevin L. Polk, Aveniros LLC; Sarah R. Powell, University of Texas at Austin; Janine T. Remillard, University of Pennsylvania; John P. Smith III, Michigan State University; Mary Kay Stein, University of Pittsburgh; Dale Truding, Arlington Heights District 25, Arlington Heights, Illinois; Judith S. Zawojewski, Illinois Institute of Technology

Note
Many people have contributed to the creation of *Everyday Mathematics.* Visit http://everydaymath.uchicago.edu/authors/ for biographical sketches of *Everyday Mathematics 4* staff and copyright pages from earlier editions.

www.everydaymath.com

Send all inquiries to:
McGraw-Hill Education
8787 Orion Place
Columbus, OH 43240

ISBN: 978-0-02-136464-0
MHID: 0-02-136464-8

Printed in the United States of America.

1 2 3 4 5 6 7 8 9 RHR 20 19 18 17 16 15

Contents

Teaching Masters and Home Link Masters

Unit 1

Unit 2

Unit 5

Unit 6

Unit 7

Unit 8

Teaching Aid Masters

Game Masters

Teaching Masters and Home Link Masters

Home Link 1-1

NAME DATE

Introduction to *Third Grade Everyday Mathematics*

Welcome to *Third Grade Everyday Mathematics.* It is part of an elementary school mathematics curriculum developed by the University of Chicago School Mathematics Project.

Several features of the program are described below to help familiarize you with the structure and expectations of *Everyday Mathematics.*

A problem-solving approach based on everyday situations
By connecting their own knowledge to their experiences both in school and outside of school, children learn basic math skills in meaningful contexts so the mathematics becomes "real."

Frequent practice of basic skills
Instead of practice presented as tedious drills, children practice basic skills in a variety of ways. Children will complete daily review exercises covering a variety of topics, find patterns on the number grid and the multiplication and division facts table, work with multiplication and division fact families in different formats, analyze visual number images, and play games that are specifically designed to develop basic skills.

An instructional approach that revisits concepts regularly
To improve the development of basic skills and concepts, children regularly revisit previously learned content and repeatedly practice skills encountered earlier. The lessons are designed to build on concepts and skills throughout the year instead of treating them as isolated bits of knowledge. Research shows repeated exposure to these concepts and skills over time develops children's abilities to recall knowledge from long-term memory.

A curriculum that explores mathematical content and practices
Everyday Mathematics provides a rich problem-solving environment, which helps children develop critical thinking skills. Children solve different kinds of problems, explore multiple solution strategies, explain their thinking to others, and make sense of other children's thinking.

Please keep this Family Letter for reference as your child works through Unit 1.

Following the recommendations of the national mathematics standards, *Third Grade Everyday Mathematics* emphasizes the following content:

Numbers and Operations in Base Ten Using place-value understanding to add and subtract multidigit whole numbers, and multiply one-digit numbers by multiples of 10; rounding numbers to the nearest 10 and 100

Number and Operations—Fractions Understanding fractions as numbers; representing fractions on a number line; recognizing equivalent fractions and comparing fractions

Operations and Algebraic Thinking Developing fluency with multiplication and division facts; exploring properties of operations and the relationship between multiplication and division; solving problems involving more than one operation; using estimation to check the reasonableness of answers

Measurement and Data Solving problems involving time, liquid volume, and mass; telling time to the nearest minute and calculating elapsed time; measuring and estimating mass in grams and kilograms and volume in liters; organizing and representing data with bar and picture graphs; measuring to the nearest $\frac{1}{4}$ inch and organizing measurement data on line plots

Geometric Measurement Measuring areas of rectangles by tiling with square units; finding area measures by counting square units and multiplying side lengths; solving problems involving areas and perimeters of rectangles

Geometry Recognizing categories of shapes with shared attributes, such as quadrilaterals; dividing shapes into equal parts and naming parts with a fraction

Everyday Mathematics provides you with many opportunities to monitor your child's progress and to participate in your child's mathematics experiences. Throughout the year, you will receive Family Letters to keep you informed of the mathematical content that your child will be studying in each unit. Each letter includes a vocabulary list, suggested Do-Anytime Activities for you and your child, and an answer guide to selected Home Link (homework) activities. You will enjoy seeing your child's confidence and comprehension soar as he or she connects mathematics to everyday life.

We look forward to an exciting year!

Unit 1: Math Tools, Time, and Multiplication

This unit reviews and extends mathematical concepts that were developed in *Second Grade Everyday Mathematics*. In Unit 1, children will . . .

- use number grids to add and subtract.
- round 2- and 3-digit numbers to the nearest tens and hundreds on open number lines.
- review math tools including clocks, rulers, and calculators.
- tell time to the nearest minute.
- measure time intervals in minutes and solve problems involving elapsed time.
- begin a yearlong Length-of-Day project that involves collecting, recording, and graphing sunrise and sunset data.
- collect and organize data on scaled bar graphs.
- analyze "Quick Looks" of equally-grouped dot patterns and arrays to develop fluency with multiplication.
- solve multiplication and division number stories using strategies based on intuitive understandings of equal groups, arrays, and sharing.
- play games, such as *Multiplication Draw*, to strengthen number skills and develop fact fluency for 2s, 5s, and 10s multiplication facts.

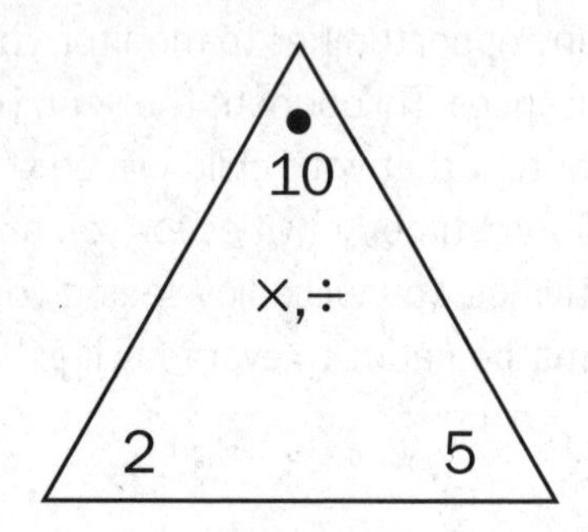

- estimate and measure mass in grams and kilograms.

Do-Anytime Activities

The following activities provide practice for concepts and skills taught in this unit.

- Discuss examples of mathematics in everyday life: road signs, recipe measurements, weight, time, and so on.
- Discuss household tools that can be used to help solve mathematical problems, such as coins, thermometers, and clocks.
- Tell time to the nearest minute on analog clocks and calculate how long daily activities take.

Vocabulary

Important terms in Unit 1:

array An arrangement of objects in a regular pattern, usually rows and columns.

elapsed time The difference in two times. For example, between 12:45 P.M. and 1:30 P.M., 45 minutes have elapsed.

equal groups Sets with the same number of elements, such as cars with 5 passengers each, rows with 6 chairs each, and boxes containing 100 paper clips each.

equal-grouping situation A situation in which a quantity is divided into equal groups. The total and size of each group are known and the number of groups is unknown. For example: How many tables seating 4 people each are needed to seat 28 people?

equal-sharing situation A situation in which a quantity is shared equally. The total quantity and the number of shares are known, and the size of each share is unknown. For example: There are 12 toys to share equally among 4 children. How many toys will each child get?

estimate An answer close to, or approximating, an exact answer.

mass A measure of the amount of matter in an object. Mass is not affected by gravity, so it is the same on Earth, the moon, or anywhere else in space. Mass is usually measured in grams, kilograms, and other metric units.

number grid A table in which numbers are arranged consecutively, usually in rows of ten. A move from one number to the next within a row is a change of 1; a move from one number to the next within a column is a change of 10.

open number line A line on which children can mark points or numbers that are useful for solving problems.

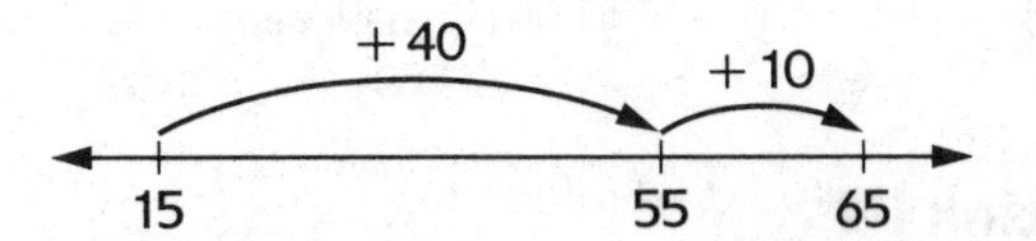

round To approximate a number to make it easier to work with or to make it better reflect the precision of the data. For example, to add $37 + 57$, one might round 37 to 40 and 57 to 60, add $40 + 60 = 100$, and then say that $37 + 57$ is a little less than 100.

As You Help Your Child with Homework

As your child brings home assignments, you may want to go over the instructions together, clarifying them as necessary. Each Family Letter will contain answers, such as the following, to guide you through the unit's Home Links. Answers to Home Links 1-2 and 1-12 are not shown.

Home Link 1-1

1. 26
2. 28
3. Sample answer: I counted by 10s from 97 to 117 and got 20. Then I counted by ones to 125 and got 8. So the answer is 28.
4. 6
5. 7
6. 13
7. 13

Home Link 1-3

1. Answers vary.
2. 8:00
3. 3:30
4. 6:15
5. 11:45
6. 7:10
7. 5:40
8. Answers vary.

Home Link 1-4

1. 90
2. 300
3. $40 + 60 = 100$; 94; Sample answer: Yes. 94 is close to my estimate of 100, so my answer is reasonable.

Home Link 1-5

1. 8:00; 8:06
2. 3:30; 3:39
3. 1:45; 1:52

Home Link 1-6

The swim meet was 2 hours and 30 minutes long.

Home Link 1-7

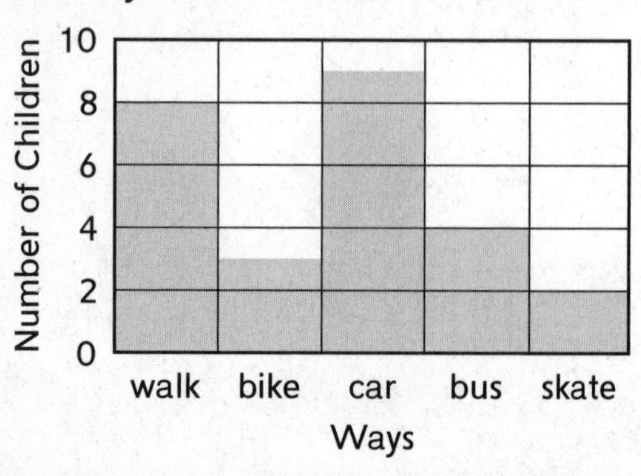

Home Link 1-8

1. 30 apples
   Sample answer: 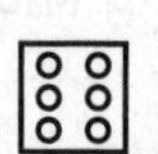
   Number model: $5 \times 6 = 30$;
   $6 + 6 + 6 + 6 + 6 = 30$
2. 24 plants
   Sample answer: ••••••••
   ••••••••
   ••••••••
   Number model: $3 \times 8 = 24$; $8 + 8 + 8 = 24$

Home Link 1-9

1. 3 baskets
2. 10 bags

Home Link 1-10

1. 10, 10, 10, 10
2. 12, 12, 12, 12
3. $2 \times 7 = 14$; $7 \times 2 = 14$; $14 \div 2 = 7$; $14 \div 7 = 2$
4. $2 \times 8 = 16$; $8 \times 2 = 16$; $16 \div 2 = 8$; $16 \div 8 = 2$
5. $5 \times 4 = 20$; $4 \times 5 = 20$; $20 \div 5 = 4$; $20 \div 4 = 5$
6. $10 \times 6 = 60$; $6 \times 10 = 60$; $60 \div 10 = 6$; $60 \div 6 = 10$

Home Link 1-11

1. 20 minutes

Home Link 1-13

1. about 500 grams
2. Sample answer: There are 1,000 grams in each kilogram, so Marco's bag has a mass of 1 kilogram. Emmi's bag has a mass of 2 kilograms, so her bag has more mass.
3. 10
4. 11
5. 12
6. 13

Finding Differences on a Number Grid

Home Link 1-1

NAME DATE

Family Note Today your child reviewed patterns on the number grid and used them to find differences between numbers. For example, one way to find the difference between 87 and 115 on the number grid is: Start at 87. Count the number of tens to 107. There are 2 tens, or 20. Count the number of ones from 107 to 115. There are 8 ones, or 8. The difference between 87 and 107 is 2 tens and 8 ones, or 28. Formal subtraction methods will be covered in the next unit.

Please return this Home Link to school tomorrow.

81	82	83	84	85	86	87	88	89	90
91	92	93	94	95	96	97	98	99	100
101	102	103	104	105	106	107	108	109	110
111	112	113	114	115	116	117	118	119	120
121	122	123	124	125	126	127	128	129	130

Use the number grid to help you solve the following problems.

① The difference between 83 and 109 is ______.

② The difference between 97 and 125 is ______.

③ Explain how you solved Problem 2. ______________________________

____

____

____

Practice

Unit
pencils

Solve.

④ $13 = 7 +$ ______

⑤ $13 = 6 +$ ______

⑥ $6 =$ ______ $- 7$

⑦ $7 =$ ______ $- 6$

Finding Differences

Lesson 1-1

NAME DATE

Use the number grid to find the difference between the pairs of numbers. Explain what you did.

① The difference between 324 and 351 is ______.

How did you use the number grid to find the difference?

____

____

____

____

② What number is 34 less than 173? ______

How did you use the number grid to find the number?

____

____

____

____

③ Do your own.

The difference between ______ and ______ is ______.

How did you use the number grid?

____

____

____

____

Number-Grid Difference

Home Link 1-2

NAME DATE

Family Note Today your child received an *Everyday Mathematics Student Reference Book.* Children can use this book to look up and review topics in mathematics. You may want to take some time to explore this book. Your child also looked up the directions for and played *Number-Grid Difference,* a game that helps to develop mental subtraction strategies. For game directions, see below or *Student Reference Book,* page 251 .

Materials

- number cards 0–9 (4 of each) If you use a regular deck of playing cards, use Jacks as 0s, Aces as 1s, and remove 10s and the other face cards.
- 1 completed number grid (see next page)
- 2 record sheets (see next page)
- 2 beans, pennies, or other counters
- calculator (optional)

Directions

1. Shuffle the cards. Place the deck number-side down on a table.
2. Both players take 2 cards from the deck and use them to make a 2-digit number. Mark both numbers with counters on the number grid.
3. Players now take turns. When it is your turn:
   - Find the difference between the 2 marked numbers. This is your score.
   - Record the 2 numbers and your score on the record sheet.
4. Continue playing until each player has recorded the scores for 5 turns.
5. Add your 5 scores. Players may use a calculator to add.
6. The player with the lower sum wins the game.

Please keep these directions and the number grid at home for future reference. Cut off and return the record sheet portion of the Home Link to school tomorrow.

Number-Grid Difference (continued)

Home Link 1-2

NAME DATE

Show someone at home how to play *Number-Grid Difference.*

−9	−8	−7	−6	−5	−4	−3	−2	−1	0
1	2	3	4	5	6	7	8	9	10
11	12	13	14	15	16	17	18	19	20
21	22	23	24	25	26	27	28	29	30
31	32	33	34	35	36	37	38	39	40
41	42	43	44	45	46	47	48	49	50
51	52	53	54	55	56	57	58	59	60
61	62	63	64	65	66	67	68	69	70
71	72	73	74	75	76	77	78	79	80
81	82	83	84	85	86	87	88	89	90
91	92	93	94	95	96	97	98	99	100
101	102	103	104	105	106	107	108	109	110
111	112	113	114	115	116	117	118	119	120

My Record Sheet				My Partner's Record Sheet			
Round	My Number	My Partner's Number	Difference	Round	My Number	My Partner's Number	Difference
1				1			
2				2			
3				3			
4				4			
5				5			

TOTAL ________ **TOTAL** ________

Copyright © McGraw-Hill Education. Permission is granted to reproduce for classroom use.

Telling Time

Home Link 1-3

NAME DATE

Family Note Today your child explored some of the math tools commonly used in third grade. We reviewed how to read a ruler to the nearest inch and centimeter, and how to tell time to the nearest hour, half hour, and 5 minutes. Help your child read each time by paying attention to the position of both the hour and the minute hands.

Please return this Home Link to school tomorrow.

SRB 184-186

① Draw the hour hand and the minute hand to show the time right now. Write the time.

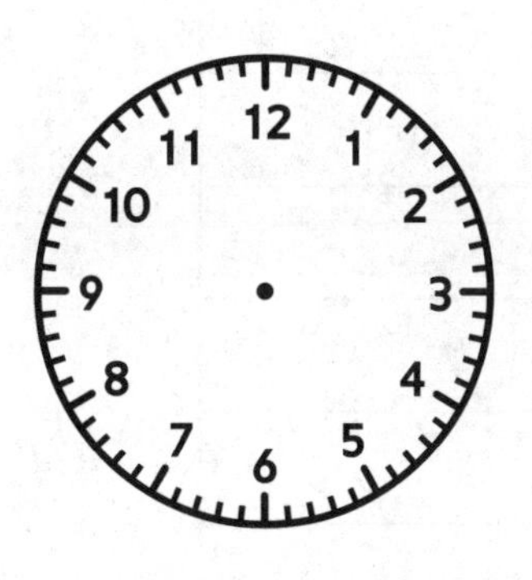

_____ : _____

Write the time shown.

②

_____ : _____

③

_____ : _____

④

_____ : _____

⑤

_____ : _____

⑥

_____ : _____

⑦

_____ : _____

⑧ Show someone how you solved the hardest problem on this page.

Calculator Puzzles

Lesson 1-4

NAME DATE

Solve the calculator puzzles. Add or subtract to find the "Change to" number. You may use a number grid to help. When you finish, compare answers with a partner. If you and your partner disagree on an answer, try the problem again.

Enter	Change to	How?
42	52	
61	51	
80	40	
58	78	
110	90	
46	146	
238	538	
78	108	

Solve the calculator puzzles. Add or subtract to find the "Change to" number. You may use a number grid to help. When you finish, compare answers with a partner. If you and your partner disagree on an answer, try the problem again.

Enter	Change to	How?
42	52	
61	51	
80	40	
58	78	
110	90	
46	146	
238	538	
78	108	

Copyright © McGraw-Hill Education. Permission is granted to reproduce for classroom use.

Rounding Numbers

Home Link 1-4

NAME DATE

Family Note Today your child used open number lines (*see Example*) to help round numbers to the nearest 10 and to the nearest 100. Rounding is one way to estimate calculations. For example, to estimate 83 − 37, your child can round 83 to 80 and 37 to 40, and then easily subtract 80 − 40 = 40, so an estimated answer for 83 − 37 is about 40. The actual answer, 46, is close to 40. Have your child explain how to use an open number line to round numbers.

Please return this Home Link to school tomorrow.

SRB 104-107

Example: What is 72 rounded to the **nearest 10**? 70

70 72 75 80

Which two multiples of 10 are closest to 72?

Round each number. Show your work on an open number line.

① What is 87 rounded to the **nearest 10**? ______

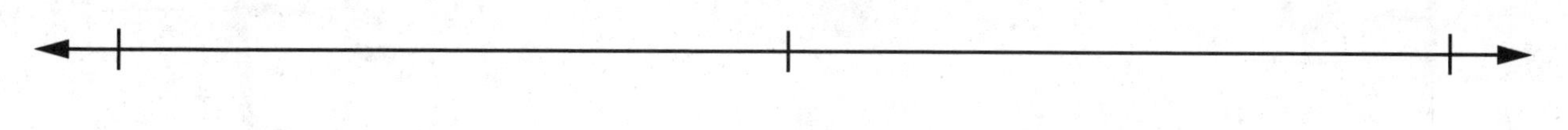

② What is 283 rounded to the **nearest 100**? ______

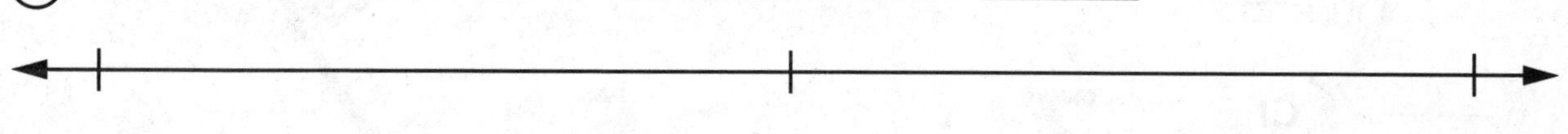

③ Round the numbers in the problem below to the nearest 10. You may sketch an open number line to help.

Unit books

Use the rounded numbers to estimate the answer. Then solve.

$$\begin{array}{r} 38 \\ +56 \\ \hline \end{array}$$

Estimate: ______ + ______ = ______

Is your answer reasonable? ______ Explain. ______

______

Five-Minute Marks

Lesson 1-5

NAME DATE TIME

① Color the hour hand red.

② Color the minute hand green.

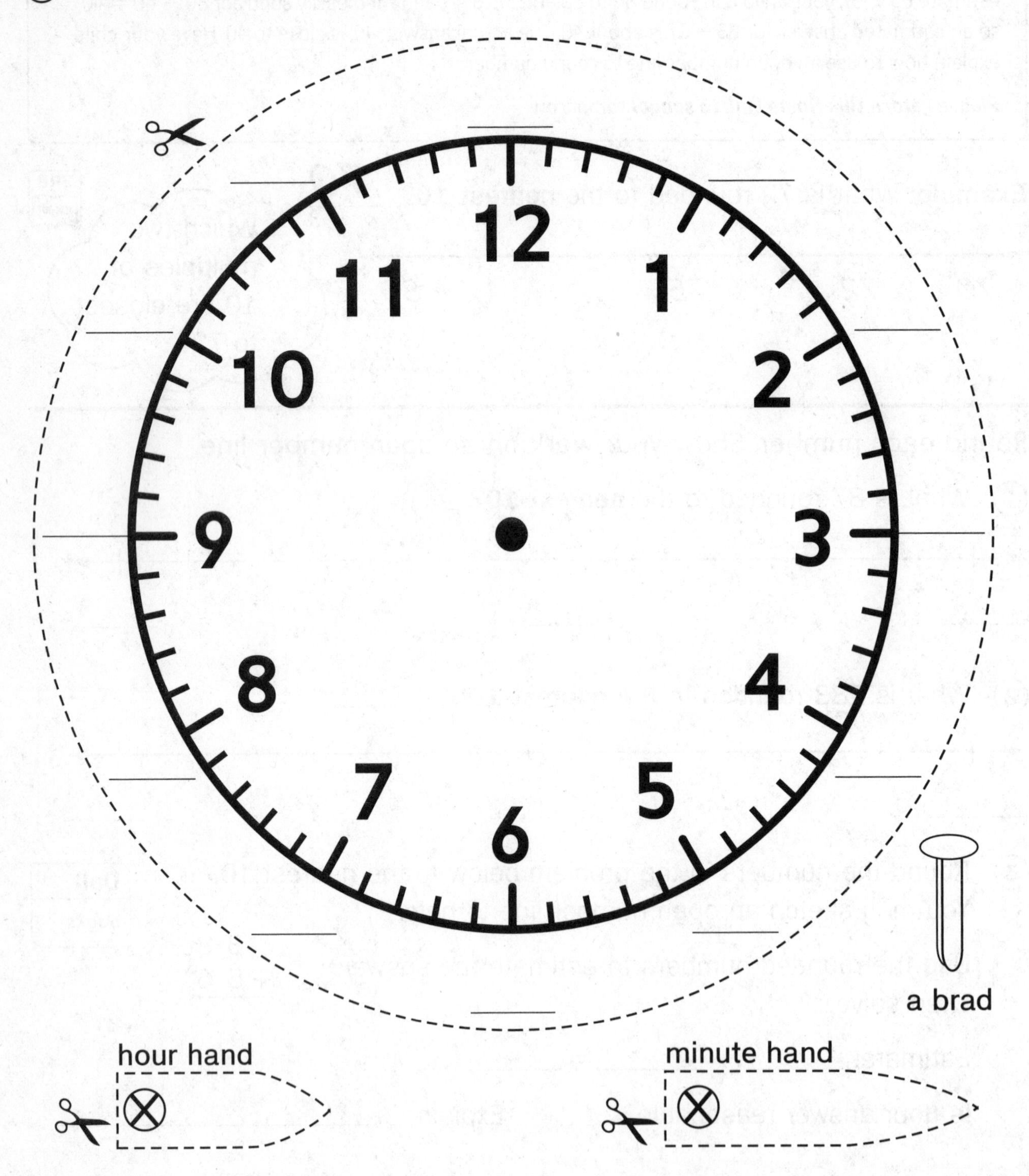

Clock Faces

Lesson 1-5

NAME DATE TIME

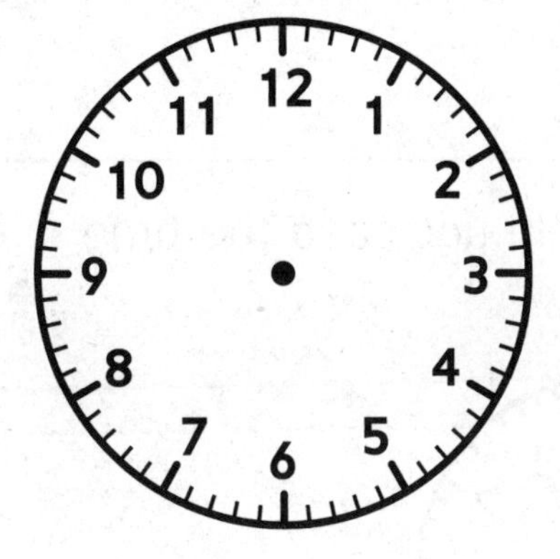

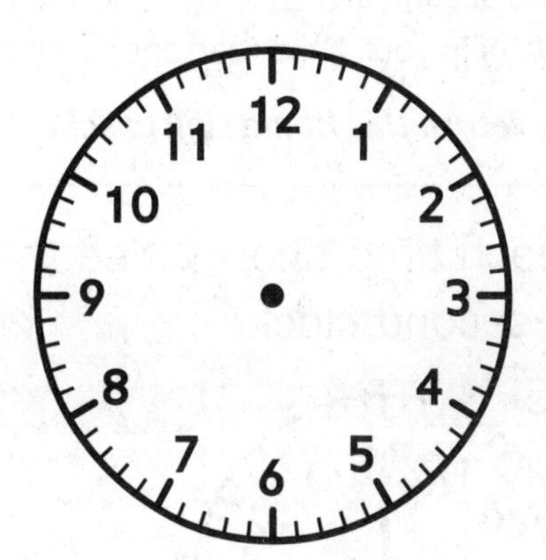

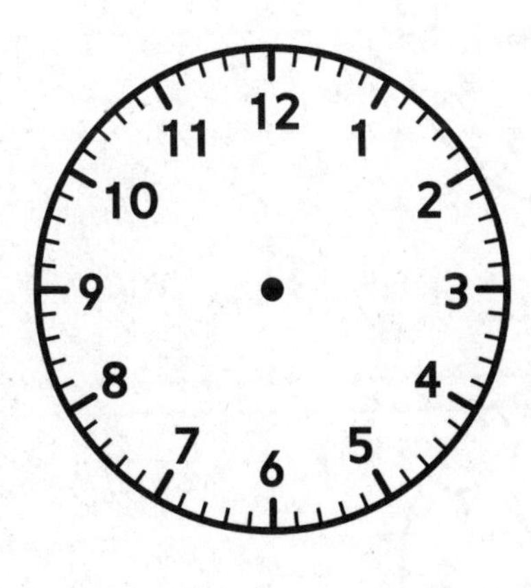

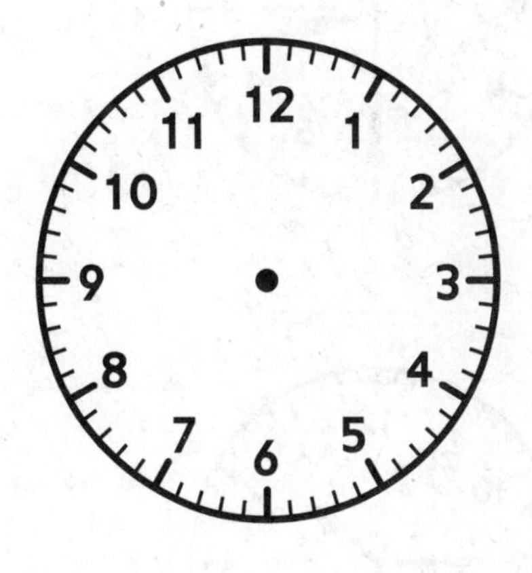

Telling Time to the Nearest Minute

Home Link 1-5

NAME DATE TIME

Family Note Today your child practiced telling time to the nearest minute on analog clocks. Children used familiar times on the hour and half hour to help them read more precise times. For example, in Problem 1 the first clock shows 8:00. Children can use 8:00 as a familiar time to help them read the second clock as 8:06. They start at 8:00 and count by 5s to 8:05 and then 1 more to 8:06. As needed, help your child read and write each time.

Please return this Home Link to school tomorrow.

Write each time shown. Use the first clock to help you read the time on the second clock.

1.  ______  ______

2.  ______  ______

3.  ______  ______

Talk about when you may need to tell time to the nearest minute.

How Long Is a Morning?

Lesson 1-6

NAME DATE TIME

Write the times:

We start at ______.

We go to lunch at ______.

Use at least one strategy to find how long you are in school in the morning. Show how you solve the problem.

How long was your morning? ______ hours and ______ minutes

Finding Elapsed Time

Home Link 1-6

NAME DATE TIME

Family Note Your child is learning how to use a model, such as a number line or clock, to determine elapsed time. Today we used an open number line like the one shown in the example below to figure out how long a morning class lasts. Have your child explain the example to you.

Please return this Home Link to school tomorrow.

Example: A swim meet started at 3:45 P.M. and ended at 6:15 P.M. Fill in familiar times on the number line and use it to answer the question.

Ava solved the problem this way:

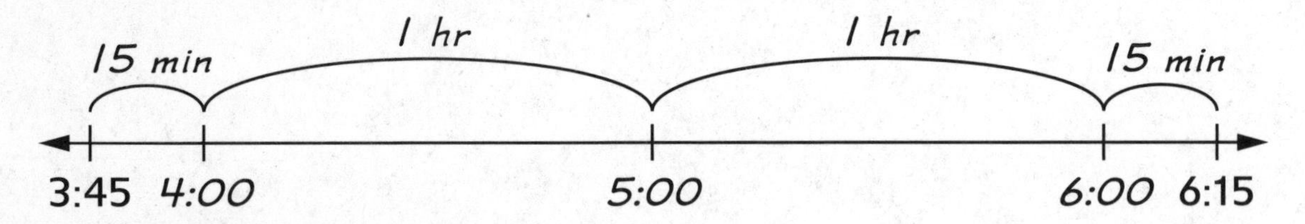

15 min + 1 hr + 1 hr + 15 min =
2 hrs and 30 min

How long was the swim meet? ______ hours and ______ minutes

1. Explain Ava's strategy to someone at home.
2. How much time do you usually have between the end of school and when you go to bed?

   I leave school at ______. I go to bed at ______.

   Make a model to help you answer the question.

   I have about ______ hours and ______ minutes after school before I go to bed.

Interpreting a Tally Chart

Lesson 1-7

NAME DATE TIME

Look at the tally chart below.

Be ready to talk about what the tally marks show.

Favorite Drinks	
Drinks	**Number of Children**
milk	𝍸 |
lemonade	𝍸
water	𝍸 ||
juice	𝍸 ||||

Interpreting a Tally Chart

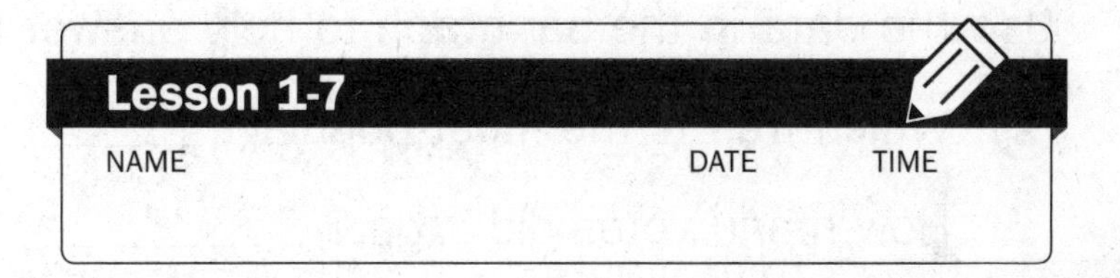
Lesson 1-7

NAME DATE TIME

Look at the tally chart below.

Be ready to talk about what the tally marks show.

Favorite Drinks	
Drinks	**Number of Children**
milk	𝍸 |
lemonade	𝍸
water	𝍸 ||
juice	𝍸 ||||

Graphing Data

Lesson 1-7

NAME DATE TIME

In a survey, each person was asked to name his or her favorite fruit.

This table shows the results.

apple	watermelon	pear	banana	grapes
𝍸	𝍸 III	III	IIII	𝍸 𝍸

① Make a bar graph that shows how many people chose each fruit.

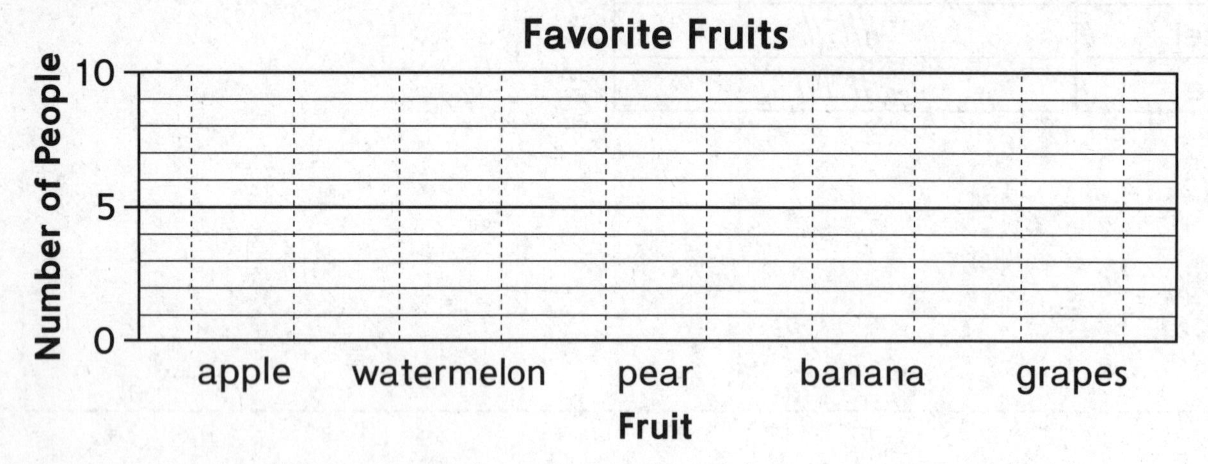

Use the data in the bar graph to help answer the questions below.

② Which fruit is the most popular? ________________

How many votes did it get? ________

③ Which fruit is the least popular? ________________

How many votes did it get? ________

④ How many more people like grapes than pears? ________

⑤ What is your favorite kind of fruit? ________________

Solving Problems in Bar Graphs

Home Link 1-7

NAME DATE TIME

Family Note Today your child collected and organized data about the number of letters in the class's first and last names into tally charts. Then children represented the data in bar graphs. Help your child read the data in the tally chart below and then represent the data on the bar graph. Note that the scale on the bar graph shows intervals of 2, so each interval represents 2 children.

Please return this Home Link to school tomorrow.

Look at the data in the tally chart.

How Bay School 3rd Graders Get to School	
Ways	**Number of Children**
walk	~~IIII~~ III
bike	III
car	~~IIII~~ IIII
bus	IIII
skate	II

Show the data in the tally chart on the bar graph. Look carefully at the scale.

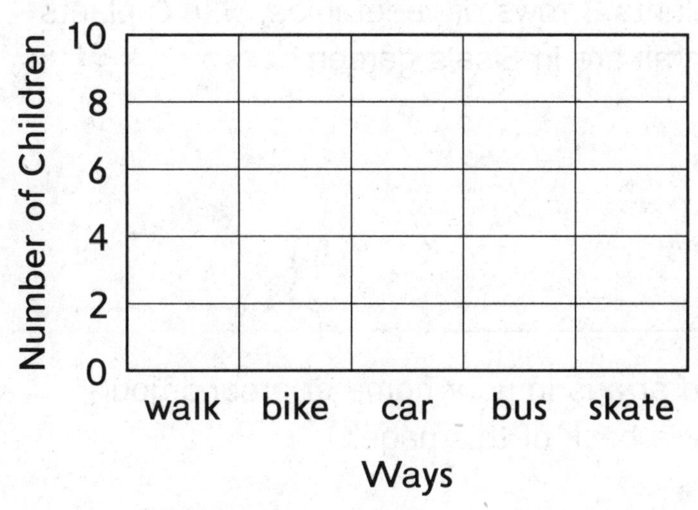

Sharing Strategies for Multiplication

Home Link 1-8

NAME DATE TIME

Family Note Today your child explored number stories that involved placing items in equal groups and organizing them into rows and columns, or arrays. (See examples below.) We used drawings and multiplication number models to help make sense of these stories. Help your child make sense of the number stories below. Note that each story can be represented by either an addition or a multiplication number model; one or the other is acceptable.

Please return this Home Link to school tomorrow.

SRB 38, 41-43

For each number story:

- Draw a picture to match.
- Solve the problem.
- Write a number model to represent the story and your answer.

1. Thaddeus buys 5 bags of apples for a picnic. There are 6 apples in each bag. How many apples does he have?

   _______ apples

   Number model: ______________________________

2. Elsa is planting a garden. She plants 3 rows of vegetables, with 8 plants in each row. How many plants in all are in Elsa's garden?

   _______ plants

   Number model: ______________________________

3. Find equal groups of objects and arrays in your home or around your neighborhood. Record them on the back of this page.

4. Write an equal-groups number story about one set of objects. Use the back of this page. Solve the number story.

Equal Groups of Cookies

Lesson 1-9

NAME DATE TIME

① Draw 12 cookies divided equally between the 2 plates.

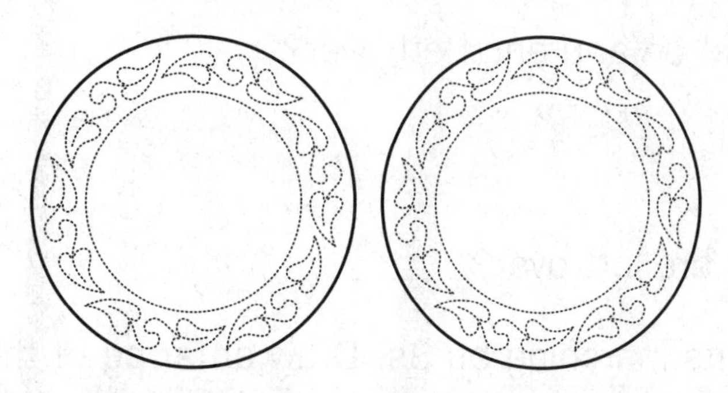

② Tom and Hannah arrive. Now 4 children share the cookies. Draw a picture to show how they share the 12 cookies.

③ Draw a picture to show how 6 children share 12 cookies.

④ There are 12 children. If each child gets 3 cookies, how many cookies do they have all together?

______________ (unit)

Draw a picture to show how you found your answer.

Exploring Remainders

Lesson 1-9

NAME DATE TIME

① There are 25 ants marching by 2s. Draw an array of the ants.

THINK: Will there be any ants left over?

How many ants are left over? _______

② There are 25 ants marching by 3s. Draw an array of the ants.

THINK: Will there be any ants left over?

How many ants are left over? _______

③ There are 25 ants marching by 4s. Draw an array of the ants.

THINK: Will there be any ants left over?

How many ants are left over? _______

④ There are 25 ants marching by 5s. Draw an array of the ants.

THINK: Will there be any ants left over?

How many ants are left over? _______

⑤ Why did Joe think of himself as a "remainder of one"?

____

____

____

Exploring Equal Shares

Lesson 1-9

NAME DATE TIME

Draw a picture to show how you shared the pennies in each problem. You may use pennies or counters to help.

Write how much money each person gets.

10¢ shared by:

① 2 people:

_______ ¢ per person

② 5 people:

_______ ¢ per person

16¢ shared by:

③ 2 people:

_______ ¢ per person

④ 4 people:

_______ ¢ per person

20¢ shared by:

⑤ 5 people:

_______ ¢ per person

⑥ 2 people:

_______ ¢ per person

Introducing Division

Home Link 1-9

NAME DATE TIME

Family Note Today your child explored ways to solve number stories using division. In the stories below the total number of objects is given, so your child needs to find either the number in each group or the number of groups. If needed, help your child count out pennies or dried beans to match the total in each story and use them to act out the story.

Please send in an unopened, 1-liter bottle of water for use in an upcoming lesson on measuring mass.

Please return this Home Link to school tomorrow.

Draw pictures to help you solve each number story.
Record your answers.

1. Connie has 18 toys to put away. She puts 6 toys in each basket. How many baskets does she use?

   _______ baskets

2. Jamal is bagging prizes for the school fair. There are 30 prizes and Jamal wants to put 3 prizes into each bag. How many bags did Jamal make?

   _______ bags

3. Think of things at home that could be shared equally by your family. Record them on the back of this page.

4. Write a number story about equally sharing one of the things you wrote for Problem 3. Use the back of this paper. Then solve your number story.

Skip Counting on the Number Grid

Lesson 1-10

NAME DATE TIME

① Start at 0 and count by 2s on the number grid. Mark an X through each number in your count.

② Start at 0 again and count by 5s on the number grid. Draw a circle around each number in your count.

									0
1	2	3	4	5	6	7	8	9	10
11	12	13	14	15	16	17	18	19	20
21	22	23	24	25	26	27	28	29	30
31	32	33	34	35	36	37	38	39	40
41	42	43	44	45	46	47	48	49	50
51	52	53	54	55	56	57	58	59	60
61	62	63	64	65	66	67	68	69	70
71	72	73	74	75	76	77	78	79	80
81	82	83	84	85	86	87	88	89	90
91	92	93	94	95	96	97	98	99	100
101	102	103	104	105	106	107	108	109	110
111	112	113	114	115	116	117	118	119	120
121	122	123	124	125	126	127	128	129	130
131	132	133	134	135	136	137	138	139	140
141	142	143	144	145	146	147	148	149	150

③ List the numbers that are marked with an X and a circle.

_____, _____, _____, _____, _____, _____, _____, _____,

_____, _____, _____, _____, _____, _____, _____, _____

A Paper-Folding Pattern

Lesson 1-10

NAME DATE TIME

Step 1 Fold a piece of paper in half 1 time to make 2 smaller rectangles. Unfold the paper.

Step 2 Fold that piece of paper in half 2 times to make 4 smaller rectangles. Unfold the paper.

Step 3 Fold the same piece of paper in half 3 times to make smaller rectangles. Do not unfold the paper.

How many smaller rectangles should you get? ______ rectangles

1. Unfold the paper to check.
   Write about any patterns that you see.

   ____

   ____

2. Predict how many rectangles you will have after folding the paper in half 4 times and 5 times.

   With 4 folds, I think I will have ______ rectangles.

   With 5 folds, I think I will have ______ rectangles.

   Try it out and check your answers.

   With 4 folds, I had ______ rectangles.

   With 5 folds, I had ______ rectangles.

3. Predict how many rectangles you will have after folding the paper in half 6 times, 7 times, and 8 times.

   6 folds: ______ 7 folds: ______ 8 folds: ______

4. Explain how your work with 1, 2, 3, 4, and 5 folds helped you to make your predictions for 6, 7, and 8 folds.

   ____

   ____

Try to fold the paper to check your predictions.

Foundational Multiplication Facts

Home Link 1-10

NAME DATE TIME

Family Note Today your child worked on developing strategies for solving 2s, 5s, and 10s multiplication facts. These facts will be used later to help solve related multiplication facts. **Fact Triangles** are the *Everyday Mathematics* version of traditional flash cards. They are better tools for building fact fluency and mental-math reflexes, however, because they emphasize fact families.

A **fact family** is a group of facts made from the same three numbers. For 6, 5, and 30, the multiplication and division fact family is $5 \times 6 = 30$, $6 \times 5 = 30$, $30 \div 6 = 5$, $30 \div 5 = 6$.

Fact Triangles arrange the three numbers such that the product is below the dot at the top and the factors are in the other two corners.

Use Fact Triangles to practice basic facts with your child. Cut out the triangles from the three attached sheets. Cover either the number below the large dot (the product) or one of the numbers in a corner (a factor).

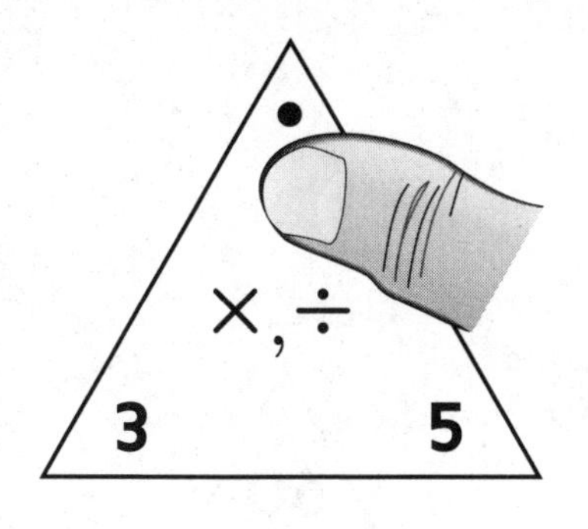

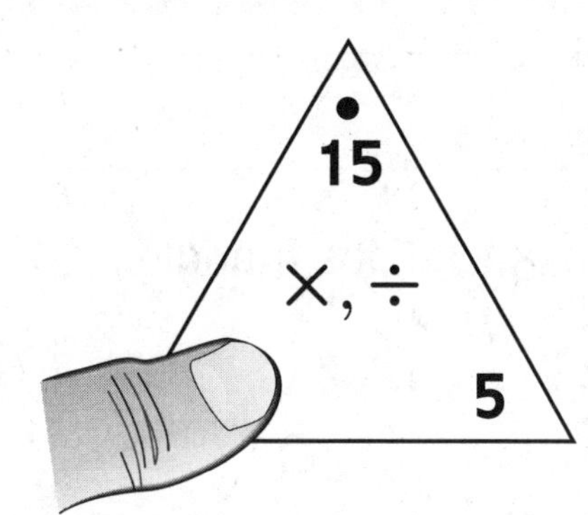

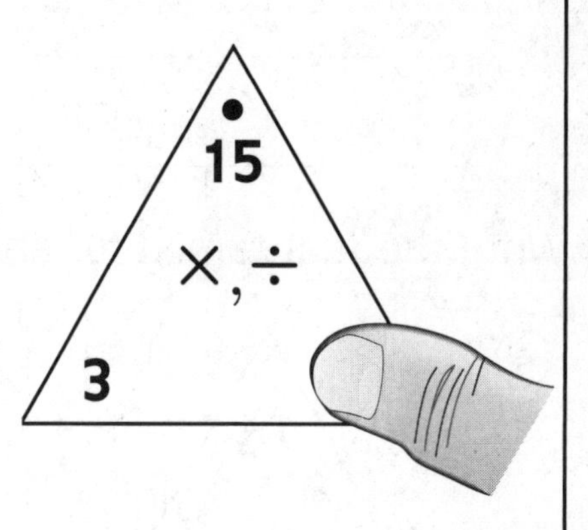

Your child may mentally solve any of the following number sentences to find the product, 15.

$3 \times 5 = ?$

$5 \times 3 = ?$

$? \div 3 = 5$

$? \div 5 = 3$

Find the factor, 3.

$15 \div 5 = ?$

$15 \div ? = 5$

$5 \times ? = 15$

$? \times 5 = 15$

Find the factor, 5.

$15 \div 3 = ?$

$15 \div ? = 3$

$3 \times ? = 15$

$? \times 3 = 15$

If your child misses a fact, flash the other two problems and then return to the fact that was missed. *Example:* Ravi can't answer $15 \div 3$. Flash 3×5, and then $15 \div 5$, and finally $15 \div 3$ a second time.

Make this activity brief and fun. Spend about 10 minutes each night for the next few weeks, or until your child learns them all. The work you do at home will support the work we are doing at school.

*Please return the **second page** of this Home Link to school tomorrow.*

Foundational Multiplication Facts (continued)

Home Link 1-10

NAME DATE TIME

Tell someone at home about multiplication/division fact families.

① The numbers 2, 5, and 10 form the following facts:

$2 \times 5 =$ ______ ______ $\div 2 = 5$

$5 \times 2 =$ ______ ______ $\div 5 = 2$

② Knowing $6 \times 2 =$ ______ and $2 \times 6 =$ ______

helps me know ______ $\div 2 = 6$ and ______ $\div 6 = 2$.

③ The numbers 2, 7, and 14 form this multiplication/division fact family:

______ ______

______ ______

Write the fact family for each ×, ÷ Fact Triangle.

④

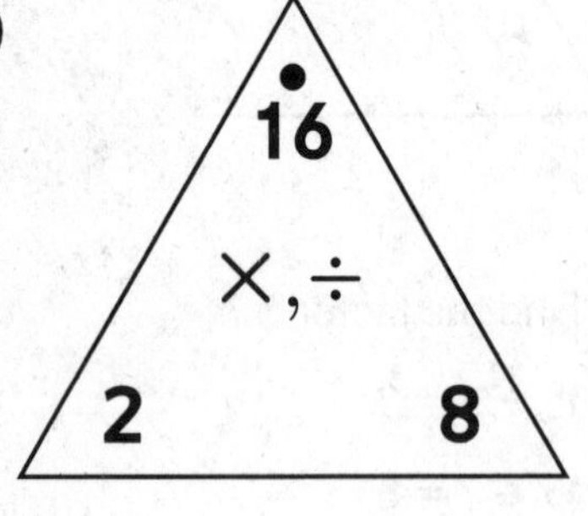

______
______
______
______

⑤

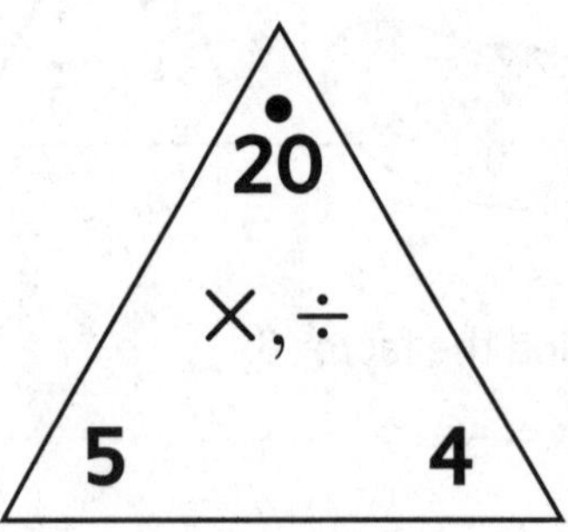

______
______
______
______

⑥

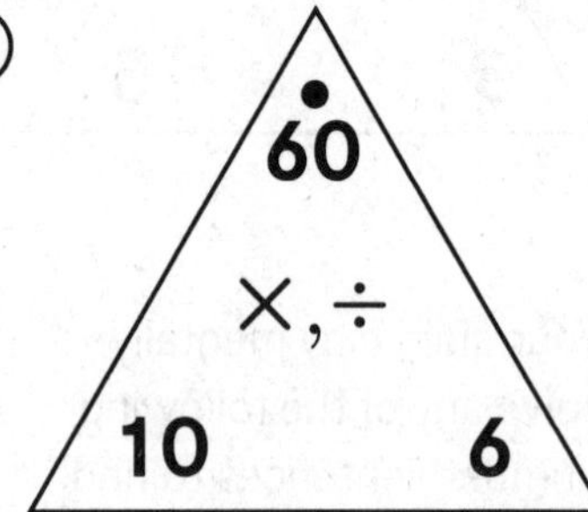

______
______
______
______

×, ÷ Fact Triangles 1: 2s, 5s, and 10s

NAME DATE TIME

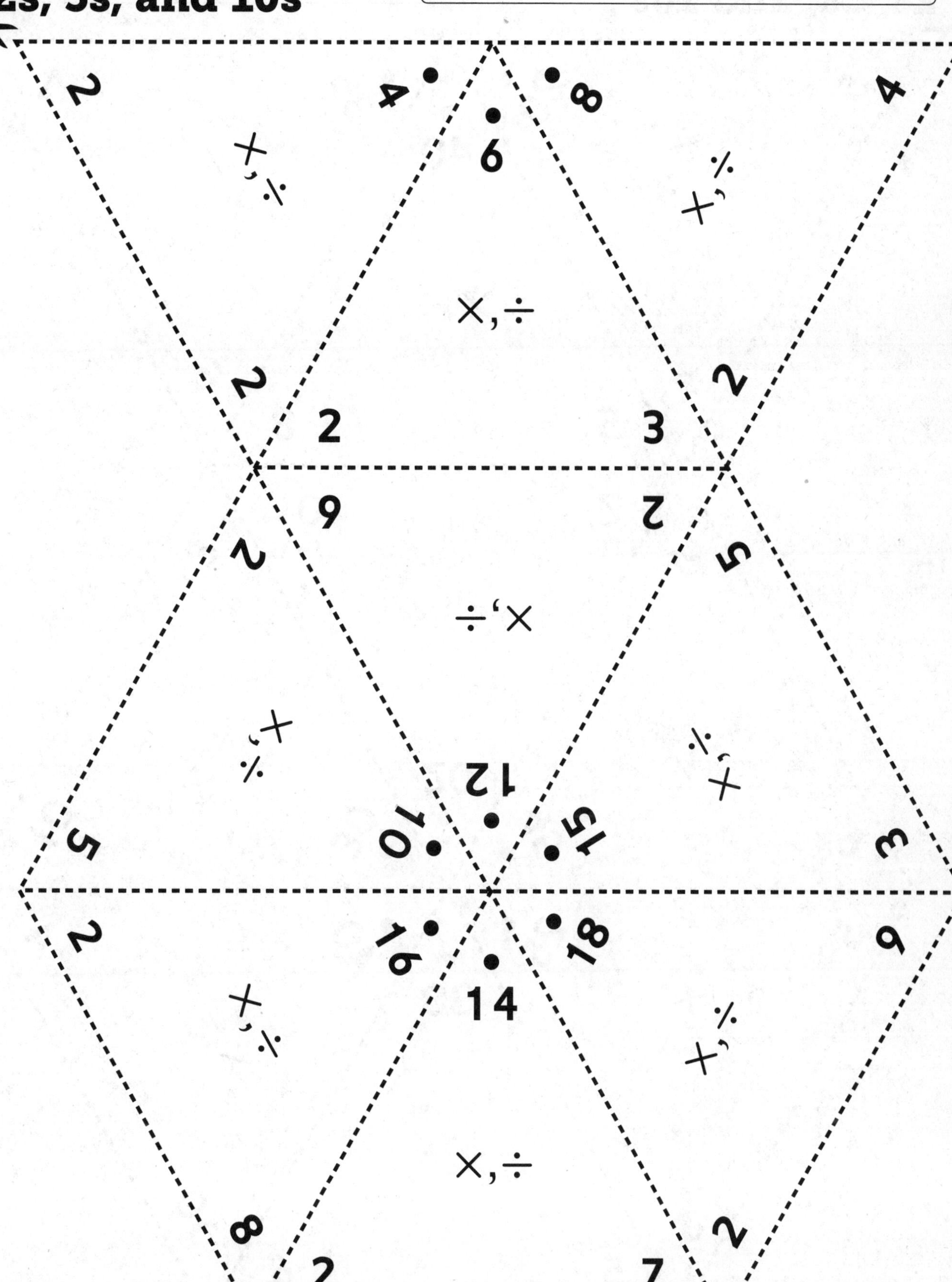

×, ÷ Fact Triangles 2: 2s, 5s, and 10s

NAME DATE TIME

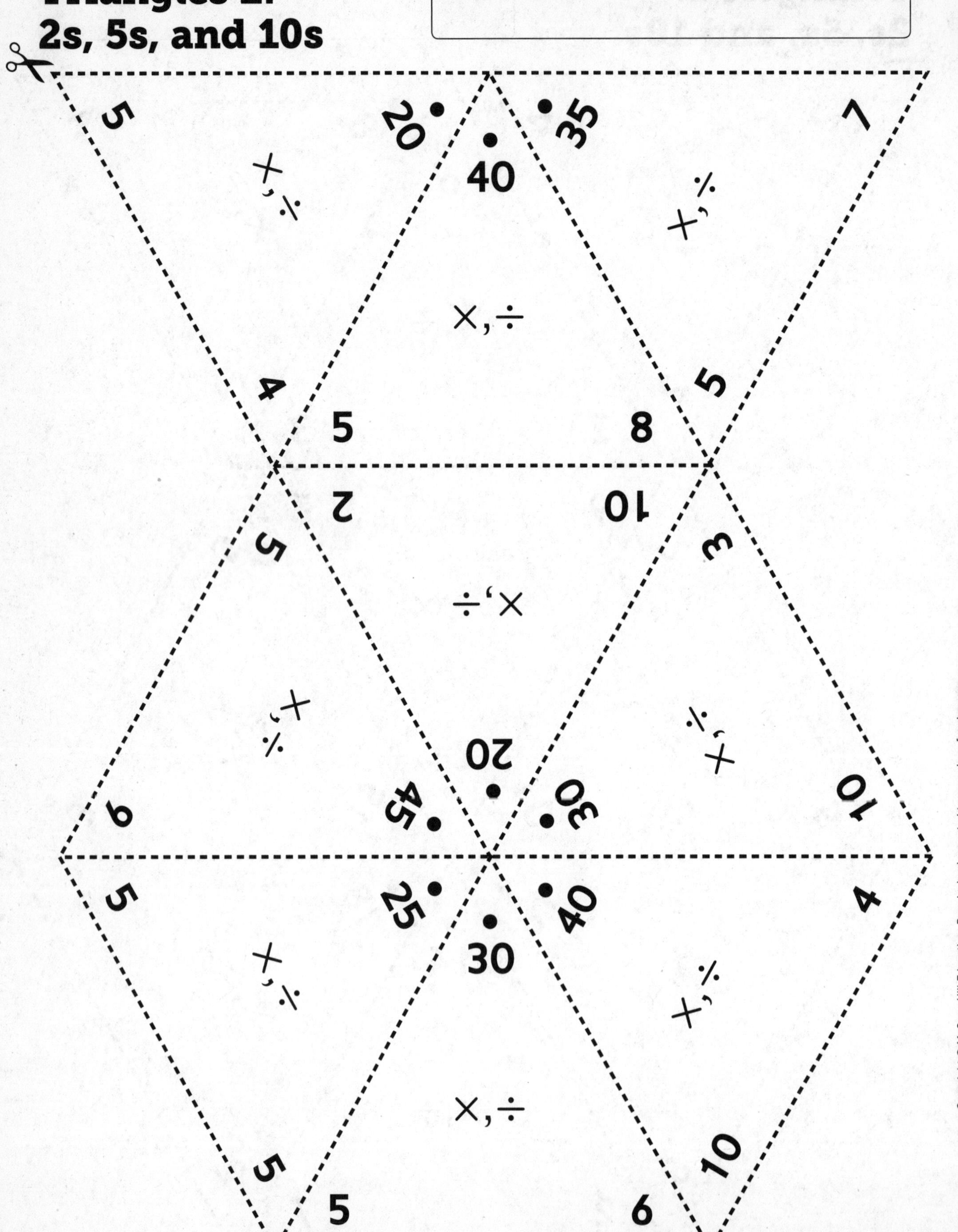

×, ÷ Fact Triangles 3: 2s, 5s, and 10s

NAME DATE TIME

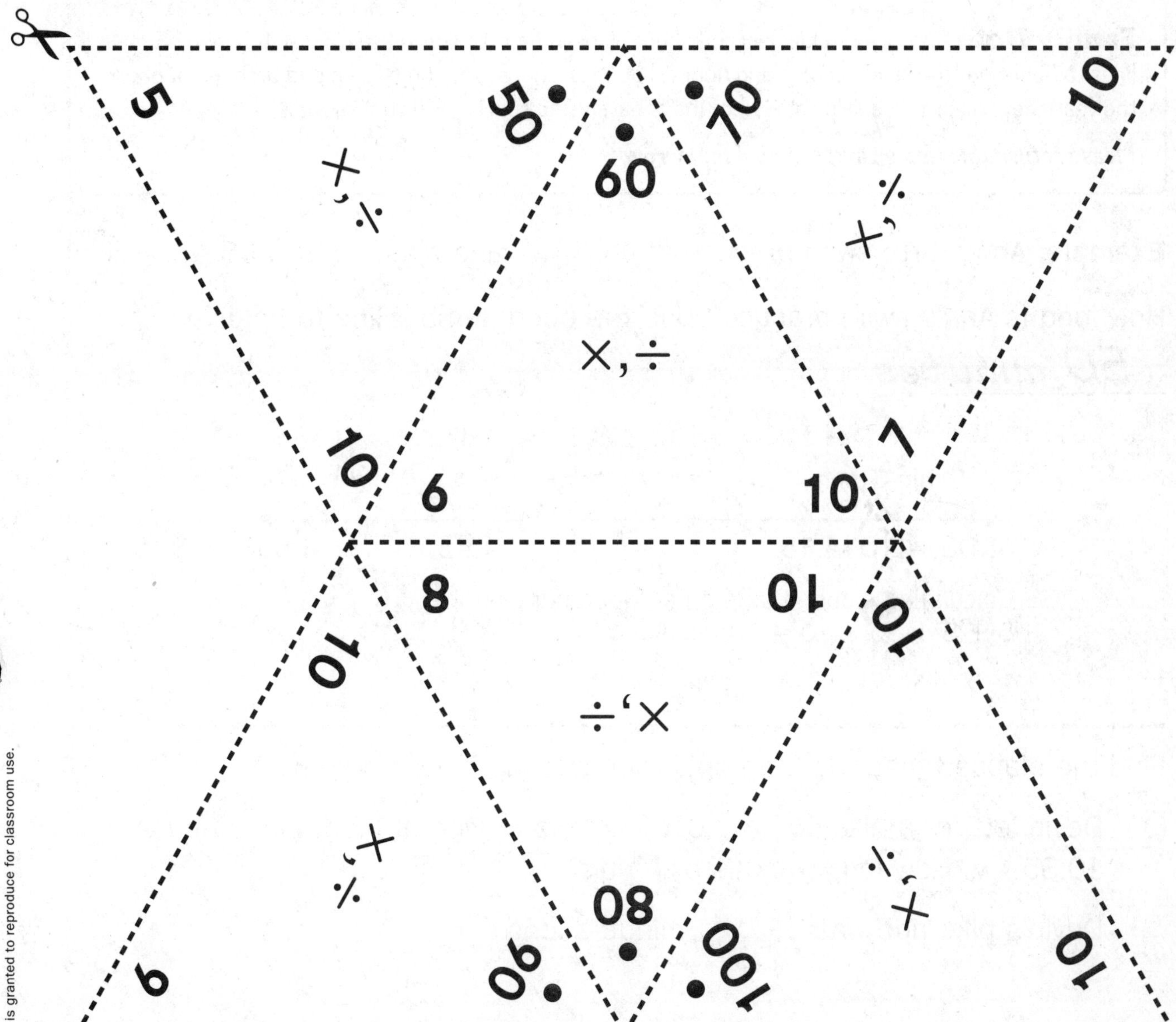

Finding Elapsed Time

Home Link 1-11

NAME DATE TIME

Family Note Today your child learned about elapsed time. Children use clocks and open number lines to figure out the total minutes and hours that pass from a start time to an end time. Throughout the year, they will practice calculating lengths of days using sunrise and sunset data.

Please return this Home Link to school tomorrow.

Example: Ann starts swim practice at 4:05 P.M. and finishes at 4:55 P.M.

How long is Ann's swim practice? Use the open number line to help.

50 minutes

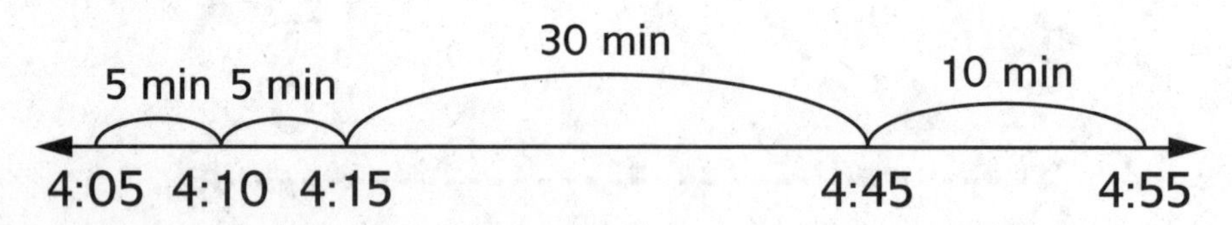

Calculating elapsed time on an open number line:
5 + 5 + 30 + 10 = 50 minutes

Find the elapsed time. Use the open number line below to help.

① Devin left for a bike ride at 10:15 A.M. He arrived at his friend's house at 10:35 A.M. How long was his bike ride?

Devin's bike ride was ______ minutes long.

Equal-Groups Riddles

Lesson 1-12

NAME DATE TIME

What Number Am I?

① If you put me into 7 equal groups with 2 in each group, 5 are left over.

What number am I? _______

Draw a picture.

② I am between 20 and 30. When you put me into 5 equal groups, there is an odd number in each group and one is left over.

What number am I? _______

Draw a picture.

③ Write your own equal-groups riddle. Give it to a partner to solve.

____

____

____

____

Finding Totals for Equal Groups

Lesson 1-12

NAME DATE TIME

You may use your calculator to help you solve the problems. Compare the strategies you and others use.

1. How many people are in my group? ______

   How many hands do the people in my group have all together? ______

   How many fingers do the people in my group have all together? ______

   Write a number model to show how you figured out the total number of fingers. ______________

2. How many tables are in the classroom? ______

   How many legs do the tables have all together? ______

   Number model: ______________

3. One flower has 5 petals.

   How many petals do 6 flowers have? ______

   Number model: ______________________________

4. Make up your own problem like the ones above. Draw a picture to help someone solve your problem.

Masses of Objects

Home Link 1-12

NAME DATE TIME

Family Note Today your child used a pan balance and grams and kilograms to compare and measure objects' masses or weights. In everyday life, mass and weight are hard to tell apart and *Everyday Mathematics* does not distinguish their differences. In later science classes your child will learn how scientists treat mass and weight.

Help your child find objects to compare at home. Below he or she will record the names of two objects that weigh about the same. Try to find objects that are different sizes or shapes.

Please remember to send an unopened 1-liter bottle of water to school with your child.

Our class is also collecting items for a Mass Museum. Help your child select an item that is 1 kilogram (2.2 pounds) or less to bring to school. Over the next several days, children will estimate and then measure the masses of objects in the museum.

Please return this Home Link to school tomorrow.

- Find objects that you can hold in one hand.
- Pick two objects and place one in each hand.
- Find two objects that feel like they have about the same mass or weight.
- Draw or write the names of the objects below.
- Tell someone how you know they have the same mass or weight.

Ask someone at home if you can bring things to school for the Mass Museum.

Estimating with Grams and Kilograms

Lesson 1-13

NAME DATE TIME

A nickel has a mass of about 5 grams.
A liter of water has a mass of about 1,000 grams, or 1 kilogram.

In Problems 1–6, circle a reasonable mass for each object.

① A dog might have a mass of about ________.

20 kilograms   200 kilograms   2,000 kilograms

② A can of soup might have a mass of about ________.

4 grams   40 grams   400 grams

③ A newborn baby might have a mass of about ________.

3 kilograms   30 kilograms   300 kilograms

④ A basketball might have a mass of about ________.

6 grams   600 grams   6,000 grams

⑤ A pencil might have a mass of about ________.

5 grams   50 grams   500 grams

⑥ A mouse might have a mass of about ________.

25 grams   250 grams   2,500 grams

⑦ Choose one of the problems above. Explain why you chose your answer.

____

____

____

Mass Museum Record Sheet

Lesson 1-13

NAME DATE TIME

Use benchmark objects to estimate the mass of an object.

Then use a pan balance and masses to measure the mass of your object.

Record your work below.

Object	Estimated Mass	Measured Mass

1. For which object was your estimate closest to your measurement?

   ____

2. For which object was your estimate farthest from your measurement?

   ____

3. Which object has the least mass? ______________________

   The greatest? ______________________

Estimating Mass

Home Link 1-13

NAME DATE TIME

Family Note Today your child explored grams and kilograms by measuring the masses of different objects with a pan balance and standard masses. Help your child solve the number stories below.

Please return this Home Link to school tomorrow.

Solve. Hint: 1 kilogram = 1,000 grams

① If a bottle of water has a mass of about 1 kilogram, about how much mass will it have after someone drinks 500 grams of water from it?

about ______ grams

② Emmi's bag has a mass of 2 kilograms. Marco's bag has a mass of 1,000 grams. Whose bag has more mass? Explain.

____

____

____

Practice

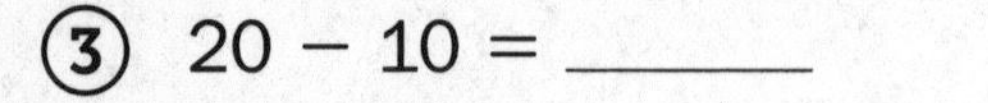

Fill in the unit box. Solve.

Unit

③ 20 − 10 = ______

④ 20 − 9 = ______

⑤ 20 − 8 = ______

⑥ 20 − 7 = ______

Unit 2: Family Letter

Home Link 1-14

NAME DATE TIME

Number Stories and Arrays

In this unit, your child will learn to make sense of and solve many number stories and explore multiplication and division. In Unit 2, children will:

- use basic facts to add and subtract larger numbers.
- solve change, comparison, and parts-and-total number stories.
- solve multistep number stories using 2 or more operations.
- understand multiplication in terms of equal groups, including multiplying by 0 and 1.
- use pictures and arrays to solve multiplication and division problems.
- understand division as sharing a quantity.
- make sense of remainders in division problems.

Understanding multiplication in terms of equal groups:

2 groups of 7 is 14
$2 \times 7 = 14$

Using pictures and arrays to divide:

15 pennies are shared equally among 3 children. How many pennies does each child get?

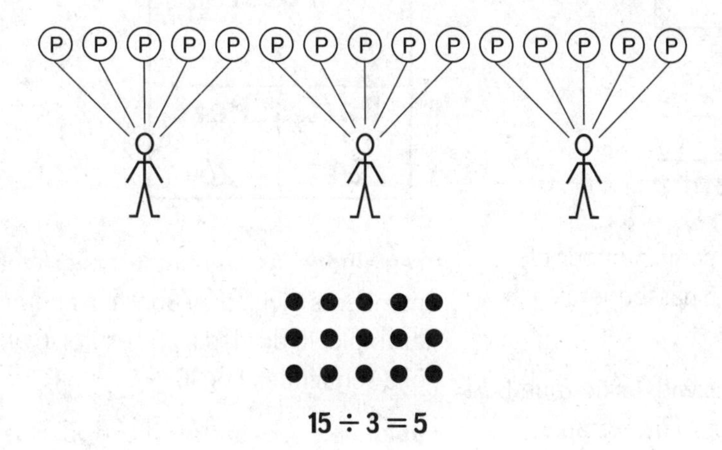

$15 \div 3 = 5$

Please keep this Family Letter for reference as your child works through Unit 2.

Vocabulary

Some important terms in Unit 2:

array An arrangement of objects in a regular pattern, usually rows and columns.

change number story A number story involving a starting quantity, a change, and an ending quantity. If the ending quantity is more than the start, it is a *change-to-more* number story. If the ending quantity is less than the start, it is a *change-to-less* number story. For example, the following is a change-to-less number story: *Rita had $28. She spent $12. How much money does Rita have now?* Change number stories can be modeled using a change diagram.

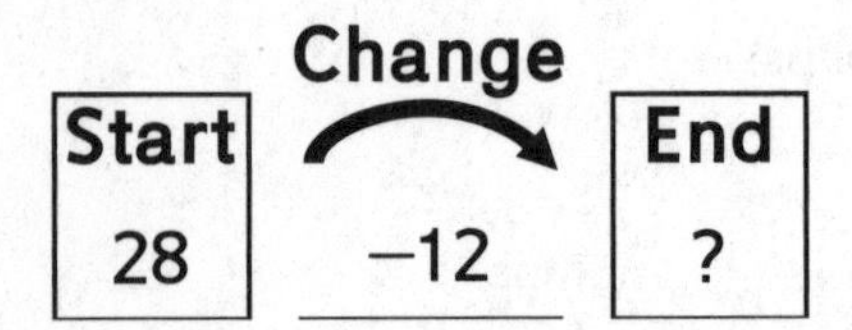

comparison number story A number story about two quantities and the difference between them. For example, *34 children ride the bus to school. 12 children walk to school. How many more children ride the bus?* Comparison number stories can be modeled using a comparison diagram.

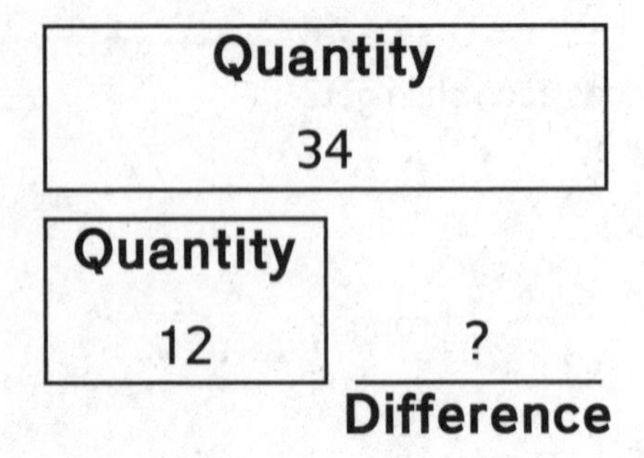

equal groups Sets with the same number of elements, such as cars with 5 passengers each and rows with 6 chairs each.

fact extensions Calculations with larger numbers that use basic arithmetic facts. For example, knowing the addition fact $5 + 8 = 13$ makes it easier to solve problems such as $50 + 80 = ?$ and $65 + ? = 73$.

Frames-and-Arrows diagrams Diagrams with connected arrows that are used to represent number patterns. Each frame contains a number, and each arrow represents a rule that determines which number goes in the next frame. There may be more than one rule, represented by different types of arrows.

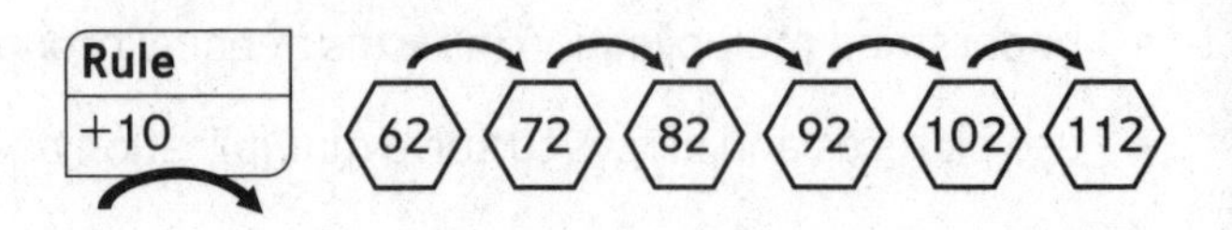

number model A number sentence or expression that models a number story or situation. For example, the story *Sally had $25, and then she earned $12* can be modeled by the number sentence $25 + 12 = 37$.

number sentence An equation such as $15 + 8 = 22$, or an inequality such as $18 > 12$.

parts-and-total number story A number story in which a whole is made up of two or more distinct parts. For example, *Leo baked 24 muffins. Nina baked 26 muffins. How many muffins in all?* Parts-and-total number stories can be modeled using a parts-and-total diagram.

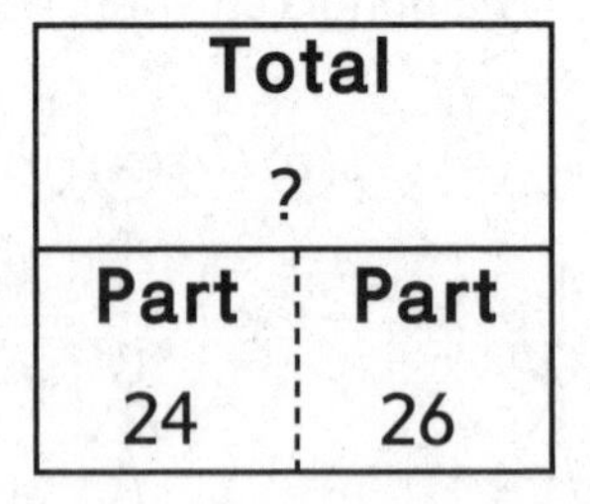

remainder An amount left over when one number is divided by another number. For example, 16 divided by 5 results in an answer of 3 remainder 1 or $16 \div 5 \rightarrow 3$ R 1.

unknown A quantity whose value is not known. An unknown is sometimes represented by a ____, a ?, or a letter. For example, in $5 + w = 13$, w is an unknown.

Do-Anytime Activities

Work with your child on some of the concepts and skills taught in this unit:

1. Practice addition and subtraction fact extensions. For example:

$6 + 7 = 13$	$13 - 7 = 6$
$60 + 70 = 130$	$23 - 7 = 16$
$600 + 700 = 1300$	$83 - 7 = 76$

2. Help your child recognize examples of equal groups, equal sharing, or array situations in everyday life. Examples: 2 boxes of granola bars with 6 bars in each box (2 equal groups of 6) gives you a total of 12 granola bars. 20 mints shared among 5 people gives each person 4 mints.
3. Continue practicing 2s, 5s, and 10s multiplication facts by using Fact Triangles and playing *Multiplication Draw* (see *Student Reference Book* for directions).

Building Skills through Games

In Unit 2 your child will practice multiplication and mental addition by playing the following games. For detailed instructions, see the *Student Reference Book.*

Array Bingo Players make a 4-by-4 array of array cards. They draw number cards and try to match them with array cards showing that number of dots. If an array card matches, they turn it facedown.

Division Arrays Players use counters to create arrays with a given total number and number of rows. They determine whether or not there are any leftovers or a *remainder* and write division number models to match their arrays.

Roll to 1,000 Players mentally add the results of dice rolls.

As You Help Your Child with Homework

As your child brings home assignments, you may want to go over the instructions together, clarifying them as necessary. The answers listed below will guide you through this unit's Home Links.

Home Link 2-1

1. 16; 26; 76; 106
2. 12; 22; 62; 282
3. 8; 28; 58; 98
4. 5; 15; 115; 475
5. 13; 130; 1,300; 13,000

Home Link 2-2

Number model: $750 - 300 = ?$; $300 + ? = 750$
450 cans; Sample answer: The unknown has to be smaller than 750. I know $3 + 4 = 7$, so 3 [100s] + 4 [100s] is 7 [100s]. 7 [100s] + 50 is 750.

Home Link 2-3

$35 + ? = 52$ or $52 - 35 = ?$; \$17; Sample answer: The answer has to be less than \$35 and \$52 because she started with \$35 and ended with \$52. The answer makes my number model true.

Home Link 2-4

1. 20 balloons **2.** 26 marbles

Home Link 2-5

1. $5 \times 2 = 10$ or $2 + 2 + 2 + 2 + 2 = 10$; $10 + 6 = 16$; 16 points

2. $4 \times 10 = 40$ or $10 + 10 + 10 + 10 = 40$; $40 + 8 = 48$; 48 pages

Home Link 2-6

1. 30 markers

2. \$0. Sample answer: If I buy 0 packs of markers, I do not buy any markers, so my cost is \$0.

3. Sample answer: I have 1 hand with 5 fingers. How many fingers do I have in all? I have 5 fingers in all.

Home Link 2-7

1. Sample answers: 1-by-12, $1 \times 12 = 12$; 12-by-1, $12 \times 1 = 12$; 3-by-4, $3 \times 4 = 12$; 4-by-3, $4 \times 3 = 12$; 2-by-6, $2 \times 6 = 12$; 6-by-2, $6 \times 2 = 12$

2. Sample answer: No. When I try to make an array with 5 rows, there are 2 left over.

Home Link 2-8

1. 8 stickers **2.** 5 cars

Home Link 2-9

1. 6 marbles for each friend, 0 marbles left over; $24 \div 4 = 6$

2. 5 flowers in each vase, 4 flowers left over; $29 \div 5 \rightarrow 5$ R 4

3. 4 rows of stamps, 6 stamps left over; $46 \div 10 \rightarrow 4$ R 6

Home Link 2-10

1. ●●●●● ●●●●● ●●●●● 0

2. ●●●●●● ●●●●●● ●●●●●● ●●●●●● ●●●●●● ●●

3. Sample answers: A package of markers, all the pillows in the house

4. 25 **5.** 8 **6.** 4 **7.** 9

Home Link 2-11

1. 15, 18, 21, 27

2. 900, 700, 600, 500

3. Rule: + 6; 30, 36, 54

4. 30 **5.** 60 **6.** 4 **7.** 4

Home Link 2-12

1. Sample answer: The volume doesn't change because even though the containers are different, the amount of liquid stays the same.

2.

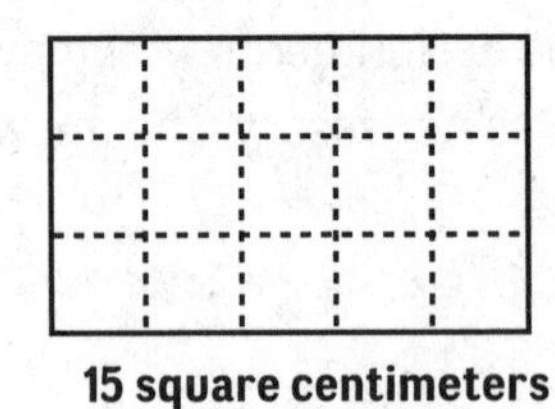

15 square centimeters

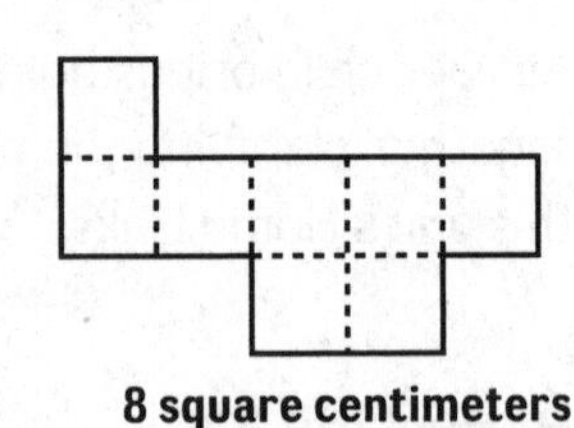

8 square centimeters

3. 12 **4.** 7 **5.** 9 **6.** 2

Fact Extensions Record Sheet

Lesson 2-1

DATE TIME

Fill in the unit box. Work with a partner. Take turns posing problems.

Unit

Start with a basic fact. Then give fact extensions. Record the problems your partner gives you and solve them below.

Extending Facts to 2-Digit Problems	Extending Facts to 3-Digit Problems
Example: Basic fact: 8 + 7 = 15 **a.** Extension: 18 + 7 = 25 **b.** Extension: 58 + 7 = 65	**Example:** Basic fact: 15 − 7 = 8 **a.** Extension: 255 − 7 = 248 **b.** Extension: 325 − 7 = 318
1. Basic fact: ______ **a.** Extension: ______ **b.** Extension: ______	**6.** Basic fact: ______ **a.** Extension: ______ **b.** Extension: ______
2. Basic fact: ______ **a.** Extension: ______ **b.** Extension: ______	**7.** Basic fact: ______ **a.** Extension: ______ **b.** Extension: ______
3. Basic fact: ______ **a.** Extension: ______ **b.** Extension: ______	**8.** Basic fact: ______ **a.** Extension: ______ **b.** Extension: ______
4. Basic fact: ______ **a.** Extension: ______ **b.** Extension: ______	**9.** Basic fact: ______ **a.** Extension: ______ **b.** Extension: ______
5. Basic fact: ______ **a.** Extension: ______ **b.** Extension: ______	**10.** Basic fact: ______ **a.** Extension: ______ **b.** Extension: ______

More Fact Extensions

Lesson 2-1

DATE TIME

Add or subtract on your calculator to complete these problems. You may use a number grid to help.

①

Enter	Change to	How?
22	30	+ 8
72	80	______
60	52	______

What combination of 10 could help you? ______

②

Enter	Change to	How?
50	34	______
940	1,000	______
700	660	______

What combination of 10 could help you? ______

③

Enter	Change to	How?
541	560	______
1,000	100	______
400	299	______

What combination of 10 could help you? ______

④ Write your own:

Enter	Change to	How?
______	______	______
______	______	______
______	______	______

What combination of 10 could help you? ______

Try This

⑤

Enter	Change to	How?
70	37	______
617	740	______
523	680	______

What combination of 10 could help you? ______

Fact Extensions

NAME DATE TIME

Family Note Today your child used basic facts to solve similar problems with larger numbers. These similar problems are known as fact extensions. For example, the basic fact $6 + 7 = 13$ helps solve the fact extension $60 + 70 = 130$. Talk to your child about the patterns in each set of problems. Help your child think of more fact extensions to complete this Home Link.

Please return this Home Link to school tomorrow.

Write the answer for each problem.

① I know: $9 + 7$ This helps me know: $19 + 7$ $69 + 7$ $99 + 7$

② I know: $8 + 4$ This helps me know: $18 + 4$ $58 + 4$ $278 + 4$

③ I know: $15 - 7$ This helps me know: $35 - 7$ $65 - 7$ $105 - 7$

④ I know: $13 - 8$ This helps me know: $23 - 8$ $123 - 8$ $483 - 8$

⑤ I know: $6 + 7$ This helps me know: $60 + 70$ $600 + 700$ $6{,}000 + 7{,}000$

Make up another set of fact extensions.

⑥ I know: This helps me know:

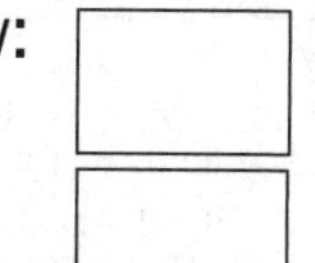

Number Stories: Drinks Vending Machine

Lesson 2-2

NAME DATE TIME

For each number story, do the following:

- Use the Drinks Vending Machine Poster on *Student Reference Book*, page 268.
- Write a number model for the problem. Use ? for the unknown.
- You may draw a diagram like those shown below or a picture to help.

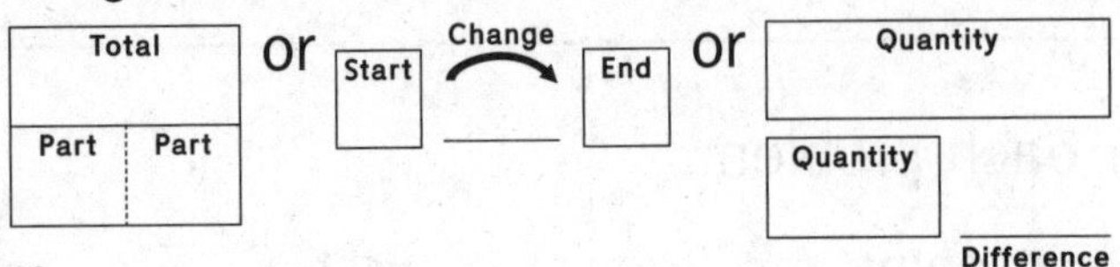

- Solve the problem.
- Write the answer.

① How much money does one orange juice and one grape juice cost together?

Number model: ______________

Answer: ______________

② How much more money does the strawberry yogurt drink cost than grape juice?

Number model: ______________

Answer: ______________

③ Use the Drinks Vending Machine Poster and write your own story.

____

____

Number model: ______________________________

Answer: ____________________________________

Check: How do you know your answer makes sense?

____

____

Number Stories

Home Link 2-2

NAME DATE TIME

Family Note Today your child reviewed parts-and-total, change, and comparison diagrams. These diagrams help organize the information in a number story. For more information, see *Student Reference Book*, page 76. Remind your child to write the unit with the answer. For example, the problem below asks about the number of cans, so the answer should include cans as the unit.

Please return this Home Link to school tomorrow.

SRB 30-31, 76

For the problem below:

- Write a number model. Use ? for the unknown.
- You may draw a diagram like the ones shown below or a picture to help.

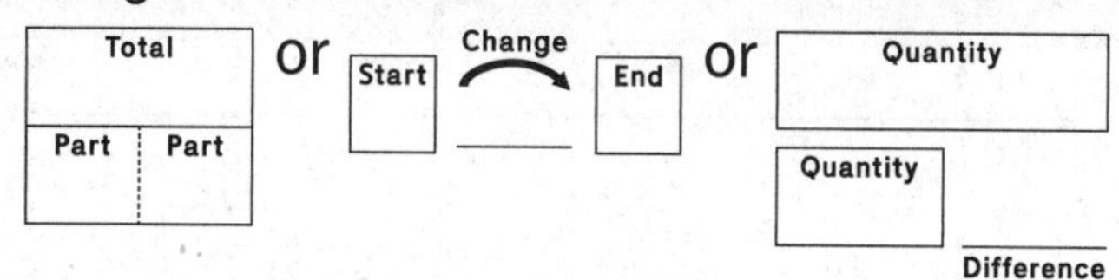

- Solve the problem and write your answer.
- Explain how you know your answer makes sense.

The second- and third-grade classes collected 750 cans to recycle. The second graders collected 300 cans. How many cans did the third graders collect?

Number model: ____________________

Answer the question: ____________________ (unit)

Check: How do you know your answer makes sense?

____

____

Changing the Calculator Display

Lesson 2-3

NAME DATE TIME

Solve the calculator problems. Use a number grid to help. Check your answers on a calculator. Write a number model to show what you did.

Example:

Change 24 to 35. *Add 10 and then add 1, which is 11.*

Number model: *24 + 10 + 1 = 35 or 24 + 11 = 35*

① Change 18 to 38. ______________________

Number model: ______________________

② Change 30 to 80. ______________________

Number model: ______________________

③ Change 21 to 63. ______________________

Number model: ______________________

④ Change 97 to 45. ______________________

Number model: ______________________

⑤ Change 100 to 62. ______________________

Number model: ______________________

More Number Stories

Home Link 2-3

NAME DATE TIME

Family Note Today your child solved more number stories using diagrams or pictures to help organize the information in the problems. Remind your child to write the unit with the answer. For example, the unit in the problem below is dollars, which can be represented by the dollar sign ($). Talk with your child about how he or she knows an answer makes sense.

Please return this Home Link to school tomorrow.

SRB 76, 30-31

For the number story below:

- Write a number model. Use a ? for the unknown.
- You may draw a diagram like those shown below or a picture to help.

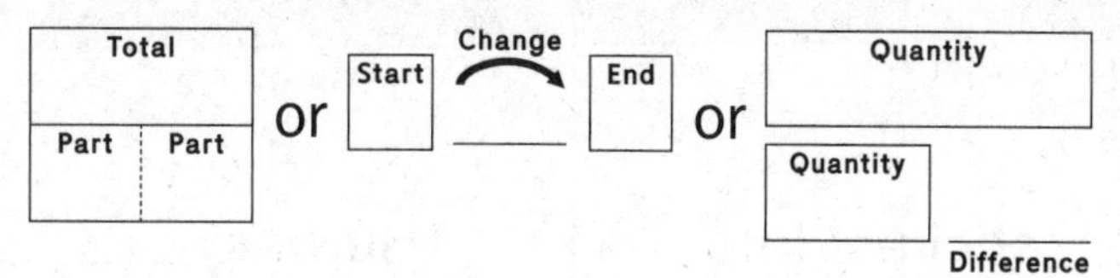

- Solve and write your answer with the unit.
- Explain how you know your answer makes sense.

Jasmine had $35. She earned some money helping her neighbors. Now she has $52. How much money did she earn?

Number model: ______________________________

Answer the question: ____________ (unit)

Check: How do you know your answer makes sense?

____

____

Sticker Stories

Lesson 2-4

NAME DATE TIME

Follow the directions on Activity Card 22.

① Jeremy has ______ pack(s) of stickers. Each pack has ______ sticker(s). He gives ______ sticker(s) away. How many stickers does he have left? He has ______ stickers left. Number models: ______ ______	② Jeremy has ______ pack(s) of stickers. Each pack has ______ sticker(s). He gives ______ sticker(s) away. How many stickers does he have left? He has ______ stickers left. Number models: ______ ______
③ Jeremy has ______ pack(s) of stickers. Each pack has ______ sticker(s). He gives ______ sticker(s) away. How many stickers does he have left? He has ______ stickers left. Number models: ______ ______	④ Jeremy has ______ pack(s) of stickers. Each pack has ______ sticker(s). He gives ______ sticker(s) away. How many stickers does he have left? He has ______ stickers left. Number models: ______ ______

Multistep Number Stories, Part 1

Home Link 2-4

NAME DATE TIME

Family Note Today your child practiced solving number stories with two or more steps. These solution strategies often combine at least two different operations (addition, subtraction, multiplication, or division). Children used drawings, words, and number models to help keep track of their thinking. Encourage your child to draw pictures or use objects to act out the stories below. Help your child make sense of each story by asking questions such as: *What do you know from the story? What do you want to find out? What is your plan? What will you do first? Next? Does your answer make sense?*

Please return this Home Link to school tomorrow.

Solve each problem. Draw pictures or use words or number models to help keep track of your thinking. Remember to write the unit.

① You have 12 red balloons and 13 blue balloons. Then 5 balloons pop. How many balloons do you have left?

Answer: ______________
(unit)

② You have 3 bags of marbles with 6 marbles in each bag. Then you find 8 more marbles. How many marbles do you have now?

Answer: ______________
(unit)

Solving a Multistep Number Story, Part 1

Lesson 2-4

NAME DATE TIME

Read the number story. Use the Snacks Vending Machine Poster on *Student Reference Book*, page 269 for prices.

Diego buys raisins and granola. If he has $1, how much change should he get back?

Study one way to solve the problem.

$100 - 45 = 55$
$55 - 40 = 15$

Diego should get 15¢ back.

Show a different way to solve the problem. Keep track of your thinking by writing number models. You may also use pictures, situation diagrams, or other representations.

Solving a Multistep Number Story, Part 2

Lesson 2-5

NAME DATE TIME

Read the number story.

Jordan and Riley have 13 scented markers. They used to have 2 full boxes of scented markers. A full box has 8 markers. How many markers did Jordan and Riley lose?

Solve the problem. Keep track of your thinking by writing number models. You may also use pictures, situation diagrams, or other representations.

Show a different way to solve the problem. Keep track of your thinking by writing number models. You may also use pictures, situation diagrams, or other representations.

Multistep Number Stories, Part 2

Home Link 2-5

NAME DATE TIME

Family Note Today your child practiced solving additional number stories with two or more steps and writing number models for each step. Help your child make sense of the stories below by asking: *What do you know from the story? What do you want to find out? What is your plan? What will you do first? Next? Have you answered the question? Does your answer make sense?*

Please return this Home Link to school tomorrow.

Solve each problem. Show your work with pictures, words, or numbers. Write number models to keep track of your thinking. Remember to write the unit.

① Each basket in basketball is worth 2 points. Cathy makes 5 baskets and scores 6 more points with free throws. How many points did she score in all?

Number models: ______________________________

____

Answer: ______________
(unit)

② Elias reads 4 chapters. Each chapter has 10 pages. Then he reads 8 more pages. How many pages does Elias read in all?

Number models: ______________________________

____

Answer: ______________
(unit)

Patterns in Multiplying by 0 and 1

Lesson 2-6

NAME DATE TIME

① Solve. What pattern do you notice?

$0 \times 4 =$ ___

$0 \times 8 =$ ___

$0 \times 5 =$ ___

Pattern: ______________________________

Explain why your pattern makes sense.

____

____

Do you think this is true for any number $\times$ 0? Explain.

____

____

② Solve. What pattern do you notice?

$1 \times 4 =$ ___

$1 \times 8 =$ ___

$1 \times 5 =$ ___

Pattern: ______________________________

Explain why your pattern makes sense.

____

____

Do you think this is true for any number $\times$ 1? Explain.

____

____

Equal-Groups Number Stories

Home Link 2-6

NAME DATE TIME

Family Note Today your child practiced using efficient ways to solve equal-groups number stories, such as using repeated addition, skip counting, or using facts he or she knows. Children also talked about what multiplying by 0 or 1 means. Encourage your child to use the number stories to explain why multiplying by 0 equals 0 and multiplying by 1 equals the number in one group.

Please return this Home Link to school tomorrow.

Solve. Show your thinking using drawings, words, or number models.

A pack of Brilliant Color Markers contains 5 markers. Each pack costs $2.

① If you buy 6 packs, how many markers will you have?

Answer: ______________ (unit)

② How much do 0 packs of Brilliant Color Markers cost?

Answer: ______________ (unit)

Explain your answer. ______________________________

____

③ Make up a number story to match the number sentence below:

$1 \times 5 = 5$

____

____

Building Arrays Record Sheet

Lesson 2-7

NAME DATE TIME

	How many rows?	How many cubes in each row?	How many cubes in all?
1			
2			
3			
4			
5			
6			

Building Arrays Record Sheet

Lesson 2-7

NAME DATE TIME

	How many rows?	How many cubes in each row?	How many cubes in all?
1			
2			
3			
4			
5			
6			

Copyright © McGraw-Hill Education. Permission is granted to reproduce for classroom use.

Representing Situations with Arrays

Home Link 2-7

NAME DATE TIME

Family Note Today your child practiced drawing arrays to represent number stories. Your child also played *Array Bingo* to practice multiplication facts with arrays and equal groups.

Please return this Home Link to school tomorrow.

SRB 41-43

① There are 12 trombone players in a parade.
Show at least 3 different ways they can be arranged into arrays.
Show your work on the dot grids below.
Write a number model for each array.

Number model: ________

Number model: ________

Number model: ________

Number model: ________

② Can you make an array with 5 rows for the 12 players? Explain. ________

____

____

Picturing Division

Lesson 2-8

NAME DATE TIME

Solve each problem.
Use pictures and words to show your thinking.

① There are 20 children in art class. If 4 children can sit at each table, how many tables do they need?

_______ tables

② There are 20 children in music class. If 6 children can sit at each table, how many tables do they need?

_______ tables

Creating Mathematical Representations

Home Link 2-8

NAME DATE TIME

Family Note Your child is learning how to create mathematical representations, such as drawings, words, and number models, to help solve division problems. In this lesson we solved division problems with and without remainders. Talk to your child about the representations he or she can use to help solve Problems 1 and 2 and how to handle the remainder in Problem 2.

Please return this Home Link to school tomorrow.

Solve. Show your thinking in a drawing or number model.

① Amit won a pack of 24 stickers in a school contest.
He put the stickers into 3 equal piles, one for himself and one each for his friends, Danny and Sue. How many stickers will each get?

Answer: Each gets _______ stickers.

② Parents are organizing a field trip to the zoo for Amit's class.
They want to take the 23 children in their cars.
If each car can carry 5 children, how many cars are needed?

Answer: _______ cars are needed for the field trip.

Dividing Strips

Lesson 2-9

NAME DATE TIME

There are 10 fruit strips to share equally among 4 children.

How many fruit strips will each child get?

Show how you would equally share the 10 fruit strips in two different ways.

One way:

Another way:

Explain how both of your answers show the same amount of fruit strips for each child.

____

____

____

____

____

Sharing Equally

Lesson 2-9

NAME ______ DATE ______ TIME ______

For each problem, write the number of counters and the number of people sharing the counters.

① I shared ______ counters equally among ______ people.

Each person got ______ counters. There were ______ counters left over.

Number model: ______________________

---

② I shared ______ counters equally among ______ people.

Each person got ______ counters. There were ______ counters left over.

Number model: ______________________

---

③ I shared ______ counters equally among ______ people.

Each person got ______ counters. There were ______ counters left over.

Number model: ______________________

---

④ I shared ______ counters equally among ______ people.

Each person got ______ counters. There were ______ counters left over.

Number model: ______________________

---

⑤ I shared ______ counters equally among ______ people.

Each person got ______ counters. There were ______ counters left over.

Number model: ______________________

Modeling with Division

Home Link 2-9

NAME DATE TIME

Family Note Today your child solved equal-sharing number stories. Sometimes when we share or divide a quantity, there are parts left over, or remainders. Your child practiced recording division number models with remainders. For example, 10 marbles shared 3 ways could be recorded as $10 \div 3 \rightarrow 3$ R 1, which can be read as "10 divided by 3 gives us 3 with a remainder of 1." Help your child solve the problems below. You may want to use counters, such as coins or dry pasta, to act out each story.

Please return this Home Link to school tomorrow.

Draw pictures to show someone at home how you can use division to solve number stories. Write a number model for each story.

1. Jamal gives 24 marbles to 4 friends. Each friend gets the same number of marbles. How many marbles does each friend get? ______ (unit)

   How many marbles are left over? ______ (unit)

   Number model: ______

2. Eliza has 29 flowers to arrange in 5 vases. She puts the same number of flowers in each vase. How many flowers does she put in each vase? ______ (unit)

   How many flowers are left over? ______ (unit)

   Number model: ______

3. A sheet of stamps has 46 stamps. A complete row has 10 stamps. How many complete rows are there? ______ (unit)

   How many stamps are left over? ______ (unit)

   Number model: ______

Exploring Equal Shares

Lesson 2-10

NAME DATE TIME

Use counters and quarter-sheets of paper to solve each problem. Record your work in the rectangles.

Example:

Nomi had 8 crayons. She gave the crayons to 4 of her friends. Each friend got the same number of crayons. Draw the number of crayons each friend got.

① Latrell shares 10 marbles with his best friend. Draw the number of marbles each child has.

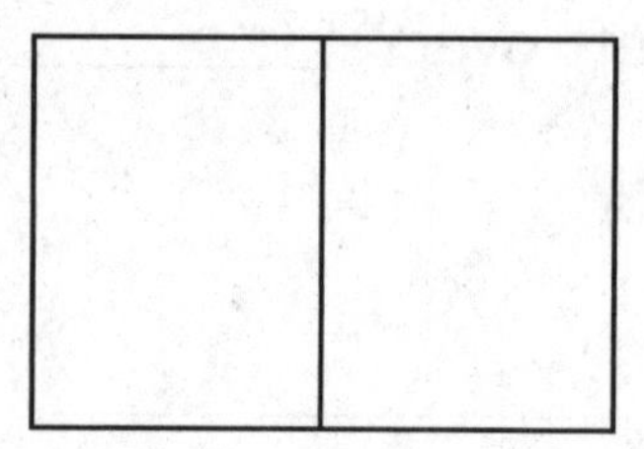

② Melissa has 6 bags of treats for her birthday party. She has a total of 12 treats in her bags. Draw the number of treats in each bag.

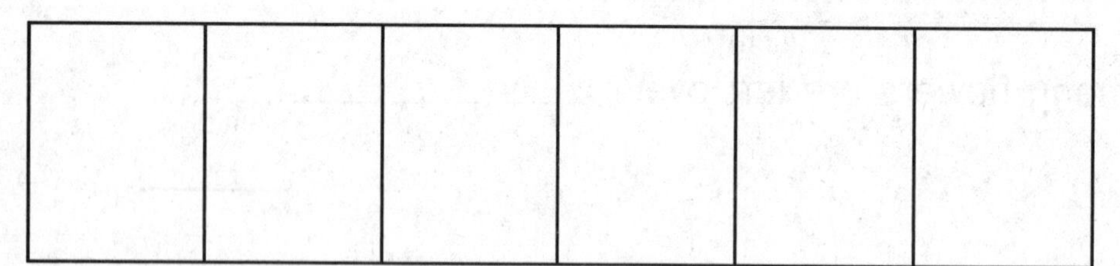

③ Make up your own problem. Draw your solution.

Division with Arrays

Home Link 2-10

NAME DATE TIME

Family Note Today your child practiced using arrays to model problems and show division with and without remainders. Children also learned a new game called *Division Arrays.*

Please return this Home Link to school tomorrow.

Use arrays to represent each division problem. If there is a remainder, show it in the Leftovers column.

	Problem	Sketch of Array Formed	Leftovers
Example	$23 \div 6$	●●● ●●● ●●● ●●● ●●● ●●●	●●●●●
1.	$15 \div 3$		
2.	$32 \div 5$		

③ List household items you could share with your family members that might have leftovers, for example, spoons, plates, and cups.

____

Practice

④ $5 \times 5 =$ _____

⑤ $40 = 5 \times$ _____

⑥ $20 \div 5 =$ _____

⑦ $45 \div$ _____ $= 5$

Two-Rule Frames and Arrows

Lesson 2-11

NAME DATE TIME

Solve the Frames-and-Arrows problems.

1. 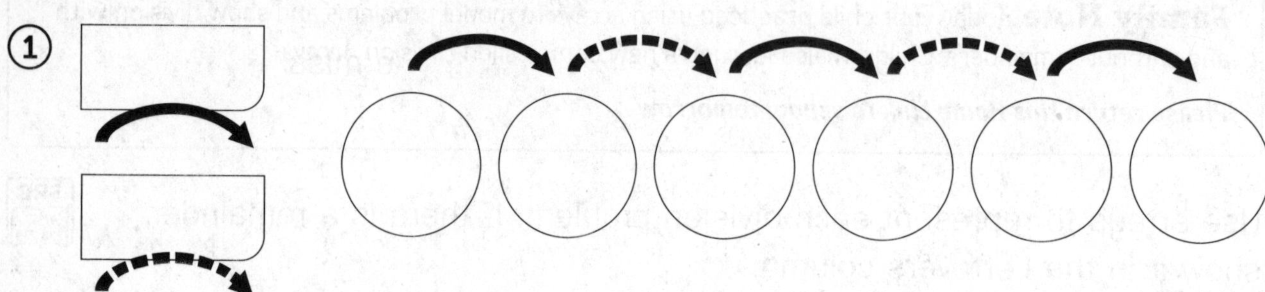

2. 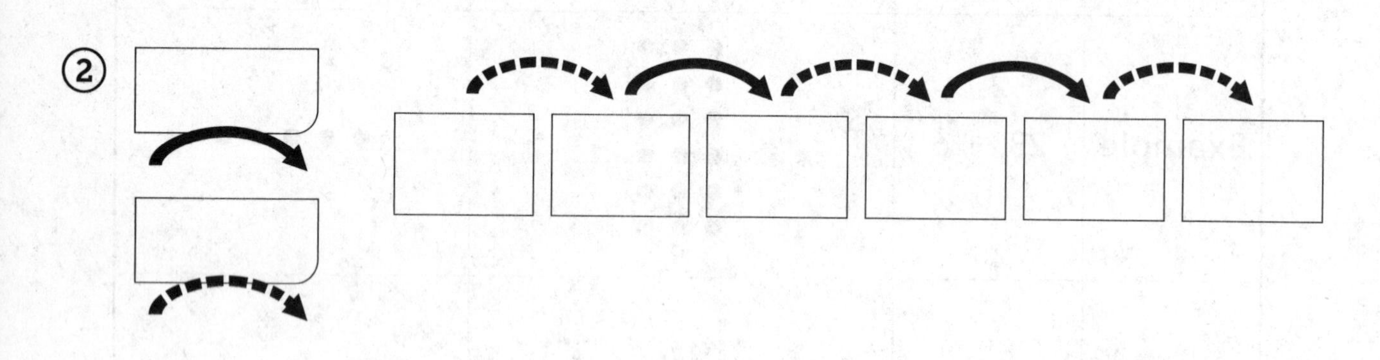

3. 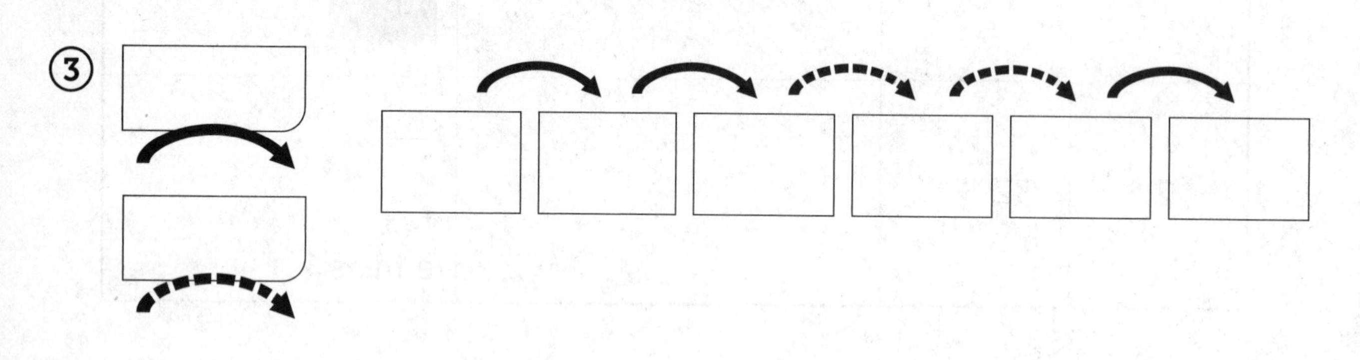

4. 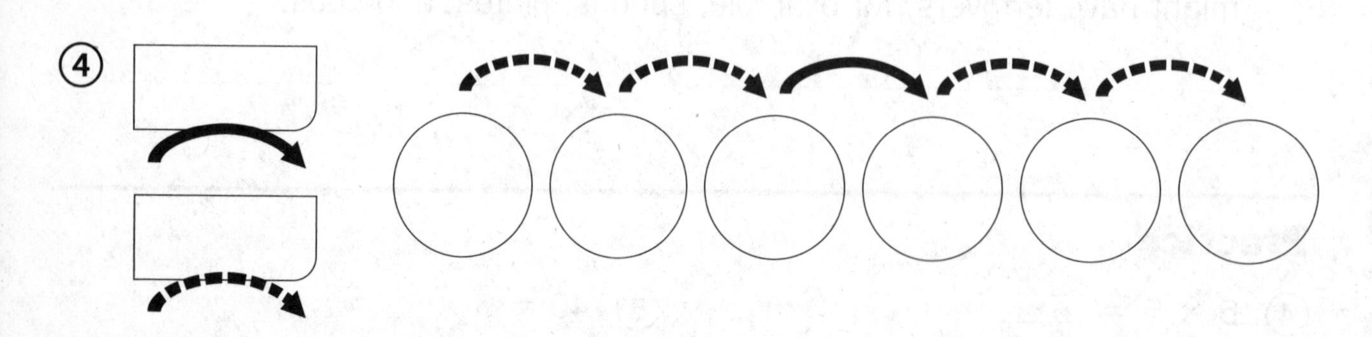

Number Patterns

Lesson 2-11

NAME DATE TIME

Math Message

For each problem, find the pattern and use it to find the missing numbers.

① 37, 40, 43, ______, ______, ______

② 27, 25, ______, 21, ______, ______

③ 1, 2, 4, 8, ______, ______

④ ______, ______, 36, 33, ______, 27

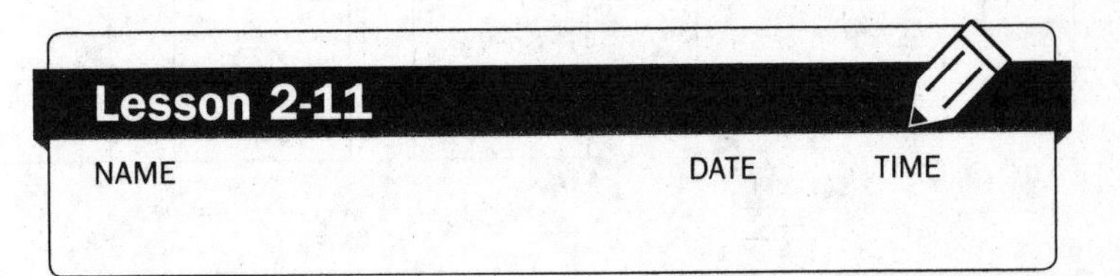

Lesson 2-11

NAME DATE TIME

Number Patterns

Math Message

For each problem, find the pattern and use it to find the missing numbers.

① 37, 40, 43, ______, ______, ______

② 27, 25, ______, 21, ______, ______

③ 1, 2, 4, 8, ______, ______

④ ______, ______, 36, 33, ______, 27

Copyright © McGraw-Hill Education. Permission is granted to reproduce for classroom use.

Frames and Arrows

Lesson 2-11

NAME DATE TIME

1. **Rule**

   × 2

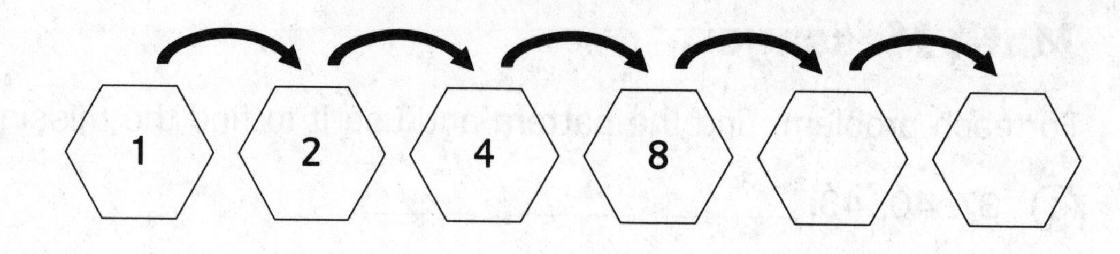

---

2. **Rule**

   − 2

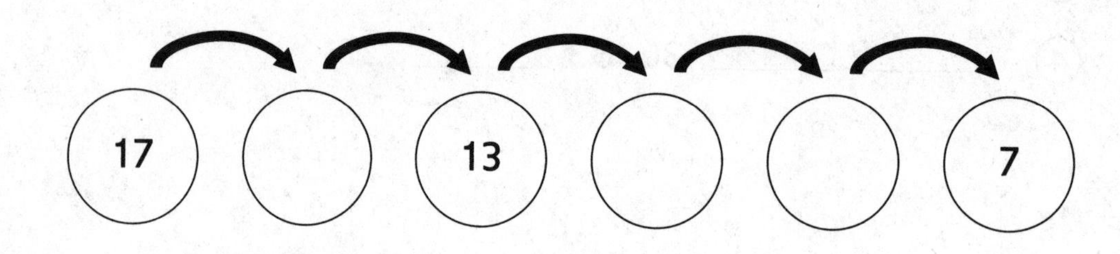

---

3. **Rule**

   ÷ 2

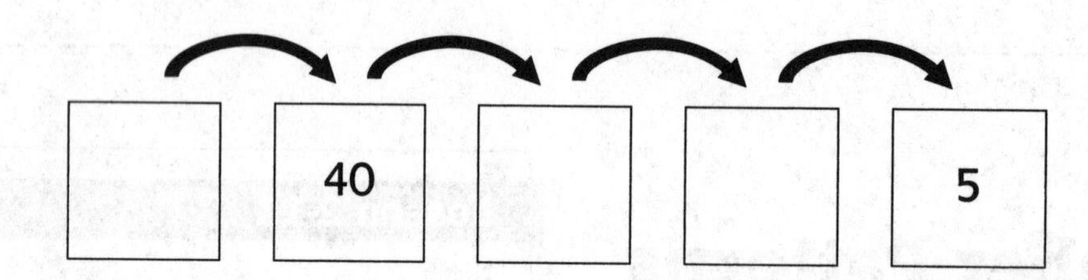

---

4. **Rule**

   − 5

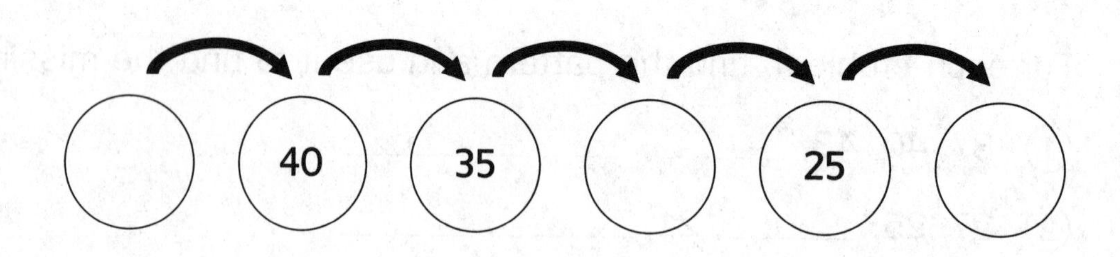

---

5. **Rule**

   + 10

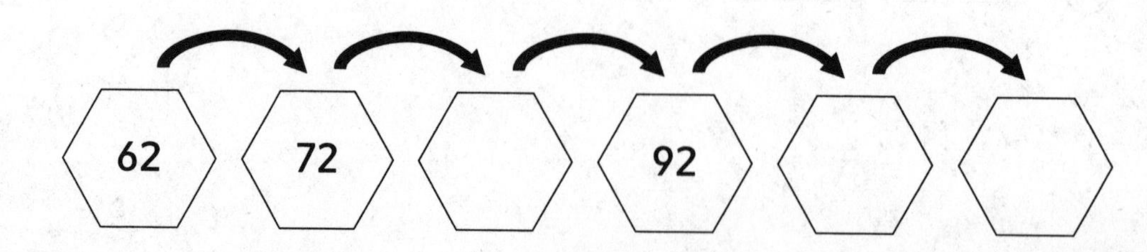

Frames and Arrows

Home Link 2-11

NAME DATE TIME

Family Note Today your child reviewed Frames and Arrows, which provide opportunities to look for addition, subtraction, multiplication, or division patterns. Your child used the patterns to fill in missing rules and blank frames.

Please return this Home Link to school tomorrow.

Show someone at home how to complete these Frames-and-Arrows diagrams.

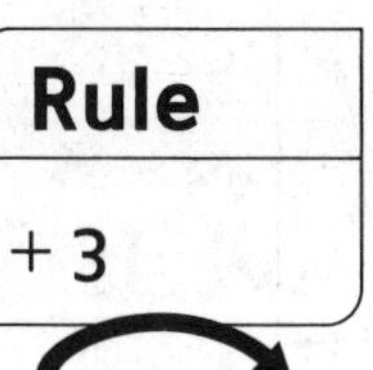

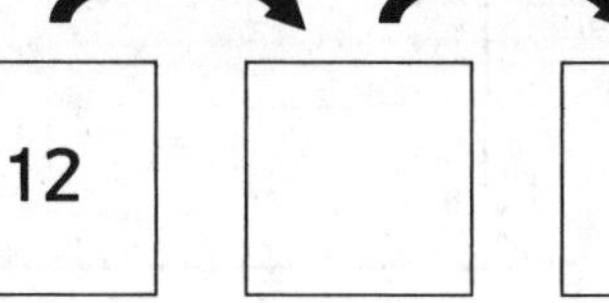

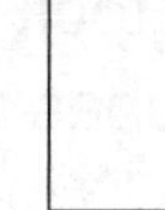

① Rule: + 3

12 → ____ → ____ → ____ → 24 → ____

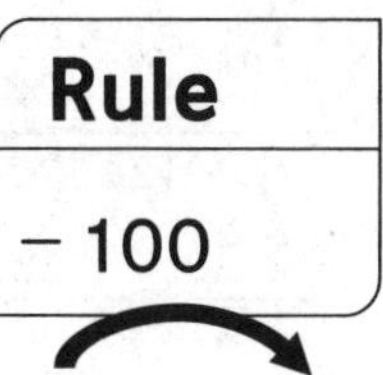

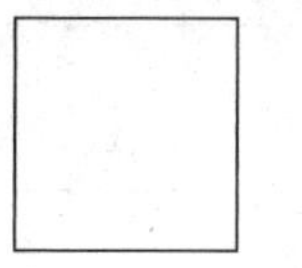

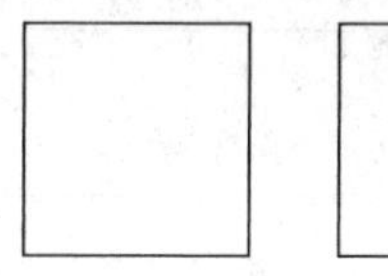

② Rule: − 100

1,000 → ____ → 800 → ____ → ____ → ____

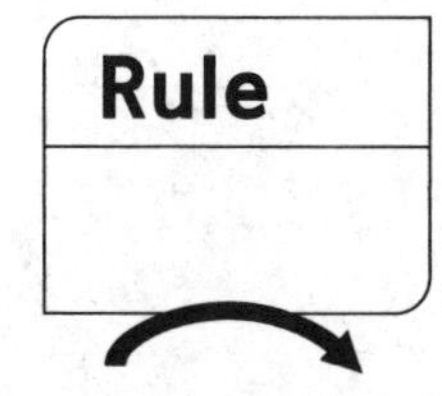

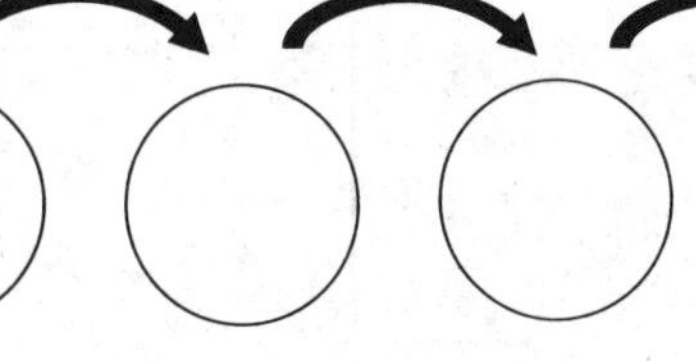

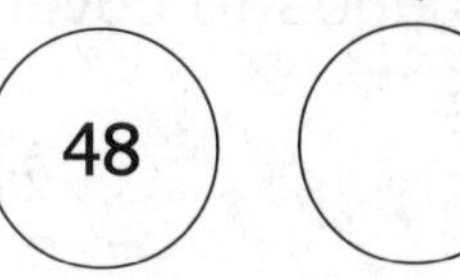

③ Rule: ____

24 → ____ → ____ → 42 → 48 → ____

Practice

Solve.

④ ______ = 6 × 5

⑤ ______ = 6 × 10

⑥ 5 × ______ = 20

⑦ 10 × ______ = 40

Estimating Area

Lesson 2-12

NAME DATE TIME

Guess how many centimeter cubes are needed to completely cover each square with no gaps or overlaps. Completely cover each square with centimeter cubes to check your guess.

① I think it will take about ___________ centimeter cubes to cover this square.

It took ___________ centimeter cubes to cover this square.

The area is ___________ sq cm.

How did you count the number of cubes used to cover this square?

____

____

② I think it will take about ___________ centimeter cubes to cover this square.

It took ___________ centimeter cubes to cover this square.

The area is ___________ sq cm.

How did you count the number of cubes used to cover this square?

____

____

Letter Areas

Lesson 2-12

NAME DATE TIME

Find the area of each letter.

① 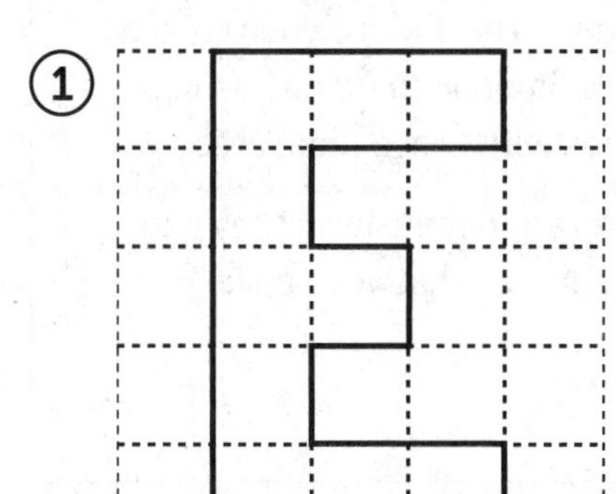

Area = ___ sq cm

② 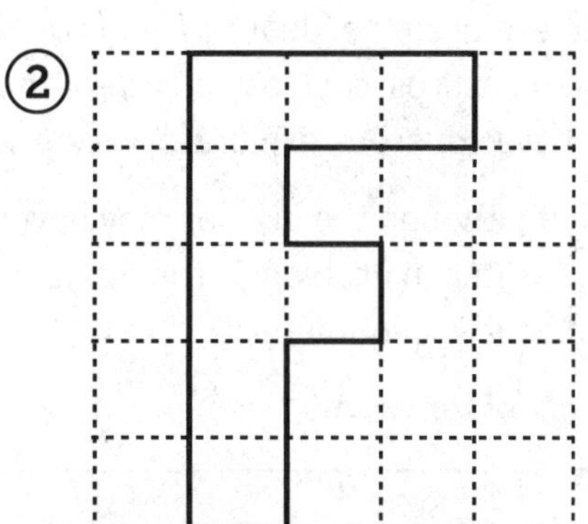

Area = ___ sq cm

③ 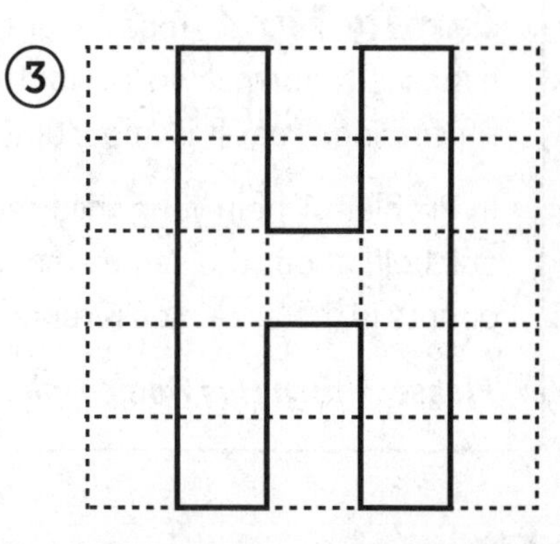

Area = ___ sq cm

④ 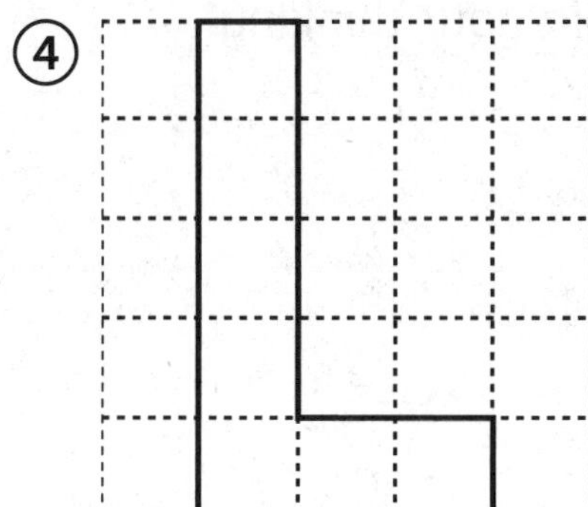

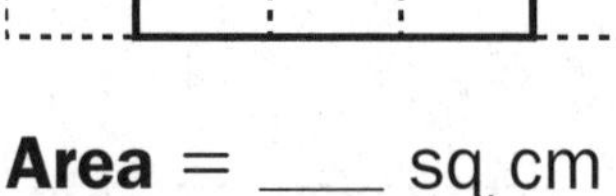

Area = ___ sq cm

⑤ 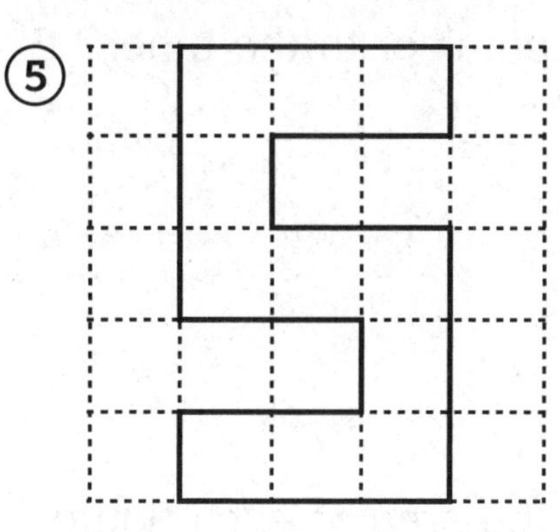

Area = ___ sq cm

⑥ 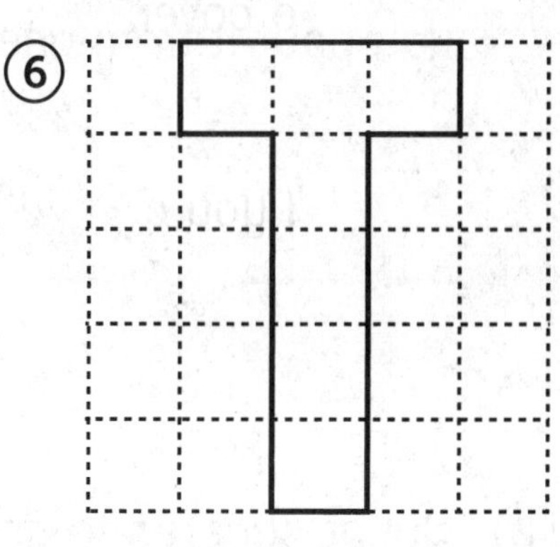

Area = ___ sq cm

Try This

An 8-by-8 checkerboard has 64 squares. Some squares on a checkerboard are white. Some are black. Squares of the same color are never next to each other.

⑦ How many white squares are in each row? ___

How many black squares? ___

⑧ If the square in one corner is black, what color is the square in the diagonal corner?

___

Liquid Volume and Area

Home Link 2-12

NAME DATE TIME

Family Note Today your child explored the ideas of *liquid volume* and *area*. Before your child is exposed to formal work with these measures (such as equivalent units of **liquid volume** or formulas for finding area), it is important to have concrete, exploratory experiences with these measures.

In Problem 1, help your child see that although the glasses may have different dimensions, they can still hold about the same amount of water. In Problem 2, the number of squares that your child counts is the area measurement in square centimeters.

Please return this Home Link to school tomorrow.

① Pour some water into a cup at home. Pour all the water from the cup into a bowl. Does the volume or amount of liquid change when you pour it from one container to the other? Explain your thinking.

② Count squares to find the area of each figure.

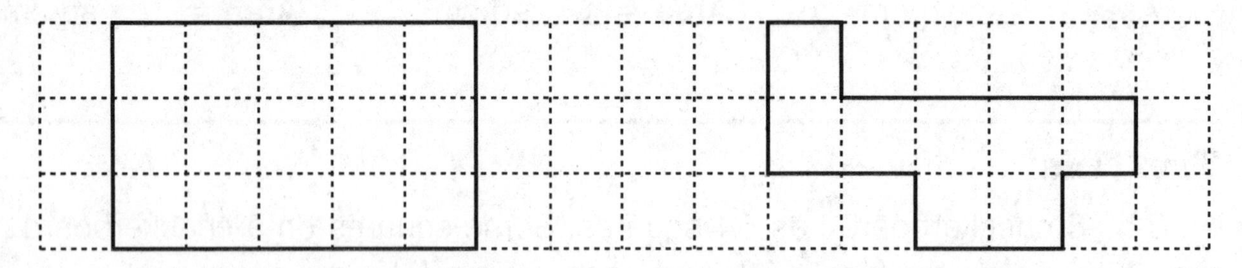

_______ square centimeters _______ square centimeters

Practice

③ $6 \times 2 =$ _______

④ $14 = 2 \times$ _______

⑤ _______ $= 18 \div 2$

⑥ $16 \div$ _______ $= 8$

Unit 3: Family Letter

Home Link 2-13

NAME DATE TIME

Operations

In Unit 3, your child will add, subtract, multiply, and divide whole numbers using a variety of problem-solving strategies and computational skills. *Everyday Mathematics* encourages children to choose from any of the methods explored in this unit, or invent their own computation methods. When children create and share their own ways of computing instead of simply learning one method, they begin to realize that problems can be solved in more than one way. They are more willing and able to take risks, think logically, and produce more reasonable answers.

In Unit 3, children will:

- Describe rules for patterns and use them to solve problems.
- Estimate to check whether their answers are reasonable.
- Add using the partial-sums and column-addition methods. Subtract using the counting-up and expand-and-trade methods.

	Tens	Ones
	6	7
+	2	5
	8	12
	9	2

$67 + 25 = 92$

column addition

$$\begin{array}{rcl} & & \quad\ \ 70 \quad 14 \\ 184 & \rightarrow & 100 + \cancel{80} + \cancel{4} \\ -\ 37 & \rightarrow & \qquad\ \ 30 + 7 \\ \hline & & 100 + 40 + 7 = 147 \end{array}$$

expand-and-trade subtraction

- Use helper facts and create arrays to solve unknown multiplication facts.
- Learn helpful rules and new groups of multiplication facts.
- Find and write equivalent names for numbers within name-collection boxes.
- Collect and organize data in scaled bar and picture graphs.

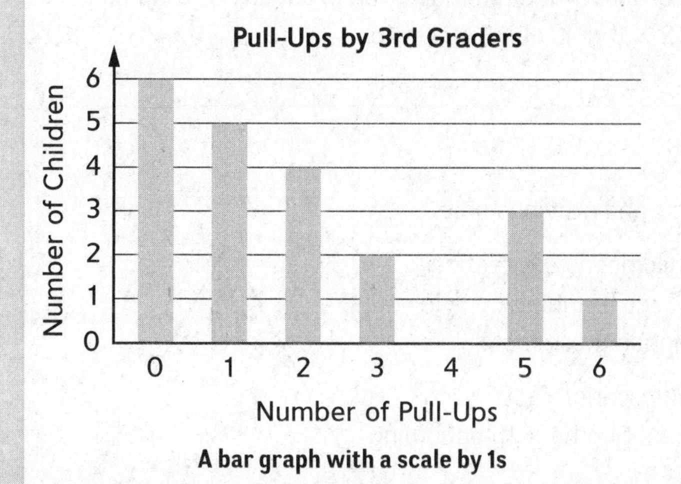

A bar graph with a scale by 1s

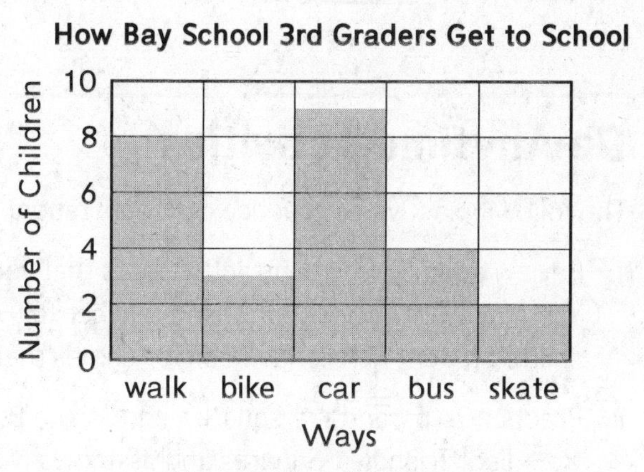

A bar graph with a scale by 2s

Please keep this Family Letter for reference as your child works through Unit 3.

Vocabulary

Important lesson components and terms used in Unit 3:

bar graph A graph with horizontal or vertical bars that represent data. Sometimes when the scale shows counts other than by 1s, the graph is called a scaled bar graph.

column addition An addition strategy in which the addends' digits are first added in each place-value column separately and then 10-for-1 trades are made until each column has only one digit. Lines may be drawn to separate the place-value columns.

estimate 1. An answer close to an exact answer. 2. To make a guess based on information you have. Some ways to estimate calculations include rounding the numbers in the problem to the nearest 10 or 100 or changing them to close-but-easier numbers. For example, to estimate the sum 245 + 92, one might calculate 200 + 100 = 300 or 245 + 100 = 345.

expanded form A way of writing a number as the sum of the values of each digit. For example, 356 is 300 + 50 + 6 in expanded form.

function machine An imaginary device that receives inputs and pairs them with outputs using a rule.

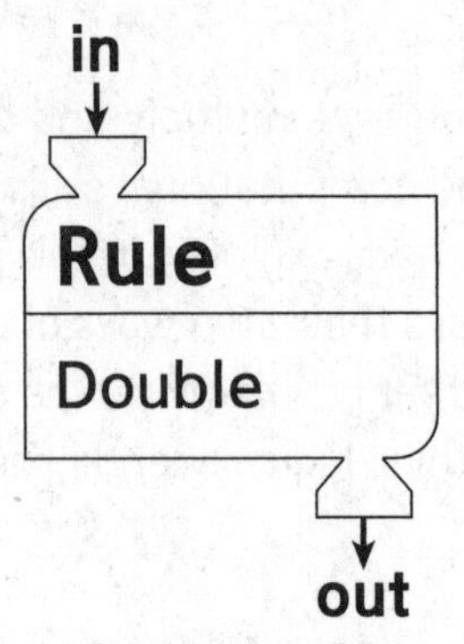

in	out
1	2
3	6
5	10
10	20
100	200

A function machine

helper facts Well-known facts used to help figure out less familiar facts.

input A number inserted into a function machine that applies a rule to pair the input with an output.

output A number paired to an input by a function machine applying a rule.

partial-sums addition An addition method in which separate sums are computed for each place value of the numbers and then added to get a final sum.

picture graph A graph constructed with picture symbols. Sometimes when a symbol represents more than one, the graph is called a scaled picture graph.

turn-around rule for multiplication A rule that says two numbers can be multiplied in either order without changing the product. For example, 2 × 8 = 16 and 8 × 2 = 16.

Do-Anytime Activities

The following activities practice concepts taught in this and previous units.

1. Review addition and subtraction facts that your child needs to practice. You may want to ask your child's teacher for +, − Fact Triangles. Look in your child's *Student Reference Book* for games that practice these facts.
2. Practice 2s, 5s, and 10s, squares, and 3s and 9s multiplication facts using ×, ÷ Fact Triangles. Squares and 3s and 9s will be sent home with upcoming Home Links.

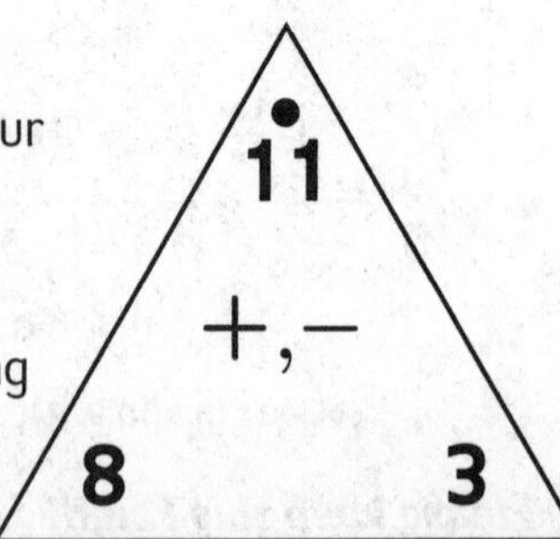

3. When adding or subtracting multidigit numbers, talk about which strategy works best for your child. Try not to impose the strategy that works best for you! Have your child make an estimate for each problem and discuss why the answer is reasonable. Here are some problems to try:

$267 + 743 =$ ________ $794 - 554 =$ ________

________ $= 851 + 697$ $840 - 694 =$ ________

Building Skills through Games

In Unit 3 your child will practice multiplication and mental addition by playing the following games. For detailed instructions, see the *Student Reference Book.*

Array Bingo Players make a 4-by-4 array of array cards. They draw number cards and try to match them to an array card with that number of dots. If there is a match, they turn the array card facedown.

8

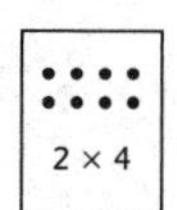

Multiplication Draw Players draw two cards and multiply the numbers on them. They add the products of 5 draws to try to get the largest sum.

Name That Number Players try to name a target number by adding, subtracting, multiplying, or dividing the numbers on two or more of five cards.

Roll to 1,000 Players mentally add the results of dice rolls.

Shuffle to 100 Players estimate to find sums close to 100.

As You Help Your Child with Homework

As your child brings home assignments, you may want to go over the instructions together, clarifying them as necessary. The answers listed below will guide you through this unit's Home Links.

Home Link 3-1

1.

in	out
14	7
7	0
12	5
15	8
10	3
21	14

2.

in	out
1	5
5	25
4	20
6	30
2	10

3.

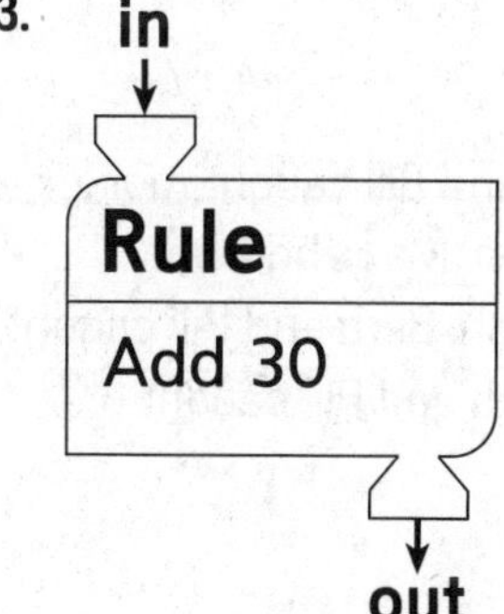

in	out
70	100
20	50
30	60
90	120
50	80
Answers vary.	

Home Link 3-2

1. Sample answer: 80 and 40; 120
2. Sample answer: 200 and 400; 400

Home Link 3-3

1. 337 2. 339 3. 562 4. 574

Home Link 3-4

Sample answer: $90 + 30 = 120$; 115

Home Link 3-5

Representations vary.

1. 202

Sample estimate: $500 - 300 = 200$

2. 122

Sample estimate: $300 - 200 = 100$

Home Link 3-6

1. 194 2. 202 3. 122 4. 206

Home Link 3-7

1. 2 2. 2 3. 4
4. Answers vary.

Home Link 3-8

1. 2 2. 8 3. 6 4. 7
5. Sample answer: Beth and Bill caught more fish than Max and Chen. Max and Chen caught 7 and 8, or 15 fish. Beth and Bill caught 11 and 10, or 21 fish. Beth and Bill caught 6 more fish.

Home Link 3-9

5. 817

Estimate: $50 + 770 = 820$

6. 954

Estimate: $360 + 600 = 960$

Home Link 3-10

1. $6 \times 2 = 12$ 2. $3 \times 5 = 15$
3. Sample answer: Each array shows the same number of dots, but the rows and columns are switched. I can also switch the numbers around when I multiply.
4. Answers vary.

Home Link 3-11

1.
```
× × × × × × ×  35
× × × × × × ×
× × × × × × ×
× × × × × × ×
× × × × × × ×
o o o o o o o
```
2. 42

Home Link 3-12

1a.
```
× × × ×
× × × ×
× × × ×
× × × ×
× × × ×
× × × ×
× × × ×
× × × ×
× × × ×
× × × × (crossed out)
```
b. 36

c. Sample answer: I know $10 \times 4 = 40$, so I start with 40 and subtract 1 group of 4. So $40 - 4 = 36$.

2. 90 3. 45 4. 40 5. 80

Home Link 3-13

1.

18 Sample answers:

$9 + 9$ 2×9

$6 + 6 + 6$ HHH HHH HHH ///

dieciocho $4 \times 5 - 2$ $36 \div 2$

number of days in two weeks + 4 days, one and a half dozens

2. 12
3. Answers vary.

"What's My Rule?"

Home Link 3-1

NAME DATE TIME

Family Note Today your child practiced adding, subtracting, multiplying, and dividing in "What's My Rule?" problems. Children were introduced to the "What's My Rule?" routine in *Kindergarten Everyday Mathematics.* In *First* and *Second Grade Everyday Mathematics,* they continued to use the routine to practice addition and subtraction. You can find an explanation of function machines and "What's My Rule?" tables on pages 74-75 in the *Student Reference Book.* Ask your child to explain how they work. Help your child fill in all the missing parts for these problems.

Please return this Home Link to school tomorrow.

SRB 74-75

Complete the "What's My Rule?" problems. Make up problems of your own for the last table. Explain to someone how you figured out the *in* and *out* numbers.

①

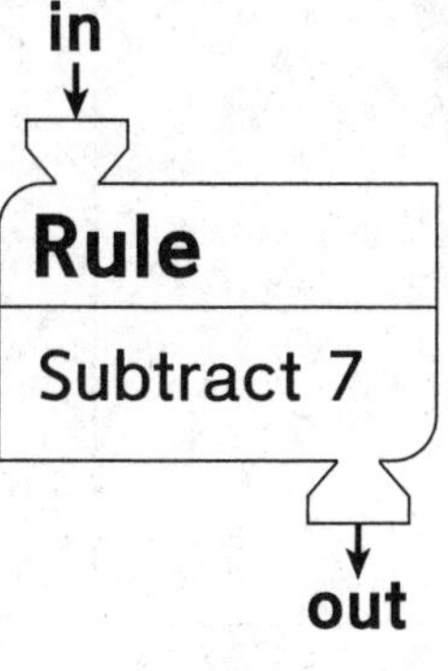

in	out
14	
7	
12	
15	
10	
21	

②

in

Rule

Multiply by 5

out

in	out
1	5
	25
4	
	30
2	

③

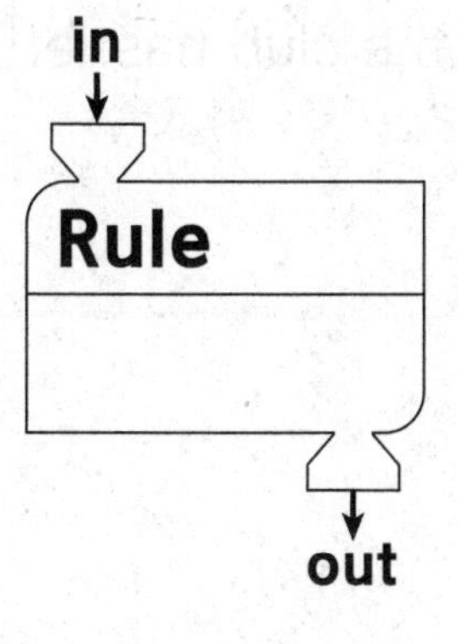

in	out
70	100
20	
	60
90	120
50	

④

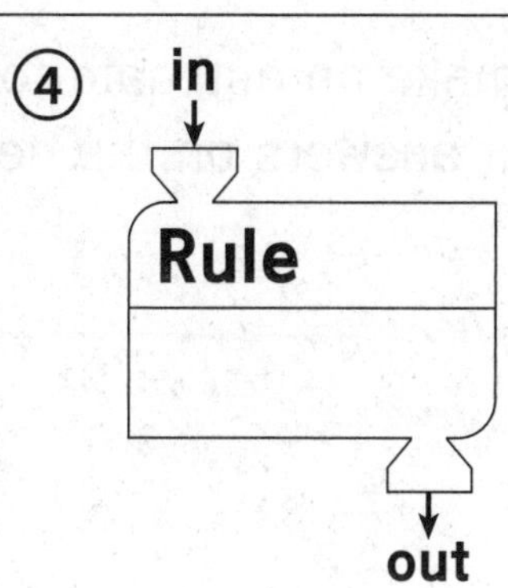

in	out

Estimating Costs

Lesson 3-2

NAME DATE TIME

Ann and her friend are walking home from school. Ann shows her friend this homework problem and her answer.

Name *Ann*

The art club has **$100** to spend on materials. They bought watercolor paints for $38 and brushes for $19.

$38

$19

How much money does the club have left over?

$67

① Suppose that you are Ann's friend and you do not have a pencil or a calculator with you.

In your head, make an estimate for how much money the club has left over. Write your answers on the next page.

Estimating Costs

(continued)

NAME DATE TIME

(2) Write your estimate for how much money the club has left over.

$_______

(3) Use your estimate to decide whether Ann's answer is reasonable. Is Ann's answer reasonable? _______

(4) Show your thinking and how you used an estimate in the thought bubble below.

Rubric for Estimating Costs

Lesson 3-2

NAME DATE TIME

Goal: Explain your mathematical thinking clearly and precisely.				
Meets Expectations	**Child 1**	**Child 2**	**Child 3**	
Shows the close-but-easier numbers used to make estimates.				
Shows how to make an estimate of how much money the art club has left over.				
Exceeds Expectations				
Explains that Ann's answer of $67 is NOT reasonable because it is not close to an estimated leftover amount of $40.				

Solving Problems with Estimation

Home Link 3-2

NAME DATE TIME

Family Note Today your child used close-but-easier numbers and estimation to solve problems. Ask your child to explain what a close-but-easier number is and when it might make sense to use an estimate rather than an exact answer. Using mental math in making estimates is important in everyday life and in *Everyday Mathematics.*

Please return this Home Link to school tomorrow.

① Use close-but-easier numbers to estimate the answer to this problem.

$$78 + 43 = ?$$

My close-but-easier numbers are ______________________.

My estimate is __________.

② At their October meeting, the school's book club set a goal for its members to read 1,000 books before the end of the year. In October the book club read 221 books, and in November they read 387 books. Without using a pencil and paper, use close-but-easier numbers to make an estimate of about how many books the club will need to read in December to reach its goal.

My close-but-easier numbers are ______________________.

The club needs to read about __________ books in December.

Partial-Sums Addition

Home Link 3-3

NAME DATE TIME

Family Note Today your child learned about adding 3-digit numbers using partial-sums addition. Your child may choose to use partial-sums addition or may prefer a different method.

Please return this Home Link to school tomorrow.

Solve each addition problem. You may want to use partial-sums addition. Use an estimate to check that your answer makes sense. Write a number model to show your estimate.

① Estimate: ______	② Estimate: ______
$\begin{array}{r} 245 \\ +\ 92 \\ \hline \end{array}$	$\begin{array}{r} 124 \\ +215 \\ \hline \end{array}$
③ Estimate: ______	④ Estimate: ______
$\begin{array}{r} 245 \\ +317 \\ \hline \end{array}$	$\begin{array}{r} 366 \\ +208 \\ \hline \end{array}$

Using Addition Strategies

Lesson 3-4

NAME DATE TIME

George and Tina used different strategies to solve 138 + 243.
See if you can figure out how they each solved the problem.

George's Strategy
138 + 243 = ?
138 + 200 = 338
338 + 40 = 378
378 + 3 = 381
138 + 243 = 381

Tina's Strategy
138 + 243 = ?
138 + 2 = 140
140 + 243 = 383
383 − 2 = 381
138 + 243 = 381

Talk to your partner about how George's strategy works.
Then talk about how Tina's strategy works.

Use either George's or Tina's strategy to solve the problems below.

255 + 139 = ?	127 + 367 = ?
Estimate: ______________	Estimate: ______________

Adding to Solve Number Stories

Lesson 3-4

NAME DATE TIME

Use the information on *Student Reference Book,* page 272 to solve each number story.

Write an addition number model for each problem. Use a ? for the unknown.

Then show your addition strategy in the space below each problem.

① Teddy drives with his mother from Oklahoma City to Des Moines and from Des Moines to Minneapolis. How many miles is their trip from Oklahoma City to Minneapolis?

Number model: ____________________

Answer: __________ miles

② Tony drives from Washington, D.C., to Cleveland on Friday. He drives back on Sunday. How many miles did Tony drive all together?

Number model: ____________________

Answer: __________ miles

③ Annabel drives with her sister from Los Angeles to Phoenix and from Phoenix to Albuquerque. How many miles is their trip from Los Angeles to Albuquerque?

Number model: ____________________

Answer: __________ miles

Multidigit Addition

Home Link 3-4

NAME DATE TIME

Family Note Today your child learned column addition, a strategy for adding multidigit numbers. Discuss the example with your child.

Please return this Home Link to school tomorrow.

Tell someone about column addition.

Example: 248 + 79 = ?

Estimate: *200 + 100 = 300*

	100s	10s	1s
	2	**4**	**8**
+		**7**	**9**
	2	11	17
	2	12	7
	3	**2**	**7**

Add each column of numbers.

There are two digits in the ones column, so trade 10 ones for 1 ten, then move 1 ten to the tens column.

There are two digits in the tens column, so trade 10 tens for 1 hundred, then move 1 hundred to the hundreds column.

248 + 79 = 327

For the problem below, estimate the sum. Then use column addition to solve. Show your work. Use your estimate to check whether your answer makes sense.

89 + 26 = ?

Estimate: ______________________

89 + 26 = __________

Subtracting on an Open Number Line

Lesson 3-5

NAME DATE TIME

Use counting-up subtraction to solve these problems. Show your work on the open number lines.

Example:

127 − 46 = ?

127 − 46 = 81

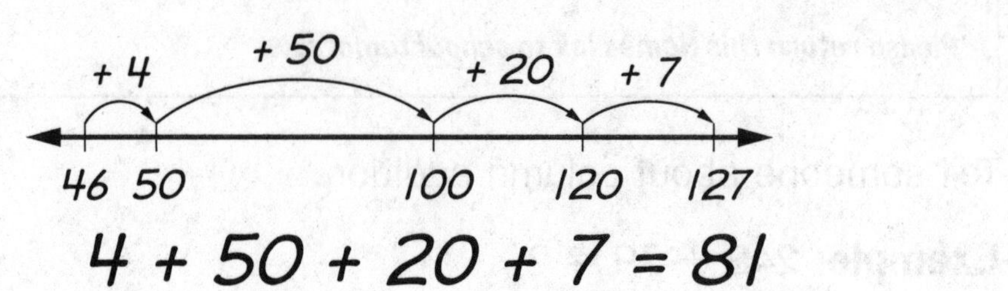

① 224 − 150 = ?

224 − 150 = ______

② ? = 346 − 97

______ = 346 − 97

③ ? = 432 − 278

______ = 432 − 278

Counting-Up Subtraction

Home Link 3-5

NAME DATE TIME

Family Note Today your child reviewed the counting-up method for subtraction. Discuss the example problem with your child.

Please return this Home Link to school tomorrow.

Explain counting-up subtraction to someone at home. Use it to solve Problems 1 and 2. Show what you did on an open number line or with number sentences. Compare your answers to your estimates to check whether your answers make sense.

Example: 468 − 274 = ?

Estimate: 500 − 300 = 200

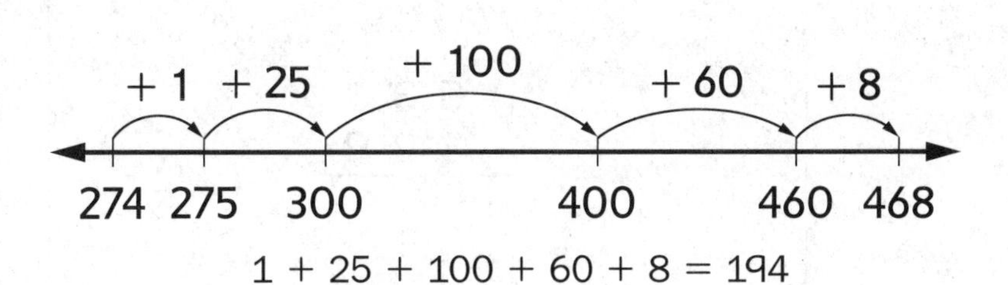

1 + 25 + 100 + 60 + 8 = 194

274 + 1 = 275
275 + 25 = 300
300 + 100 = 400
400 + 60 = 460
460 + 8 = 468

1 + 25 + 100 + 60 + 8 = 194

468 − 274 = 194

① 531 − 329 = ?	② 331 − 209 = ?
Estimate: ______	Estimate: ______
531 − 329 = ______	331 − 209 = ______

Expand-and-Trade Subtraction

Home Link 3-6

NAME DATE TIME

Family Note Today your child used expand-and-trade subtraction to find differences between 3-digit numbers. This method reinforces children's understanding of place value. Learning different strategies helps children think flexibly and apply strategies that make sense to them.

Please return this Home Link to school tomorrow.

Fill in the unit. Estimate and then solve the problems. You may use any strategy you like. Use your estimates to check that your answers make sense. On the back of this Home Link, explain how you solved one of the problems.

Unit

① Estimate: ______

$$\begin{array}{r} 468 \\ -\ 274 \\ \hline \end{array}$$

② Estimate: ______

$$\begin{array}{r} 531 \\ -\ 329 \\ \hline \end{array}$$

③ Estimate: ______

$$\begin{array}{r} 331 \\ -\ 209 \\ \hline \end{array}$$

④ Estimate: ______

$$\begin{array}{r} 653 \\ -\ 447 \\ \hline \end{array}$$

Making a Scaled Graph

Lesson 3-7

NAME DATE TIME

The table below shows the results of a survey about favorite zoo exhibits.

Birds	Mammals	Reptiles	Ocean Life
////	~~////~~ ~~////~~	~~////~~	~~////~~ ////

① Fill in the bar graph to show how many third graders chose each exhibit.

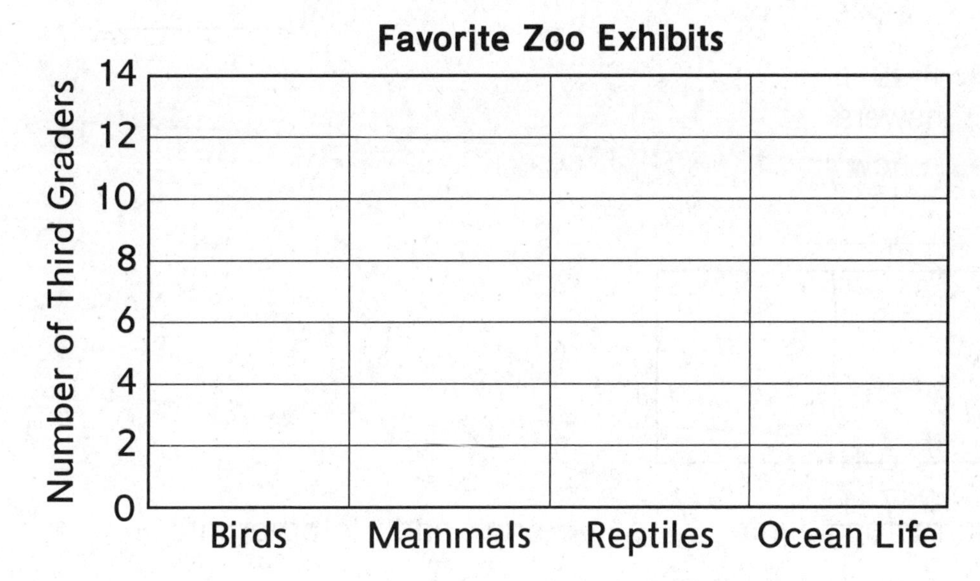

② Which exhibit is the most popular? ______________

How many votes did it get? ______________

③ Which is the least popular? ______________

How many votes did it get? ______________

④ How many more third graders like mammals than birds? ______________

⑤ How many more third graders like ocean life than reptiles? ______________

⑥ What is your favorite kind of exhibit? ______________

Partitioning Polygons

Lesson 3-7

NAME DATE TIME

Follow these directions for each shape in Problems 1 and 2.

- Draw a line segment to make two rectangles.
- Partition each rectangle into squares about this size: ☐
- Record the number of squares in each rectangle.
- Find the total number of squares in the shape.

Example:

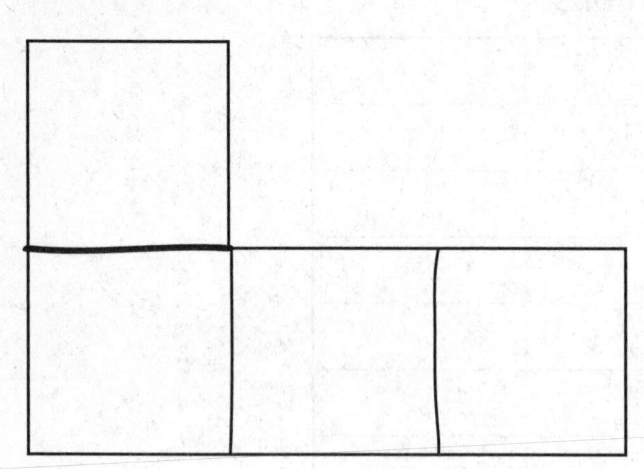

I partitioned one rectangle into __1__ square unit and the other into __3__ square units.

I partitioned the whole shape into __4__ square units.

①

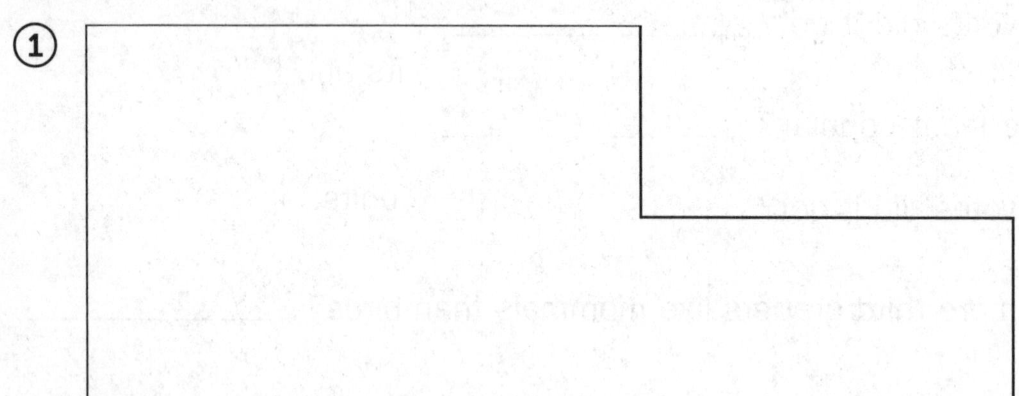

I partitioned one rectangle into ___ square units and the other into ___ square units.

I partitioned the whole shape into ___ square units.

Partitioning Polygons (continued)

Lesson 3-7

NAME DATE TIME

②

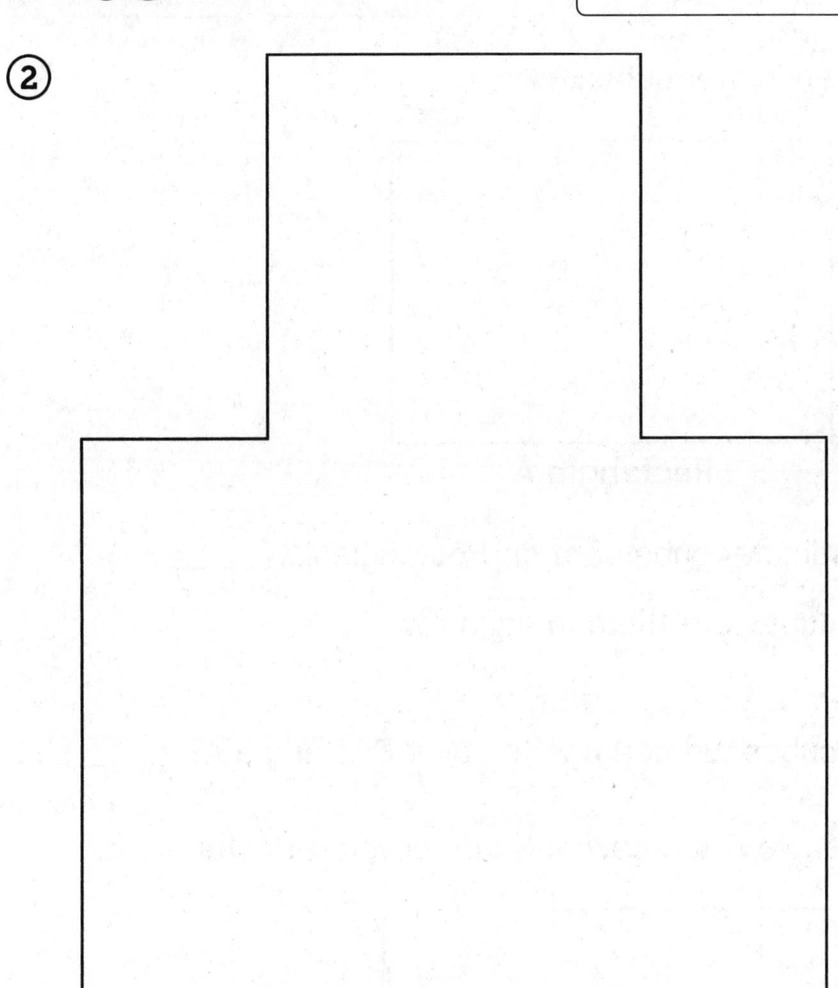

I partitioned one rectangle into _______ square units and the other into _______ square units.

I partitioned the whole shape into _______ square units.

Exploration C: Partitioning Rectangles

Lesson 3-7

NAME DATE TIME

① Tile Rectangle A with centimeter cubes.

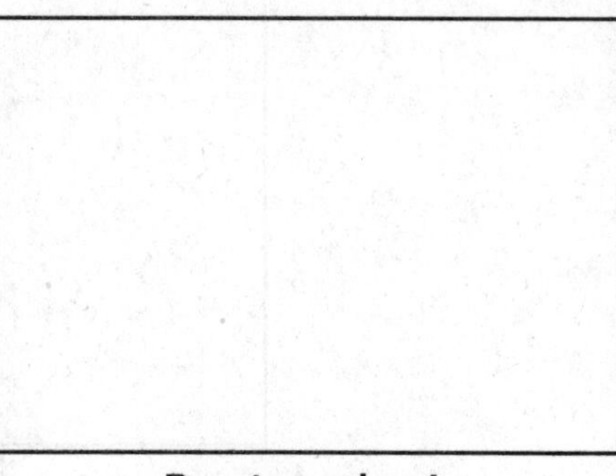

Rectangle A

How many rows of centimeter cubes are on Rectangle A? ______

How many centimeter cubes are there in each row on Rectangle A? ______

How many centimeter cubes did you use to cover Rectangle A? ______

② Draw squares on Rectangle B to show how you covered Rectangle A.

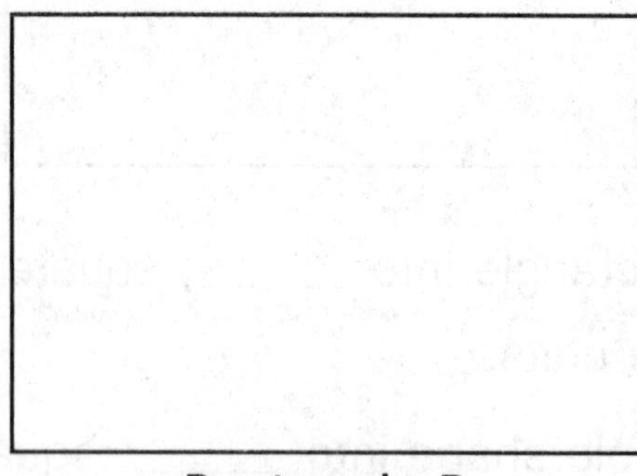

Rectangle B

How many squares did you draw? ______

Exploration C: Partitioning Rectangles (continued)

Lesson 3-7

NAME DATE TIME

③ Draw lines to partition the rectangle into 5 rows with 6 same-size squares in each row. You may use a square pattern block to help.

How many squares cover the rectangle? ______

Talk to a partner.

④ How did you figure out the total number of squares? ______

______

⑤ How are the rectangles in Problems 2 and 3 like arrays? ______

______

______

Scaled Bar Graph

Home Link 3-7

NAME DATE TIME

Family Note Today your child sorted pattern blocks and created a bar graph with a scale of more than 1 to represent the data. Scales on bar graphs should have equally spaced intervals to represent data, such as below, where the scale is marked in intervals of 2.

Please return this Home Link to school tomorrow.

Talk to someone at home about the data shown on the bar graph below. Then use the information shown on the graph to answer the questions.

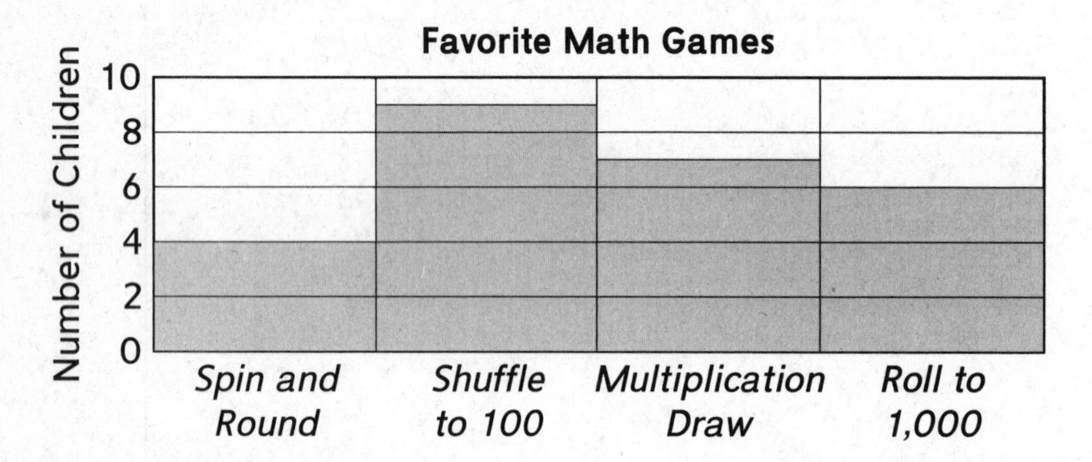

① How many more children chose *Shuffle to 100* than *Multiplication Draw*? ____

② How many more children chose *Roll to 1,000* than *Spin and Round*? ____

③ How many fewer children chose *Shuffle to 100* than the combined total of children who chose *Roll to 1,000* and *Multiplication Draw*? ____

④ Write your own question about the graph. Then write the answer.

____

____

____

____

Completing a Picture Graph

Lesson 3-8

NAME DATE TIME

Valeria's classmates counted the number of milk cartons they recycled each day during the week. Their results are shown in the tally chart.

Day of Week	Number of Milk Cartons
Monday	𝍸
Tuesday	|||
Wednesday	𝍸 |||
Thursday	||
Friday	𝍸 |

Use the data in the tally chart to finish the picture graph.

Number of Milk Cartons

Monday	
Tuesday	
Wednesday	
Thursday	
Friday	

Key: Each □ = 1 milk carton

1. Write something you know from the picture graph.

____

____

2. Write a question that can be answered using the data in the picture graph. Then answer your question.

____

____

Drawing a Picture Graph

Lesson 3-8

NAME DATE TIME

Justin's scout troop counted the number of white-tailed deer they saw each day during their camping trip. Their results are shown in the tally chart.

Day of Week	Number of Deer
Monday	////
Tuesday	//
Wednesday	~~////~~
Thursday	~~////~~ /
Friday	///

Use the data in the tally chart to finish the picture graph.

Number of Deer the Scout Troop Saw

Key: ◯ = 2 deer

Monday	
Tuesday	
Wednesday	
Thursday	
Friday	

Use information from the picture graph to write and answer two questions.

Interpreting a Picture Graph

Home Link 3-8

NAME DATE TIME

Family Note Today your child learned to read and draw picture graphs with a scale of more than one. The key on a picture graph shows a symbol that represents the scale.

Please return this Home Link to school tomorrow.

The picture graph shows how many fish each child caught on a fishing trip.

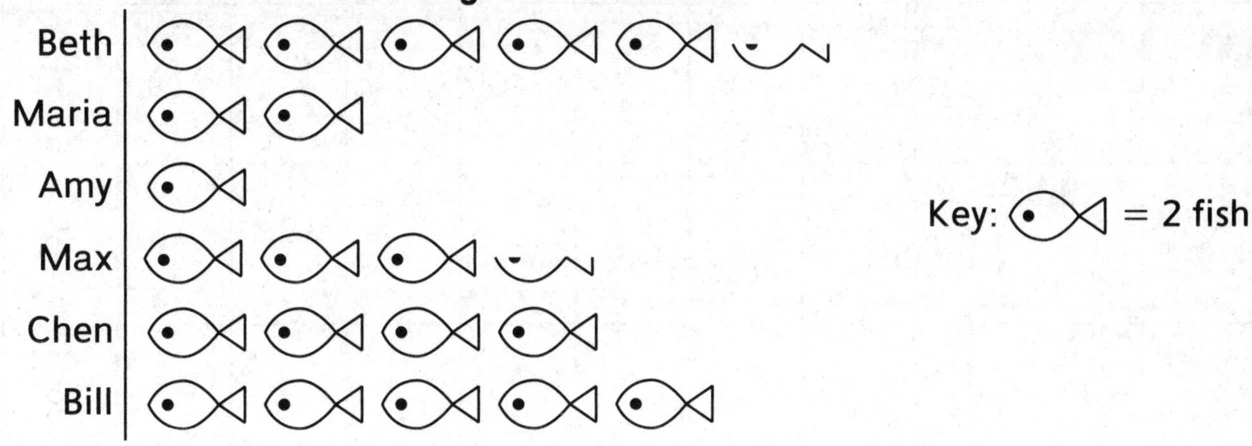

Use the graph to answer the questions.

1. How many fish did Amy catch? ____ fish
2. How many fish did Chen catch? ____ fish
3. How many more fish did Bill catch than Maria? ____ fish
4. Maria catches 3 more fish. Now how many has she caught in all? ____ fish

Revise the picture graph to show the number of fish Maria caught in all.

5. Did Chen and Max or Beth and Bill catch more fish? How many more? Explain your answer.

____

____

____

____

Rolling and Recording Squares Record Sheet

Lesson 3-9

NAME DATE TIME

1	4	9	16	25	36	49	64	81	100

Multiplication Squares

Home Link 3-9

NAME DATE TIME

Family Note Today your child learned about multiplication squares, such as $3 \times 3 = 9$ and $7 \times 7 = 49$. Help your child practice multiplication squares by completing the Rolling and Recording Squares activity below. If you don't have a 10-sided die, you will need a set of cards numbered 1 through 10, preferably two or more of each. You can use a regular deck of playing cards 2–10, using the aces as 1s.

Continue to help your child practice multiplication with the included Fact Triangles.

Please return this Home Link to school tomorrow.

Rolling and Recording Squares

Directions

1. Work with a family member.
2. Roll a 10-sided die (or draw a card) and make a multiplication square using that number as both factors.
3. Figure out the product. Shade the first open box above that product.
4. Take turns until one column is filled. (If drawing cards, reuse them.)

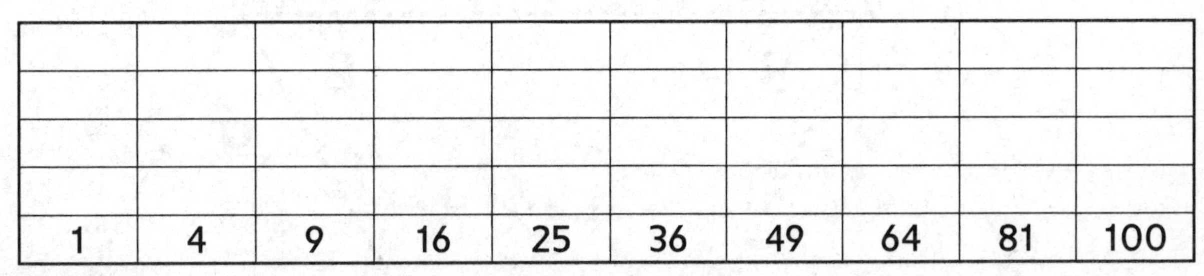

1	4	9	16	25	36	49	64	81	100

Practice

Fill in the unit box. Write these problems on the back of this page and show your work there. Write a number sentence for your estimate. Use any method you wish to solve each problem.

5. Estimate: ______________________

   49 + 768 = ______

6. Estimate: ______________________

   356 + 598 = ______

×, ÷ Fact Triangles: Multiplication Squares

NAME DATE TIME

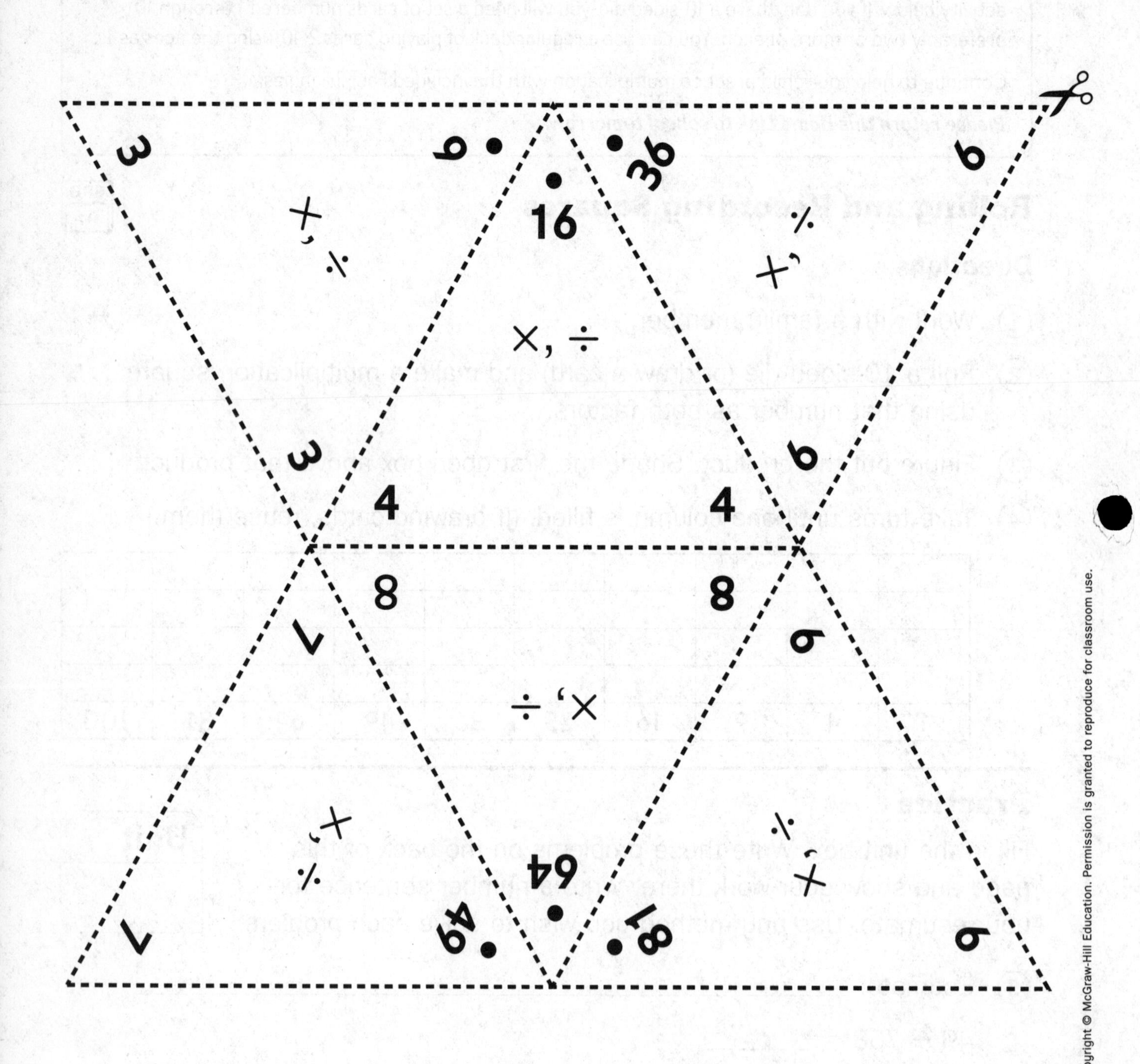

Exploring the Turn-Around Rule

Lesson 3-10

NAME DATE TIME

① Kelly says that $8 \div 2$ has the same answer as $2 \div 8$.

Matt says they do not because the turn-around rule does not work for division.

Who is correct? Why?

Use pictures and words to help explain your thinking.

② Does division have a turn-around rule? Explain your answer. ____________

____

____

____

____

____

Showing the Turn-Around Rule on a Facts Table

Lesson 3-10

NAME DATE TIME

Multiplication/Division Facts Table

+,÷		2	3	4	5	6	7		9	10
1	1	2	3	4	5	6	7		9	
2	2	4	6	8	10		14	16	18	20
3	3	6	9	12	15	18	21	24	27	30
4	4		12	16	20	24	28	32	36	40
5	5	10	15	20	25		35	40	45	50
6	6	12		24	30	36	42		54	60
		14	21	28		42	49	56	63	70
8	8	16	24	32	40	48	56	64	72	80
9	9	18	27	36	45	54	63	72	81	
10	10	20	30	40	50		70	80	90	100

The Turn-Around Rule for Multiplication

Home Link 3-10

NAME DATE TIME

Family Note Today your child explored the *turn-around rule for multiplication,* which says two numbers may be multiplied in either order and the product will remain the same. *For example:* $2 \times 5 = 10$ and $5 \times 2 = 10$. Knowing this rule can help children multiply more easily. Children also took inventory of the facts they can solve quickly and easily and those they still need to practice.

Please return this Home Link to school tomorrow.

Sketch an array to match each fact. Then sketch that array turned around. Record a number sentence to match the second array.

① $2 \times 6 = 12$ Number sentence: ______________

② $5 \times 3 = 15$ Number sentence: ______________

③ Use Problems 1 and 2 to tell someone why the turn-around rule works.

④ Choose a multiplication fact you need to practice. Write a strategy you can use to figure it out.

My fact: ________ × ________ = ________

Strategy: ______________________________

____

Solving Problems by Adding a Group

Lesson 3-11

NAME DATE TIME

You need number cards 0–10. Make two piles. Place 0, 2, 5, and 10 into one pile. Place the other cards in another pile.

For each problem, choose one card from each pile.

Use the 0, 2, 5, or 10 to fill in the first blank in the number story.

Use the other card to fill in the other blank.

Draw arrays next to the problems to show your work.

Solve to fill in the last blank.

1. **a.** Emma has _______ packages of markers.

   Each package has _______ markers.

   How many markers does Emma have? _______

   Number model: ______________________________

   b. Add one more package of markers.

   How many markers does Emma have now? _______

   Number model: ______________________________

2. **a.** Charlie has _______ bags of marbles.

   Each bag has _______ marbles.

   How many marbles does Charlie have? _______

   Number model: ______________________________

   b. Add one more bag of marbles.

   How many marbles does Charlie have now? _______

   Number model: ______________________________

Adding a Group

Home Link 3-11

NAME DATE TIME

Family Note Today your child learned another strategy for solving multiplication facts. Children used familiar facts, or helper facts, including the 2s, 5s, and 10s facts, to figure out facts they didn't know. Today your child learned the adding-a-group strategy. Children added a group to helper 2s and 5s facts to solve other facts. Eventually children will know all their multiplication facts, but in the meantime, practicing strategies such as adding a group helps them figure out facts they do not know and also supports their understanding of multiplication and its properties.

Please return this Home Link to school tomorrow.

Solve.

1. Jamila has 5 shelves of books with 7 books on each shelf. How many books does she have? Draw an array to show Jamila's books.

$5 \times 7 =$ ______

2. Jamila's sister gives her 7 more books to fill a new shelf. Now she has 6 rows of 7 books. Add a row of books to your array above. Then figure out how many books Jamila has now.

$6 \times 7 =$ ______

3. How did 5×7 help you figure out 6×7?

____

____

Solving Problems by Subtracting a Group

Lesson 3-12

NAME ______ DATE ______ TIME ______

You need number cards 0–10. Make two piles. Place 0, 2, 5, and 10 into one pile. Place the other cards in another pile.

For each problem, choose one card from each pile.

Use the 0, 2, 5, or 10 to fill in the first blank in the number story.

Use the other card to fill in the next blank.

Draw arrays next to the problems to show your work.

Solve to fill in the last blank.

1. **a.** Elliot has ______ packages of erasers.

   Each package has ______ erasers.

   How many erasers does Elliot have? ______

   Number model: ____________________

   b. Subtract one package of erasers.

   How many erasers does Elliot have now? ______

   Number model: ____________________

2. **a.** Elena has ______ packages of pens.

   Each package has ______ pens.

   How many pens does Elena have? ______

   Number model: ____________________

   b. Subtract one package of pens.

   How many pens does Elena have now? ______

   Number model: ____________________

Subtracting a Group

Home Link 3-12

NAME DATE TIME

Family Note Today your child learned the subtracting-a-group strategy for solving multiplication facts. Children subtracted groups from 5s and 10s helper facts to solve other facts. For example, to solve 4×6, they might start with $5 \times 6 = 30$ and subtract a group of 6: $30 - 6 = 24$, so $4 \times 6 = 24$. Continue to help your child practice multiplication with the included Fact Triangles.

Please return this Home Link to school tomorrow.

① Use 10×4 and the array below to help figure out 9×4.

$9 \times 4 = ?$

Helper fact: $10 \times 4 = 40$

× × × ×
× × × ×
× × × ×
× × × ×
× × × ×
× × × ×
× × × ×
× × × ×
× × × ×
× × × ×

a. Draw on the array above to show how to use 10×4 to figure out 9×4.

b. Solve. $9 \times 4 =$ ______

c. How did knowing 10×4 help you figure out 9×4?

____

____

Practice

② $9 \times 10 =$ ___

③ $9 \times 5 =$ ___

④ ___ $= 8 \times 5$

⑤ ___ $= 8 \times 10$

×, ÷ Fact Triangles: 3s and 9s

Home Link 3-12

NAME DATE TIME

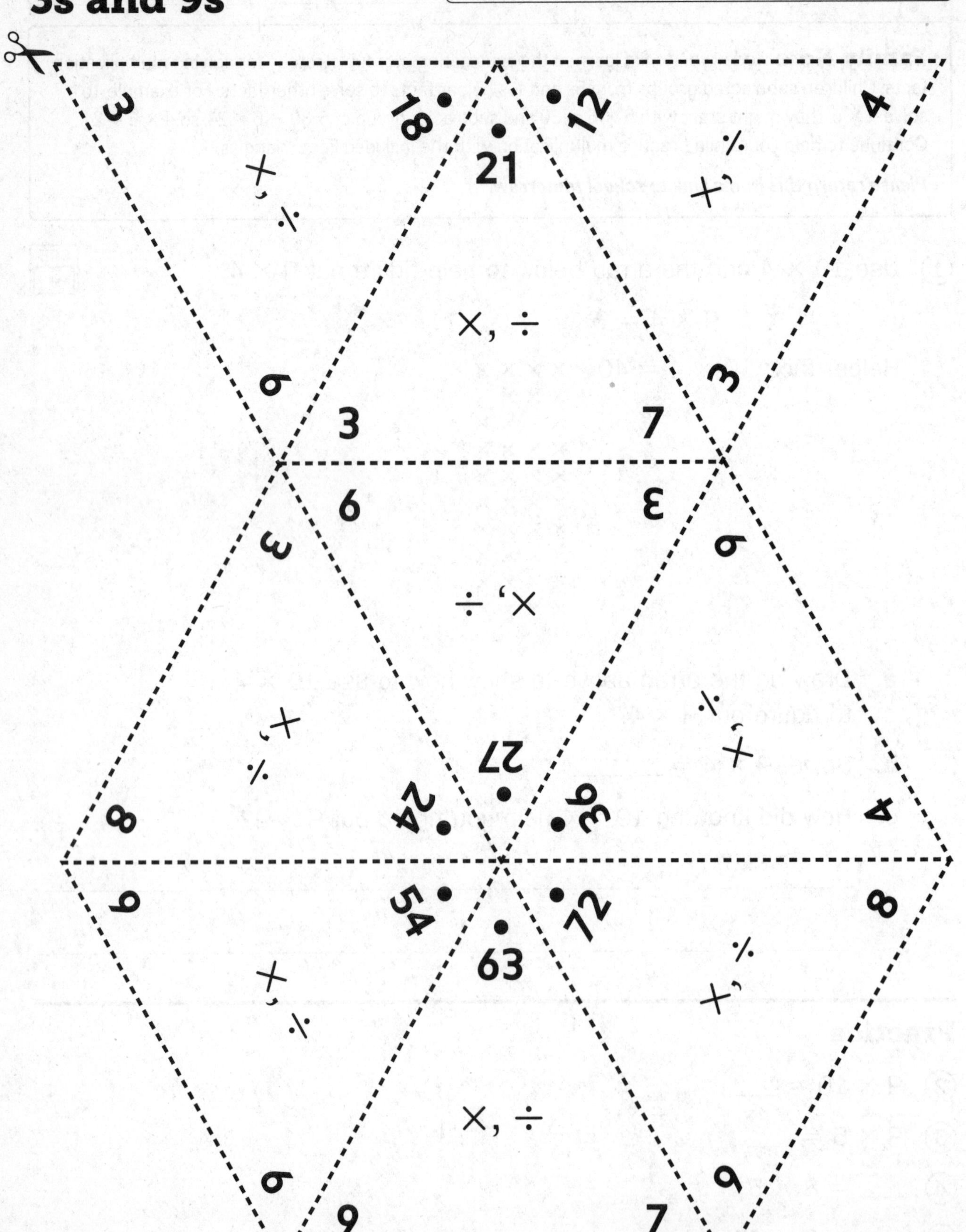

Writing Equivalent Names

Lesson 3-13

NAME DATE TIME

Cross out the names that do not belong. Write at least 8 more names. Include at least one of each type of name: addition, subtraction, multiplication, and division.

60

$6 \times 5 \times 2$ $100 - 40$ $74 - 8$

$40 + 18$ ||||||| $980 - 900$

Choose one of the names that does not belong in the name-collection box.

Explain why it does not belong. ______________________________

____

Frames and Arrows

Lesson 3-13

NAME DATE TIME

Fill in the missing numbers and rule. For Problem 2, create your own rule and fill in the diagram with the number pattern that fits your rule.

①

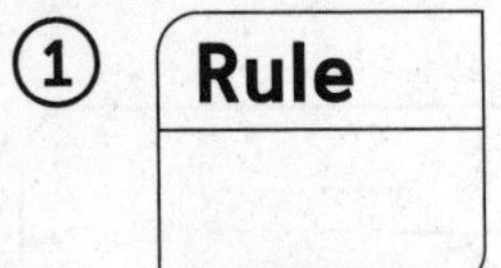

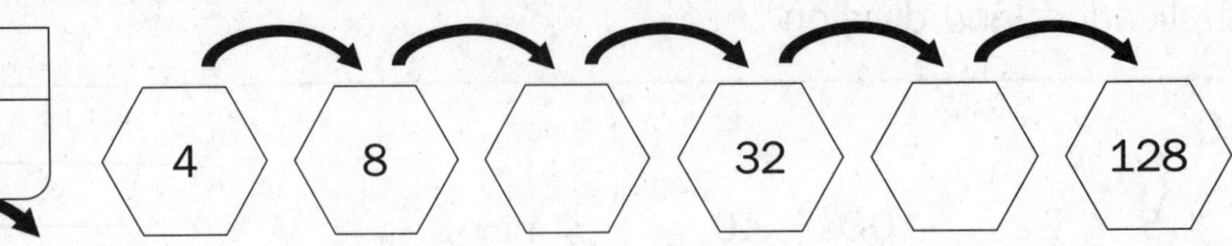

②

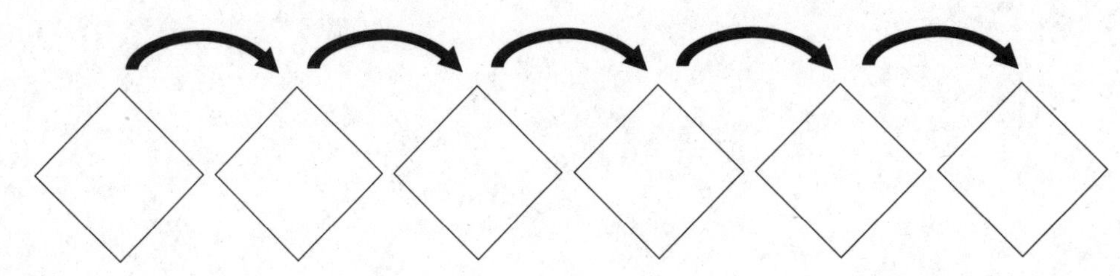

Use two rules to fill in the empty frames.

③

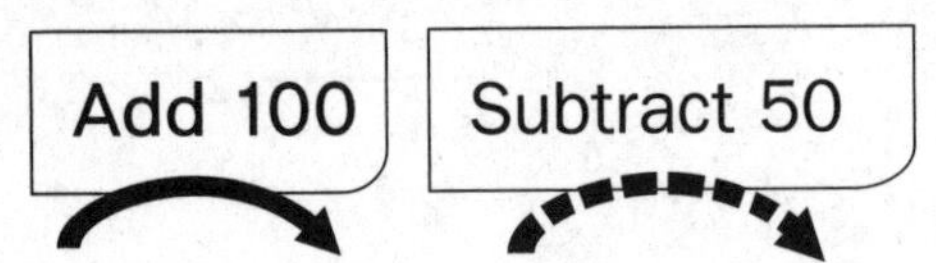

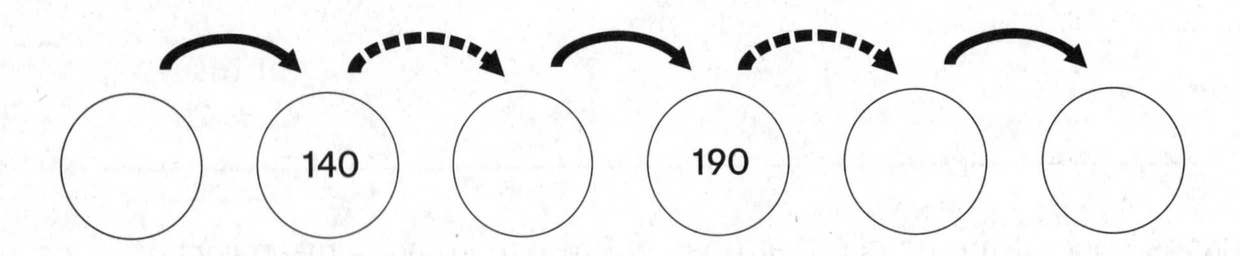

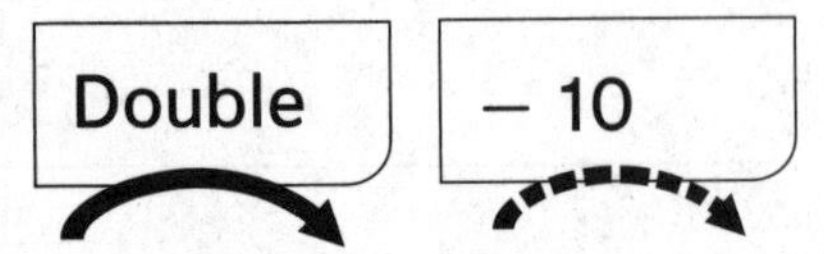

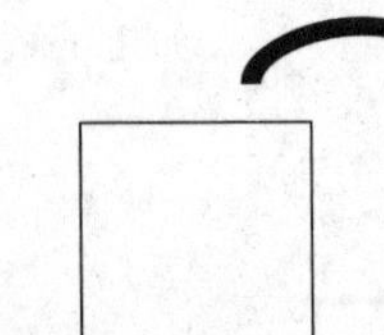

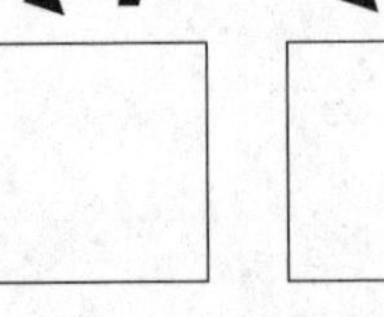

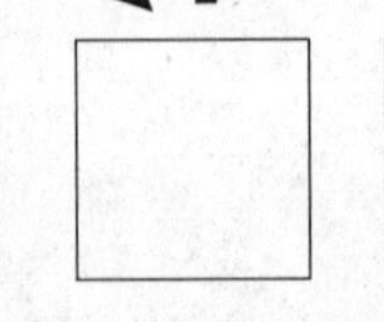

180

Name-Collection Boxes

Home Link 3-13

NAME DATE TIME

Family Note Today your child discussed and wrote equivalent names for numbers in name-collection boxes. You can find an explanation of name-collection boxes on pages 96-97 in the *Student Reference Book*.

Please return this Home Link to school tomorrow.

SRB 96-97

① Write at least 10 names for the number 18 in the name-collection box. Then explain to someone at home how the box works. Have that person add another name for 18.

18

② Three of the names do not belong in this box. Cross them out. Then write the name of the box on the tag.

~~||||~~ ~~||||~~ one dozen
7 + 5
number of months in 1 year
15 − 3 10 + 2
18 − 4 9 − 3

③ Make up a problem like Problem 2. Do not write the name of the box on the tag. Write 4 names for the number and 2 names that are not names for the number.

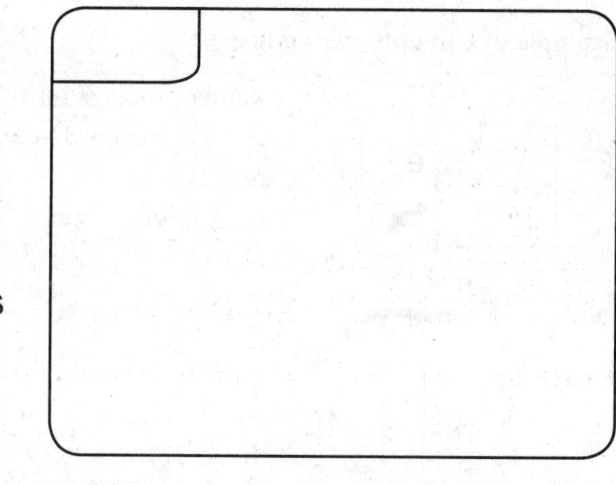

To check whether the problem makes sense, ask someone at home to tell you which 2 names do not belong. Then have that person write the name of the box on the tag.

Unit 4: Family Letter

Measurement and Geometry

In this unit children learn to make more precise measurements as they measure lengths, including perimeters, to the nearest half inch. Children will generate measurement data by measuring their shoe lengths and body parts, and they will represent the data on line plots. Building on their experiences from second grade, they will further explore attributes of polygons that help define shape categories such as quadrilaterals. Children develop an understanding of the area of rectangles and square units. They find areas by counting unit squares, repeatedly adding composite units, and multiplying side lengths. Through solving real-world and abstract problems, children will explore ways to find the perimeters of polygons and calculate the areas of rectilinear figures.

In Unit 4, children will:

- Measure to the nearest centimeter and $\frac{1}{2}$ inch.
- Generate and represent measurement data on a line plot.
- Review characteristics of polygons.
- Sort quadrilaterals into categories based on defining attributes.
- Measure perimeters of rectangles.
- Distinguish between perimeter as a measure of distance around and area as a measure of the amount of surface within the boundaries of a 2-dimensional shape.
- Find the areas of rectangles using composite units.
- Write multiplication number sentences that show how to find areas of rectangles.
- Develop strategies for finding area and perimeter.
- Find the areas of real-world rectilinear figures by partitioning figures into rectangles.

9 $9\frac{1}{2}$ 10 $10\frac{1}{2}$ 11

Shoe Lengths to Nearest $\frac{1}{2}$ inch

Line plot with stick-on notes

Example of a Rectilinear Figure

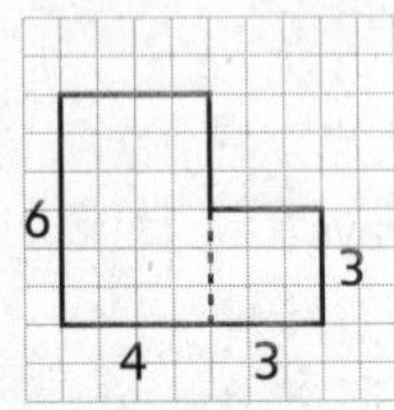

Key: □ = 1 square yard

Number models for finding areas of rectangles:
$6 \times 4 = 24$; $3 \times 3 = 9$
$24 + 9 = 33$
Area of whole shape: 33 square yards

Please keep this Family Letter for reference as your child works through Unit 4.

Vocabulary

Important terms in Unit 4:

angle A figure that is formed by two rays or line segments that have the same endpoint.

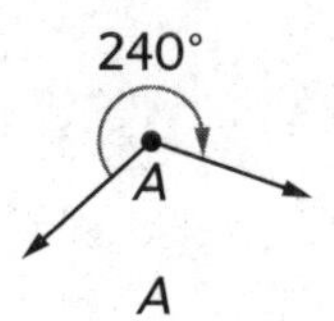

area The amount of surface inside a shape. Area is usually measured in square units such as square inches or square centimeters.

composite unit A unit made up of a group of units. A row made up of square units is a composite unit that can be used to find area.

data Information that is gathered by counting, measuring, questioning, or observing.

kite A quadrilateral that has two nonoverlapping pairs of adjacent, equal-length sides.

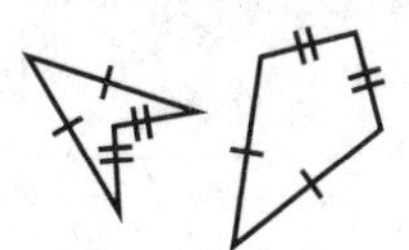

length The distance between two points.

line plot A sketch of data that uses Xs, checks, or other marks above a number line to show how many times each value appears in a set of data.

parallel line segments Segments that are always the same distance apart. They never meet or cross, even when extended.

parallelogram A trapezoid that has two pairs of parallel sides.

partition To divide a shape into smaller shapes.

perimeter The distance around a 2-dimensional figure.

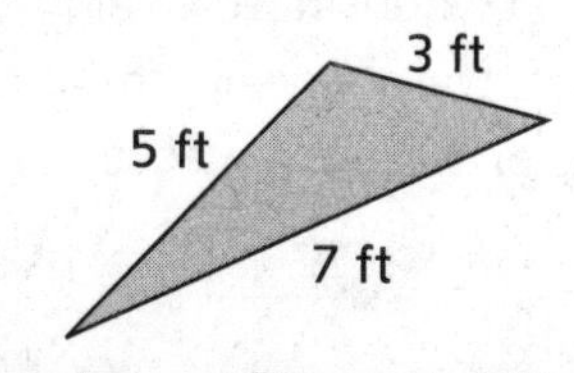

perimeter = 5 ft + 3 ft + 7 ft = 15 ft

polygon A 2-dimensional figure formed by line segments (sides) joined end to end to make one closed path. The sides may not cross one another.

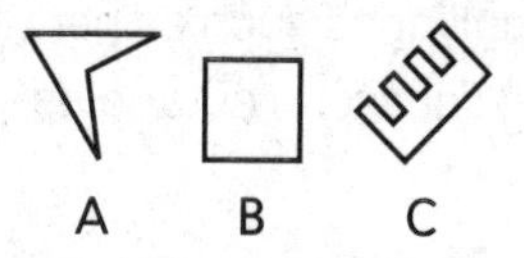

quadrilateral A 4-sided polygon. Polygons A and B above are quadrilaterals.

rectilinear figure A polygon with a right angle at each vertex.

rectangle A parallelogram with four right angles.

rhombus A parallelogram with four equal-length sides.

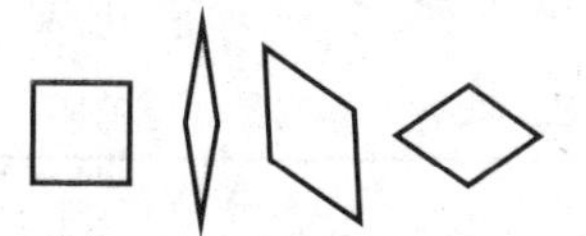

right angle A 90° angle. The sides of a right angle form a square corner.

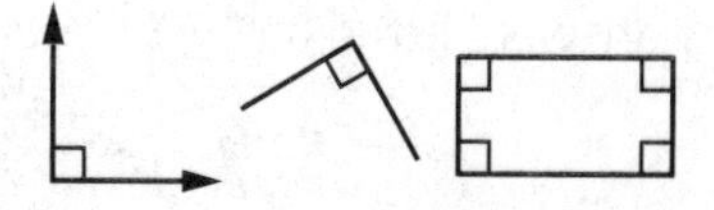

scale of a graph The unit interval, or distance between numbers, on graphs.

square A rectangle with 4 equal-length sides.

square unit A unit used to measure area.

vertex The point at which the rays of an angle, the sides of a polygon, or the edges of a polyhedron meet.

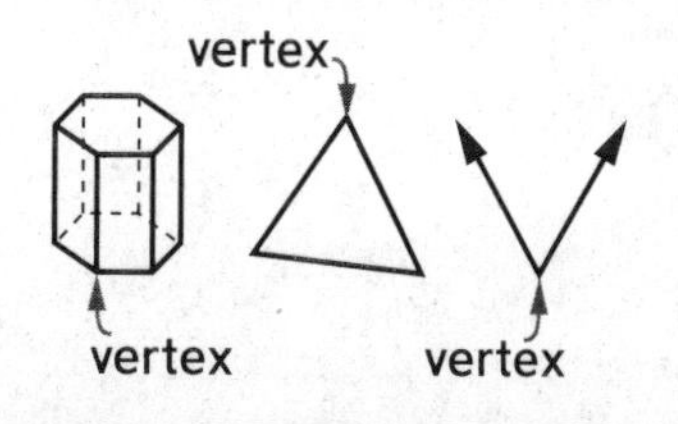

Do-Anytime Activities

The following activities provide practice for concepts taught in this and previous units.

1. Together read *Spaghetti and Meatballs for All!* by Marilyn Burns (Scholastic, 2008).
2. Help your child measure objects to the nearest centimeter or $\frac{1}{2}$ inch.
3. Ask your child to identify polygons such as pentagons, hexagons, and octagons, as well as quadrilaterals, including squares, rectangles, parallelograms, rhombuses, kites, and trapezoids.
4. Ask your child to think of situations in which knowing how to find perimeter and area can help with solving problems. Such situations include purchasing carpeting, painting walls, and building fences.

Building Skills through Games

In Unit 4 your child will practice calculating area and perimeter as well as identifying quadrilaterals by playing the following games. For detailed instructions, see the *Student Reference Book*.

The Area and Perimeter Game Children score points by finding the perimeters and areas of rectangles.

What's My Polygon Rule? Children sort polygons into categories based on their similarities and differences. They recognize additional characteristics of polygons.

As You Help Your Child with Homework

As your child brings home assignments, you may want to go over the instructions together, clarifying them as necessary. The answers listed below will guide you through this unit's Home Links.

Home Link 4-1

1. 438 **2.** 168

Home Link 4-2

1. 6 **2.** 2 **3.** 43

4. 14; 21

× × × × × × ×
× × × × × × ×
○ ○ ○ ○ ○ ○ ○

5. 30; 24

× × × × × ×
× × × × × ×
× × × × × ×
× × × × × ×
~~× × × × × ×~~

Home Link 4-3

1. Sample answers: Tape measure, toolkit ruler, 12-inch ruler, string, yardstick

3. 35; 42 **4.** 60; 54

5. 40; 48 **6.** 70; 63

Home Link 4-4

1. The third and fifth shapes should be crossed out. Sample answers: Polygons have straight sides that do not cross. The shapes I crossed out have curved sides or sides that cross.

Home Link 4-5

1. Sample answers: square; rhombus; Both shapes have 4 same-length sides. A square has to have 4 right angles. A rhombus doesn't have to have all 4 right angles.
2. Sample answers: rhombus; rectangle; Both shapes have 2 pairs of equal opposite sides. A rhombus has all 4 equal-length sides. A rectangle has 4 right angles.

Home Link 4-6

1. Sample answer: $2 + 2 + 1 + 1 = 6$; about 6 inches
2. Sample answer: $2\frac{1}{2} + 1 + 1 + 1 = 5\frac{1}{2}$; about $5\frac{1}{2}$ inches
3. Sample answer: $4 \times 5 = 20$; 20 meters
4. Sample answer: $12 + 12 + 5 + 5 = 34$; 34 centimeters

Home Link 4-7

1. I agree. The perimeter is the total length of sides, so $2 + 6 + 2 + 6 = 16$. The area is the number of squares inside the rectangle.

Home Link 4-8

1.

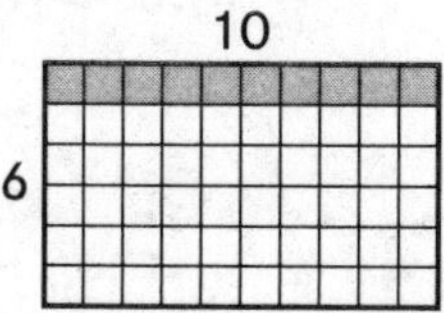

Sarah needs 60 tiles.

2.

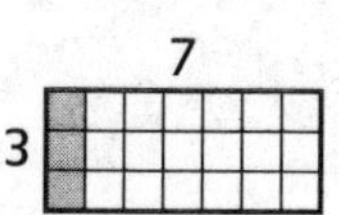

21 square yards

3. Sample answer: I made a composite unit of a column of 3 squares and counted by 3s seven times to get 21.

Home Link 4-9

1. 7; 5; $7 \times 5 = 35$; 35 square units
2. 6; 7; $6 \times 7 = 42$; 42 square units
3. $4 \times 8 = 32$
4. $9 \times 5 = 45$

Home Link 4-10

1. Area: 16; Perimeter: 20
2. Area: 12; Perimeter: 14
3. Sample answer: I multiplied two side lengths to find the area: $3 \times 4 = 12$. I added the side lengths to find the perimeter: $3 + 4 + 3 + 4 = 14$.
4. Area: 30; Perimeter: 22; Sample answer: I found the area by multiplying $5 \times 6 = 30$. I added two side lengths and doubled the total to find the perimeter: $5 + 6 = 11$ and $11 \times 2 = 22$.

Home Link 4-11

1. Perimeter = 6 feet; Perimeter = 24 feet
2. 2 quarts; Sample answer: I drew a rectangle for the wall and marked off 10 feet on one side and 8 feet on the other side. I drew in squares so that it looked like an array with 8 rows of 10 squares. I skip counted by 10s to find the total number of squares. There were 80 squares, so the area is 80 square feet. Sue needs to buy 2 quarts, but she will have some paint left over.

Home Link 4-12

1–3. Sample answers:

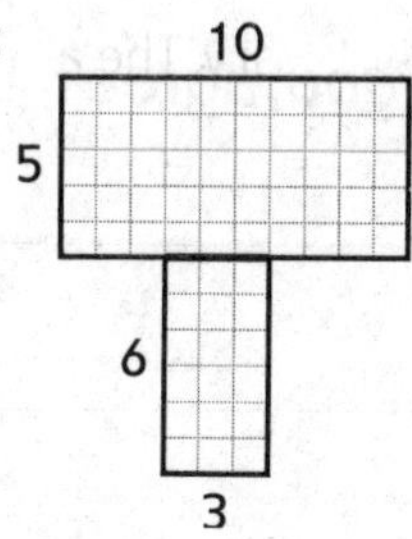

4. Sample number sentences: $5 \times 10 = 50$, $6 \times 3 = 18$; $50 + 18 = 68$
   Area: 68 square units
5. Sample answer: I can find the area of each rectangle and then add the two together to get the area of the whole shape.

Measuring with Different Rulers

Lesson 4-1

NAME DATE TIME

① Mahima says she can use this ruler to accurately measure the length of her pencil.

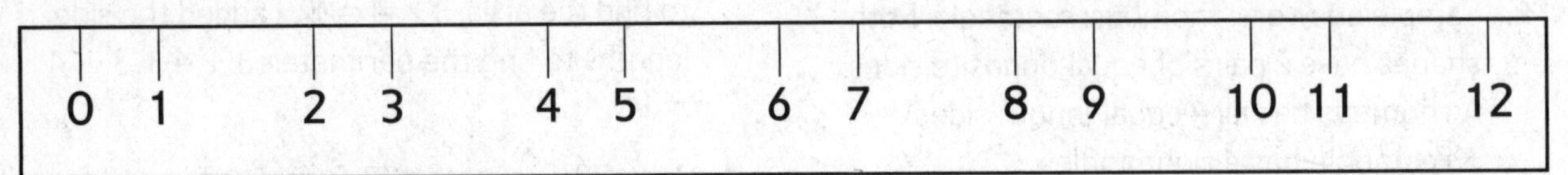

Do you agree? Explain your answer.

② Rod says he can use this ruler to accurately measure the length of his book, even though it does not begin at 0 and there are missing numbers.

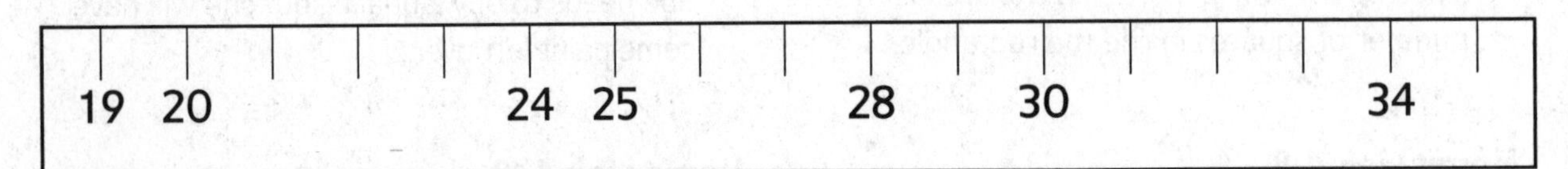

Do you agree? Explain your answer.

Measuring Objects

Lesson 4-1

NAME DATE TIME

Measure the length of each object with Ruler A and with Ruler C.

Record your measurements to the nearest $\frac{1}{2}$ inch.

Name or Sketch of Object	Ruler A Nearest $\frac{1}{2}$ Inch	Ruler C Nearest $\frac{1}{2}$ Inch
	about _______ inches	about _______ inches
	about _______ inches	about _______ inches
	about _______ inches	about _______ inches
	about _______ inches	about _______ inches
	about _______ inches	about _______ inches

Body Measures

Home Link 4-1

NAME DATE TIME

Family Note Today your child measured to the nearest half inch. Help your child measure an adult at home. Use a tape measure if available, or mark lengths on a piece of string and then measure the string with a ruler.

Please return this Home Link to school tomorrow.

Measure an adult at home to the nearest $\frac{1}{2}$ inch. Fill in the information below:

Name of adult: ______________

Height: about ______ inches

Length of shoe: about ______ inches

Around neck: about ______ inches

Around wrist: about ______ inches

Distance from waist to floor: about ______ inches

Forearm: about ______ inches	Hand span: about ______ inches	Arm span: about ______ inches
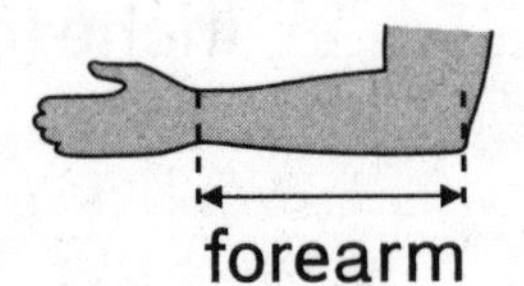 forearm	hand span	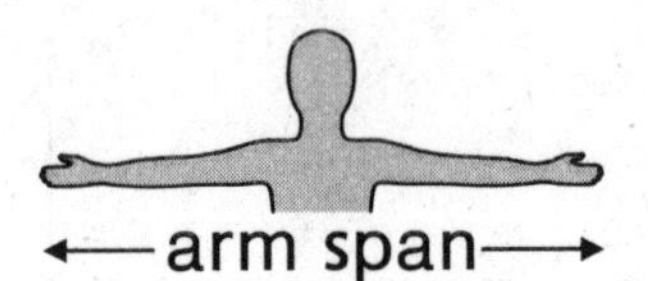 ←arm span→

Practice

Fill in the unit box. Solve. Show your work in the space below.

Unit

① ______ = 293 + 145

② 326 − 158 = ______

Plotting Plant Heights

Lesson 4-2

NAME DATE TIME

The third-grade class is growing bean plants. Use the data below to complete the line plot.

Heights of Bean Plants	Number of Plants
6 inches	//
7 inches	////
8 inches	//
9 inches	~~////~~ /

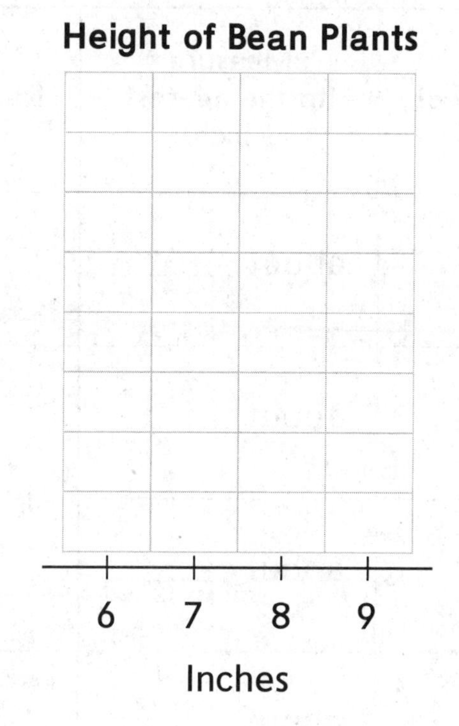

Write four things you know from your line plot.

____

____

____

____

Measuring Objects for a Line Plot

Lesson 4-2

NAME DATE TIME

Name of Object	Measure (to the nearest $\frac{1}{2}$ inch)	Name of Object	Measure (to the nearest $\frac{1}{2}$ inch)
	about ______		about ______
	about ______		about ______
	about ______		about ______
	about ______		about ______
	about ______		about ______
	about ______		about ______

Measures from shortest to longest:

______, ______, ______, ______, ______, ______, ______, ______,

______, ______, ______, ______

Describing Data

Home Link 4-2

NAME DATE TIME

Family Note Today your child represented shoe-length measures on a line plot. Help your child answer questions about the line plot below.

Please return this Home Link to school tomorrow.

Children in the Science Club collected pill bugs. The tally chart shows how many they collected. Use the data from the tally chart to complete the line plot.

Number of Pill Bugs	Number of Children
0	
1	
2	///
3	~~////~~
4	
5	//
6	//

Number of Children

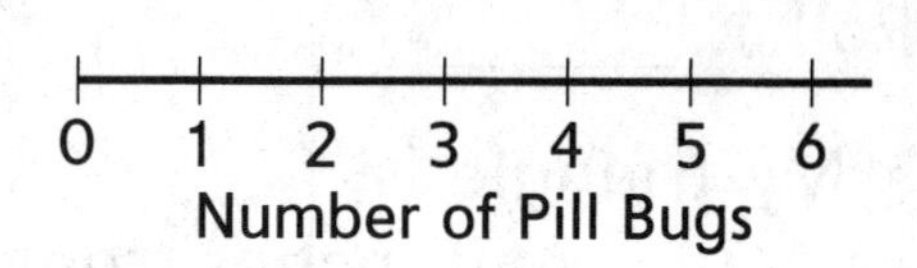

Use the information in the line plot to answer the questions.

① What is the greatest (maximum) number of pill bugs found? ______

② What is the least (minimum) number of pill bugs found? ______

③ How many pill bugs were collected all together? ______

Practice

Think how the first fact can help you solve the second. Draw an array to show your strategy. Then solve.

④ $2 \times 7 =$ ______

$3 \times 7 =$ ______

⑤ $5 \times 6 =$ ______

$4 \times 6 =$ ______

Traveling Along a Ruler Record Sheet

Lesson 4-3

NAME DATE TIME

Follow directions on Activity Card 55. Record the distance your triangle moves.

Example

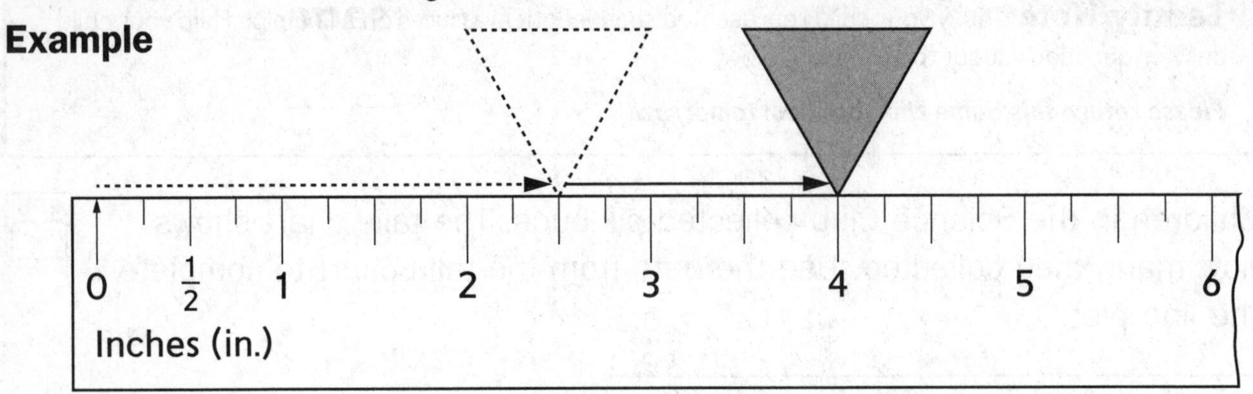

Roll 1: 5 My triangle is 5 half-inches from 0.

Roll 2: 3 Now my triangle is 8 half-inches from 0.

Round 1	
Roll 1: ______ • My triangle is ______ half-inches from 0.	Roll 2: ______ • Now my triangle is ______ half-inches from 0.
Round 2	
Roll 1: ______ • My triangle is ______ half-inches from 0.	Roll 2: ______ • Now my triangle is ______ half-inches from 0.
Round 3	
Roll 1: ______ • My triangle is ______ half-inches from 0.	Roll 2: ______ • Now my triangle is ______ half-inches from 0.
Round 4	
Roll 1: ______ • My triangle is ______ half-inches from 0.	Roll 2: ______ • Now my triangle is ______ half-inches from 0.

Head and Wrist Measurements

Lesson 4-3

NAME DATE TIME

Math Message

Name ______________________

Work with a partner.

Measure, to the nearest $\frac{1}{2}$ inch, the distance around:

Your head **Your wrist**

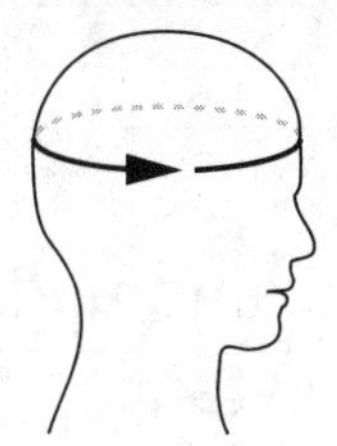

______ (inches) ______ (inches)

Math Message

Name ______________________

Work with a partner.

Measure, to the nearest $\frac{1}{2}$ inch, the distance around:

Your head **Your wrist**

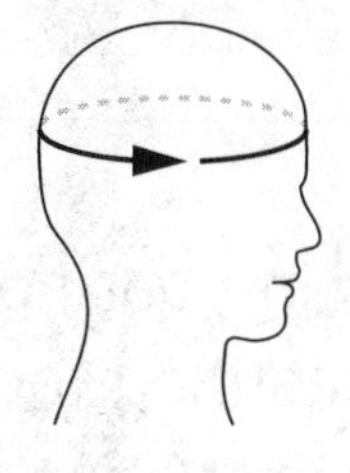

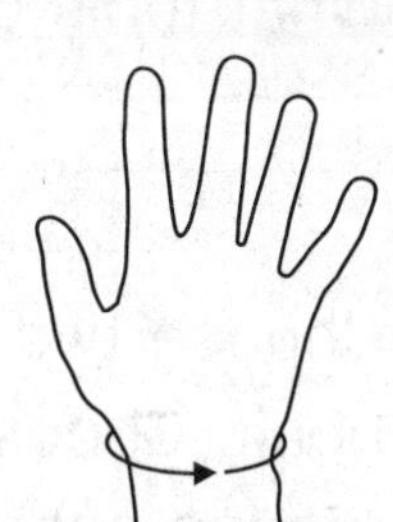

______ (inches) ______ (inches)

Math Message

Name ______________________

Work with a partner.

Measure, to the nearest $\frac{1}{2}$ inch, the distance around:

Your head **Your wrist**

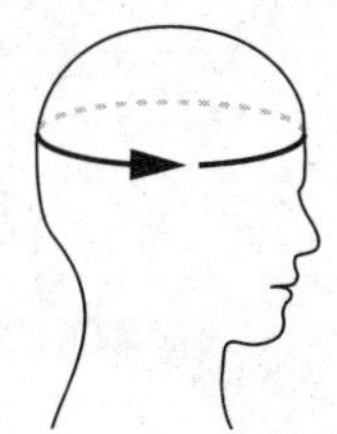

______ (inches) ______ (inches)

Math Message

Name ______________________

Work with a partner.

Measure, to the nearest $\frac{1}{2}$ inch, the distance around:

Your head **Your wrist**

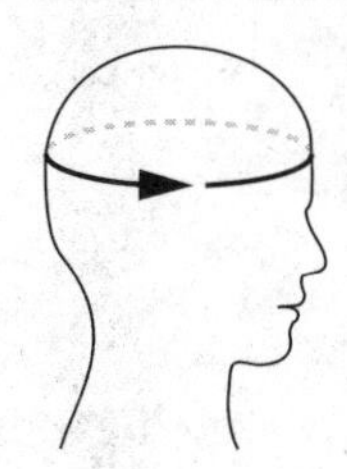

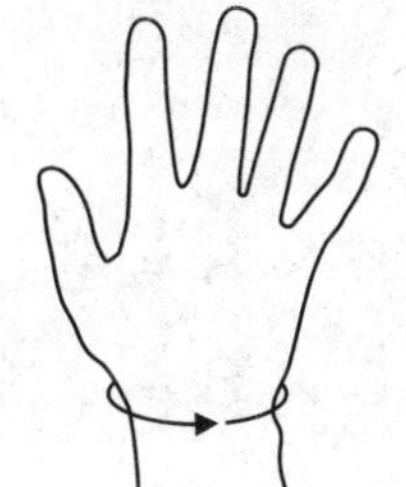

______ (inches) ______ (inches)

Measuring Distances Around Objects

Home Link 4-3

NAME DATE TIME

Family Note Today your child measured the distance around his or her head and wrist, as well as around different objects in the classroom. Finding the distance around objects helps children understand perimeter, which is formally introduced in Lesson 4-6. It is also good practice for measuring to the nearest $\frac{1}{2}$ inch.

Please return this Home Link to school tomorrow.

Talk to someone at home about finding the distances around objects.

① What tools can be used to measure the distance around an object?

____

② Choose two objects in your home, such as a small picture frame and a book. Choose a measuring tool and use it to measure the distance around each object to the nearest $\frac{1}{2}$ inch.

Object: ______________ Measurement: about ________ inches

Object: ______________ Measurement: about ________ inches

What measuring tool did you use? ____________________

Practice

Think of how the first fact can help you solve the second. Then solve both. You may draw arrays to help.

③ $5 \times 7 =$ ______
$6 \times 7 =$ ______

④ $10 \times 6 =$ ______
$9 \times 6 =$ ______

⑤ $5 \times 8 =$ ______
$6 \times 8 =$ ______

⑥ $10 \times 7 =$ ______
$9 \times 7 =$ ______

Polygons

Home Link 4-4

NAME DATE TIME

Family Note Today your child learned the names of different polygons. A polygon is a 2-dimensional shape with only straight sides that meet end to end to make one closed path. The sides may not cross one another. Polygons are named by the number of sides they have. Polygons are all around us. For example, a stop sign is an octagon, an 8-sided polygon, and this Home Link page is a rectangle, a 4-sided polygon with 4 right angles (square corners).

Please return this Home Link to school tomorrow.

① Cross out the shapes that are not polygons.

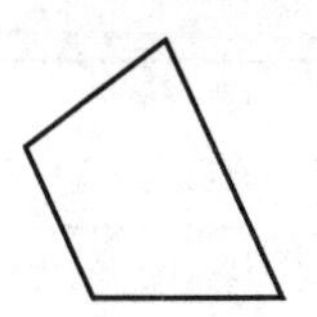 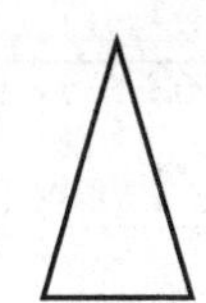 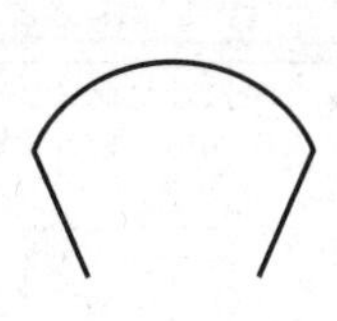  

How do you know which shapes are not polygons?

____

____

② Cut out pictures of shapes from newspapers and magazines to match each of the descriptions below. Tape or glue your pictures on the front or back of this page.

equal-length sides	parallel sides
at least one right angle	quadrilateral

Exploring Quadrilaterals in Tangrams

Lesson 4-5

NAME DATE TIME

① What quadrilaterals are used in the tangrams below?

② Explain why these quadrilaterals work well in the designs created with the tangrams.

Cut out the pieces below. Choose one of the tangram puzzles in *Grandfather Tang's Story* to solve.

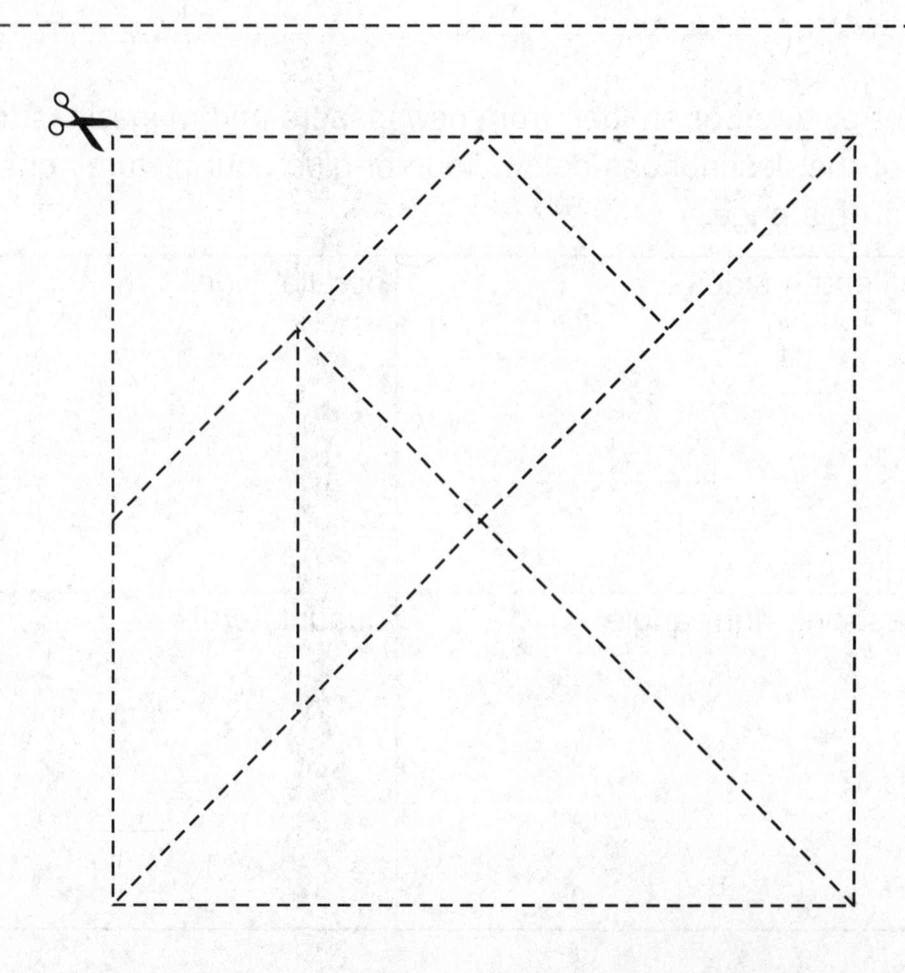

Special Quadrilaterals

Home Link 4-5

NAME DATE TIME

Family Note Today your child learned about six different categories of quadrilaterals, which are polygons that have four sides: squares, rectangles, rhombuses, parallelograms, trapezoids, and kites. Although these categories have specific definitions, a particular shape may fall into more than one category.

Please return this Home Link to school tomorrow.

SRB 216-217

1. Name the two special quadrilaterals below.

_______________ _______________

How are these two shapes alike? _______________

____

How are they different? _______________

____

2. Name the two special quadrilaterals below.

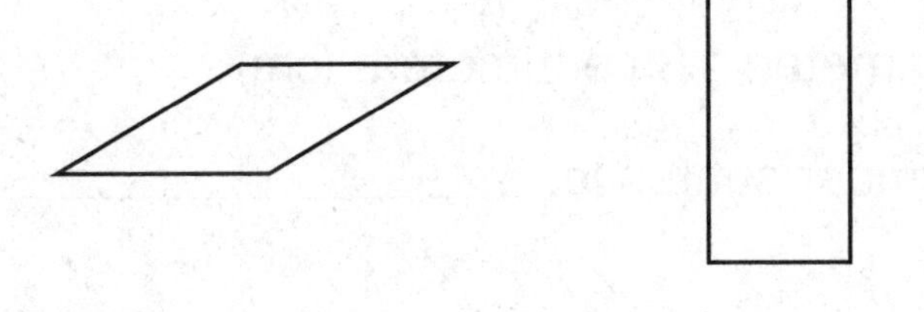

_______________ _______________

How are these two shapes alike? _______________

____

How are they different?

____

____

Finding the Perimeters of Polygons

Lesson 4-6

NAME DATE TIME

Find the perimeter of each shape. Show your thinking with number sentences.

① 3 m, 3 m, 3 m, 3 m, 3 m, 3 m

Perimeter: _______ meters (m)

Number sentence: ____________________

② 6 ft, 4 ft, 4 ft, 4 ft

Perimeter: _______ feet (ft)

Number sentence: ____________________

Use the perimeter and side lengths to find the other sides' lengths. Show how you can check your answers by adding your side lengths.

③

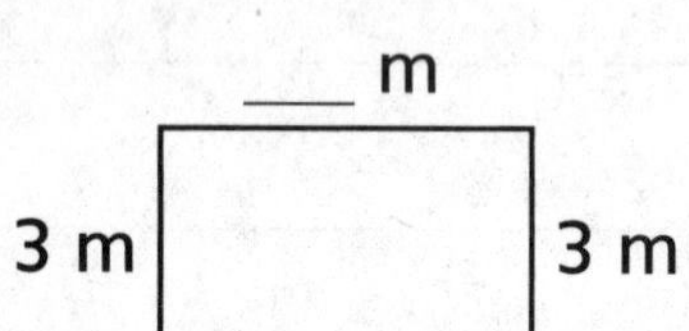

Perimeter: 16 meters (m)

Number sentence: ____________________

④

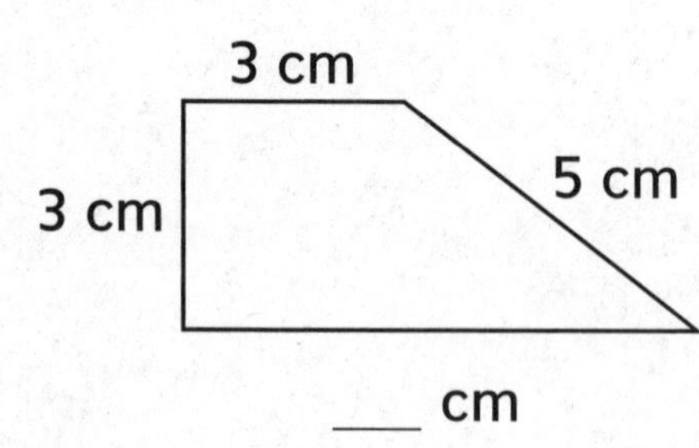

Perimeter: 18 centimeters (cm)

Number sentence: ____________________

⑤ Explain how you found the length of the unknown side in Problem 4.

____

____

____

Perimeter

NAME DATE TIME

Family Note Today your child found the perimeters of several polygons. Perimeter is the distance around a 2-dimensional shape. Finding perimeters gives your child practice measuring to the nearest $\frac{1}{2}$ inch and the nearest whole centimeter.

Please return this Home Link to school tomorrow.

If you do not have a ruler at home, cut out and use the 6-inch ruler on the next page. Measure the sides of each polygon to the nearest $\frac{1}{2}$ inch. Use the side lengths to find the perimeter of each polygon.
Write a number sentence to show how you found the perimeter.

1.

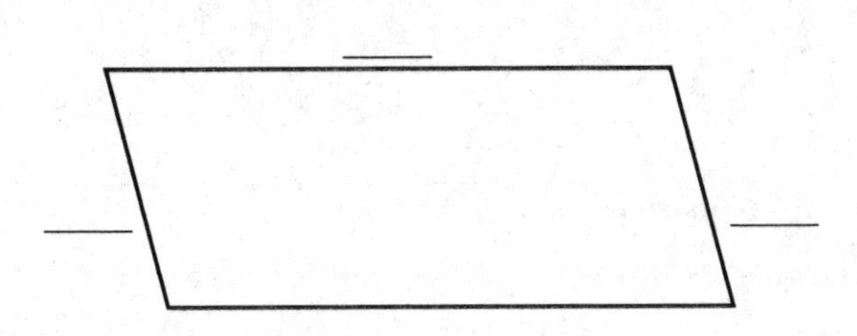

Number sentence: ______________________________

Perimeter: about ____ inches

2.

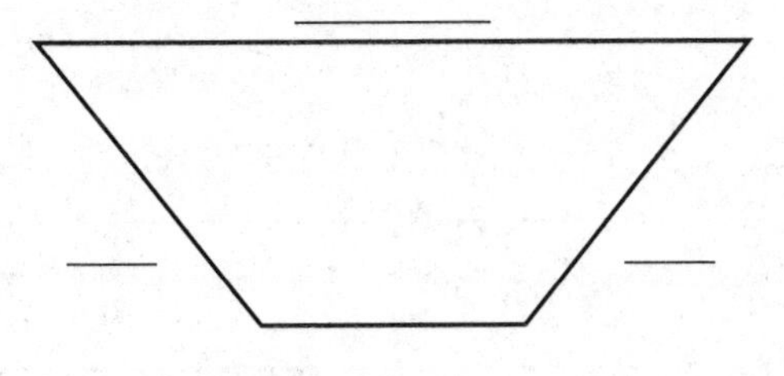

Number sentence: ______________________________

Perimeter: about _______ inches

Perimeter (continued)

Home Link 4-6

NAME DATE TIME

Find the perimeters of the square and the rectangle below.

③ 5 m

Number sentence: ______________________________

Perimeter: ______ meters (m)

④ 5 cm

12 cm

Number sentence: ______________________________

Perimeter: ______ centimeters (cm)

⑤ Draw a quadrilateral below. Find the perimeter to the nearest $\frac{1}{2}$ inch.

____

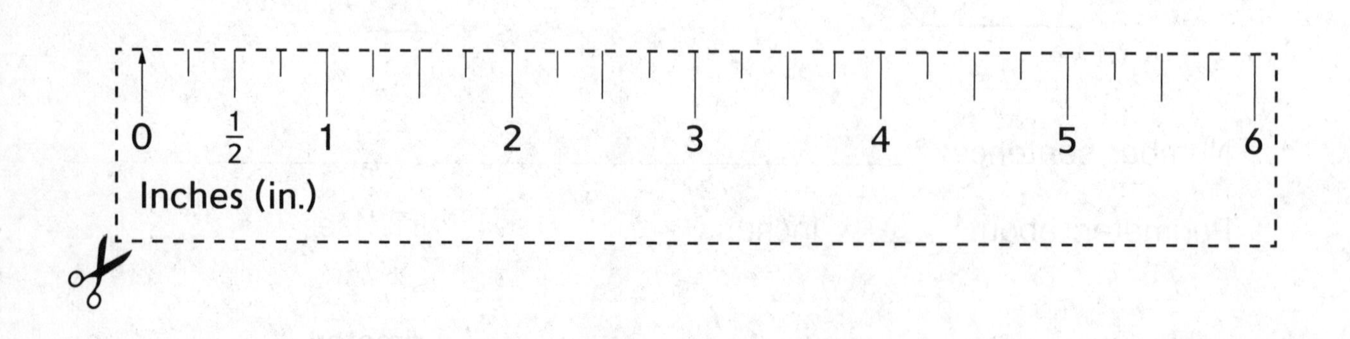

Perimeter and Area

Home Link 4-7

NAME DATE TIME

Family Note Today your child compared measuring perimeter to measuring area using 1-foot squares. *Perimeter* is the distance around a shape. It can be measured in units of length, such as centimeters, inches, feet, and so on. *Area* is the measure of surface space inside the boundary of a shape. It can be measured in square units, such as square centimeters, square inches, square feet, and so on. To measure perimeter, children used the edges of 1-foot squares as their units. To measure area, they used the area of 1-foot squares as their units.

Please return this Home Link to school tomorrow.

Trace the boundary of the rectangle with a crayon to show where you measure the perimeter.

With a different colored crayon, shade the surface inside the rectangle to show where you measure area.

① Key: ☐ = 1 square foot

Dale said the perimeter of this rectangle is 16 feet and the area is 12 square feet. Do you agree? Explain.

____

____

____

Your pace is the length of one of your steps.

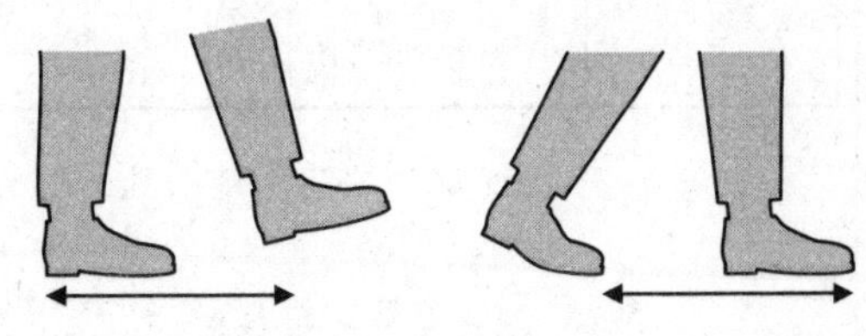

② Find the perimeter, in paces, of your bedroom. Walk along each side and count the number of paces.

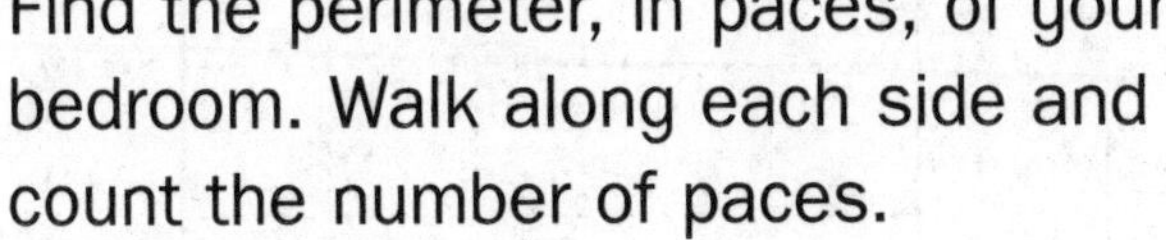

The perimeter of my bedroom is about ______ paces.

③ Which room in your home has the largest perimeter? Use your estimating skills to help you decide.

The ____________________ has the largest perimeter.

Its perimeter is about ______ paces.

Using Squares to Find Area and Perimeter

Lesson 4-8

NAME DATE TIME

Get a blue and a red colored pencil.
Shade the squares inside the rectangles blue.
Count the blue squares to find the area.
Outline the rectangles with red. Count the red edges to find the perimeter.

①

Area: ____ square centimeters (sq cm)

Perimeter: ____ centimeters (cm)

②

Area: ____ square centimeters (sq cm)

Perimeter: ____ centimeters (cm)

③

Area: ____ square centimeters (sq cm)

Perimeter: ____ centimeters (cm)

④

Area: ____ square centimeters (sq cm)

Perimeter: ____ centimeters (cm)

Exploring Area with Composite Units

Lesson 4-8

NAME DATE TIME

Partition, shade, and use a different composite unit to find each area.

① 8 by 7 Area: _______ square units	② 8 by 7 Area: _______ square units
③ 8 by 7 Area: _______ square units	④ 8 by 7 Area: _______ square units

⑤ How does the same number of squares cover each rectangle even though the composite units are different? ______________________________

____

⑥ Which composite units are easier and harder to use when finding the areas? Explain. ______________________________

____

____

Measuring Area with Composite Units

Lesson 4-8

NAME DATE TIME

Use the shaded composite units to find the area of each rectangle.

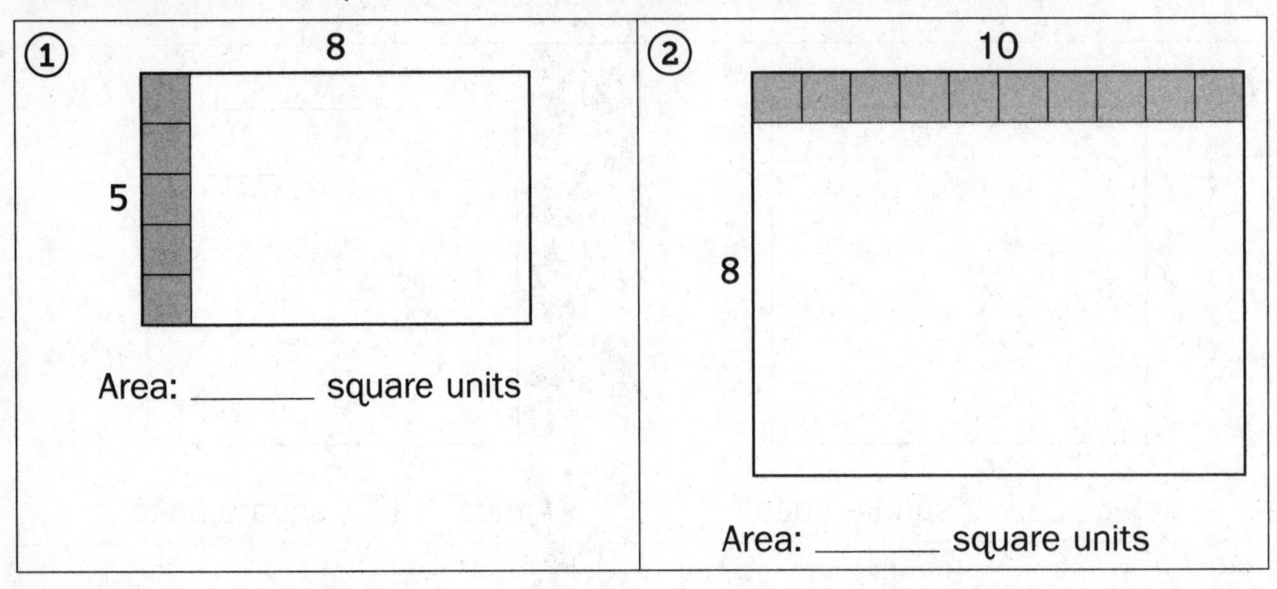

3 Shade a composite unit and use it to help find the area of the rectangle.

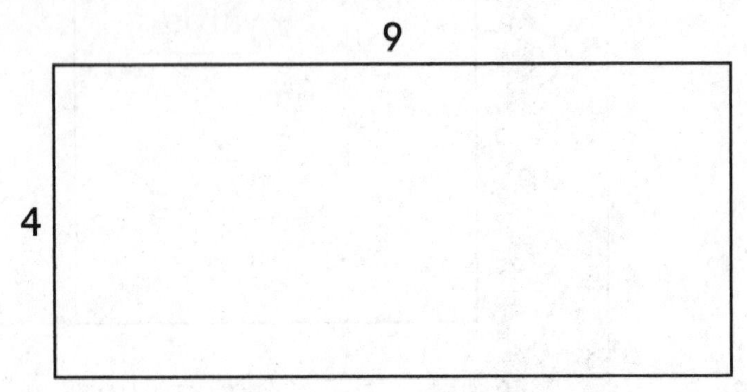

Area: _______ square units

Why did you choose the composite unit? ____________________

____

4 How did you use your composite unit in Problem 3 to find the area of the rectangle?

____

____

____

Areas of Rectangles

Home Link 4-8

NAME DATE TIME

Family Note Today your child found areas of rectangles using composite units. Composite units are made up of two or more square units. Using composite units to find area helps children see a rectangle as having a row-by-column structure, and it helps them measure area more efficiently.

Please return this Home Link to school tomorrow.

SRB 178

① Sarah tiled her floor with square tiles. This is a drawing of her floor:

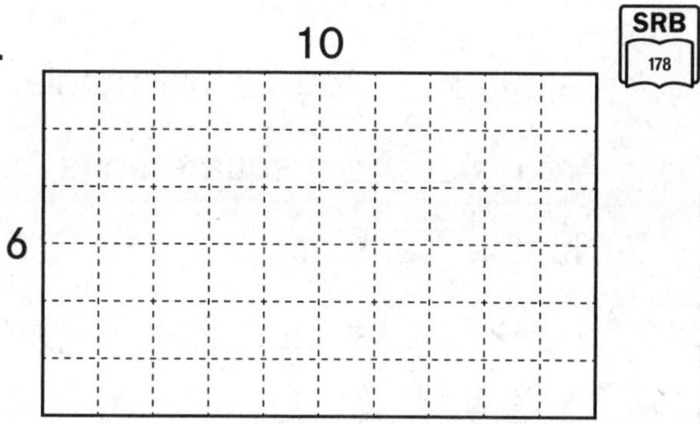

Shade a composite unit made of 10 squares. Use the composite unit to figure out the number of tiles Sarah needs.

Sarah needs _______ tiles.

② Alejandro painted a wall that is 3 yards tall and 7 yards long. This is a drawing of the wall:

7 yards

3 yards

Partition the rectangle to show 3 rows with 7 squares in each row. Shade a composite unit made of 3 squares. Then figure out the area of the wall.

How many square yards did Alejandro paint? _______ square yards

③ Explain how you found the area of the wall in Problem 2.

____

____

____

Finding Areas of Rectangles

Lesson 4-9

NAME DATE TIME

Fill in the blanks.

① 3 units

6 units

This is a ____-by-____ rectangle.

Area = ________ square units

Number sentence:

____ × ____ = ________

② 3 units

4 units

This is a ____-by-____ rectangle.

Area = ________ square units

Number sentence:

____ × ____ = ________

③

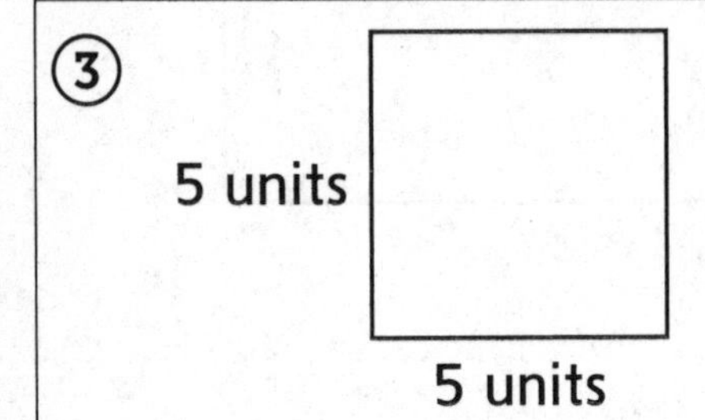

5 units

5 units

This is a ____-by-____ rectangle.

Area = ________ square units

Number sentence:

____ × ____ = ________

④

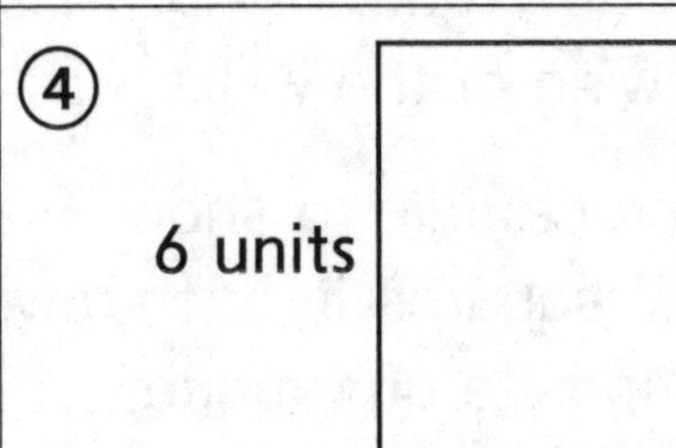

6 units

5 units

This is a ____-by-____ rectangle.

Area = ________ square units

Number sentence:

____ × ____ = ________

Areas of Rectangles

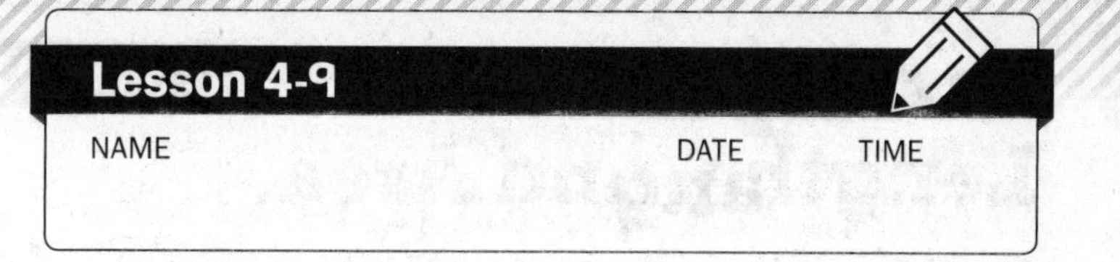

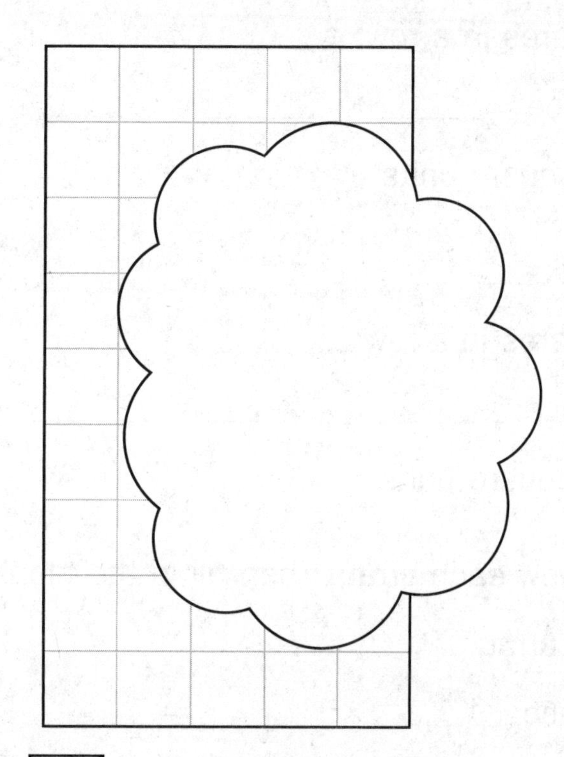

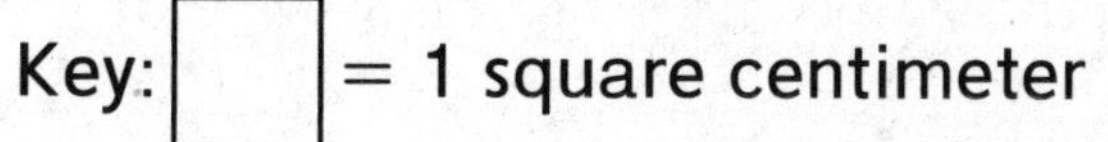

Arrays, Side Lengths, and Area

Home Link 4-9

NAME ______ DATE ______ TIME ______

Family Note Today your child learned that side lengths of rectangles correspond to the number of square units in the rectangles' rows and columns. Just as rows and columns in arrays can be multiplied to find total numbers of objects, side lengths can be multiplied to find areas of rectangles.

Please return this Home Link to school tomorrow.

Make a dot inside each small square in one row. Then fill in the blanks.

① Number of rows: ______

Number of squares in a row: ______

Number sentence: ______ × ______ = ______

Area: ______ square units

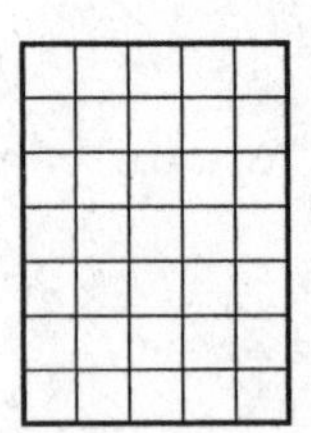

② Number of rows: ______

Number of squares in a row: ______

Number sentence: ______ × ______ = ______

Area: ______ square units

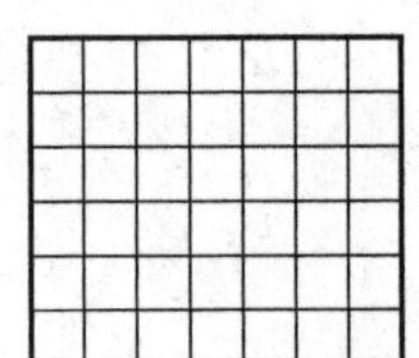

Mark the dots to show each array. Then fill in the blanks.

③ Make a 4-by-8 array.

Number sentence: ______ × ______ = ______

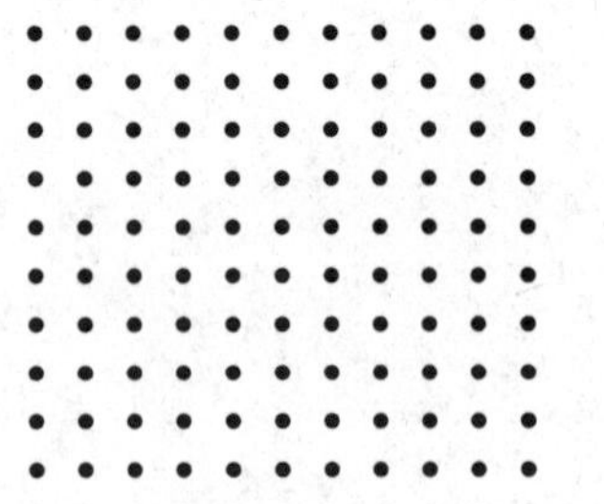

④ Make a 9-by-5 array.

Number sentence: ______ × ______ = ______

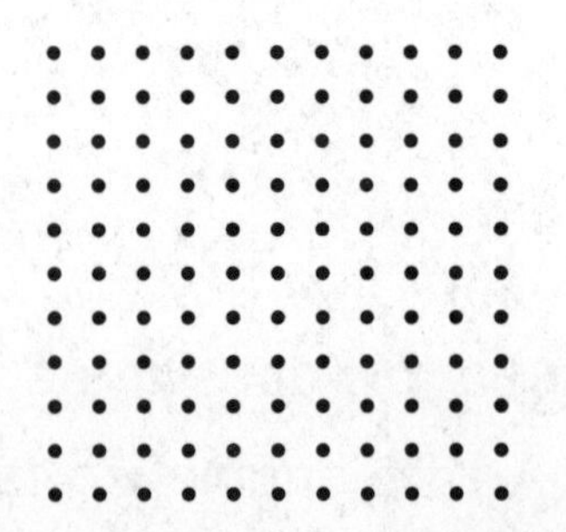

Finding and Comparing Areas

Lesson 4-10

NAME DATE TIME

- Cut out the square.
- Find its area in square inches. You may tile the area with square pattern blocks or use a ruler to measure the side lengths.
- Cut the square into two equal-size triangles.
- Find the area of each triangle in square inches.

Name: ______________________________

① Area of square: ______________ (unit)

② Area of first triangle: about ______________ (unit)

③ Area of second triangle: about ______________ (unit)

④ How do the areas of the two triangles compare?

____

⑤ How does the area of one triangle compare to the area of the original square?

____

⑥ How does the area of the two triangles together compare to the area of the original square? ______________________________

Area and Perimeter

Home Link 4-10

NAME DATE TIME

Family Note Today your child learned how to play *The Area and Perimeter Game* to practice finding the areas and the perimeters of rectangles.

Please return this Home Link to school tomorrow.

Find the area and the perimeter of each rectangle.

① This is a 2-by-8 rectangle.

Area: _______ square units

Perimeter: _______ units

② This is a 3-by-4 rectangle.

Area: _______ square units

Perimeter: _______ units

③ What strategies did you use to solve Problem 2?

____

____

Try This

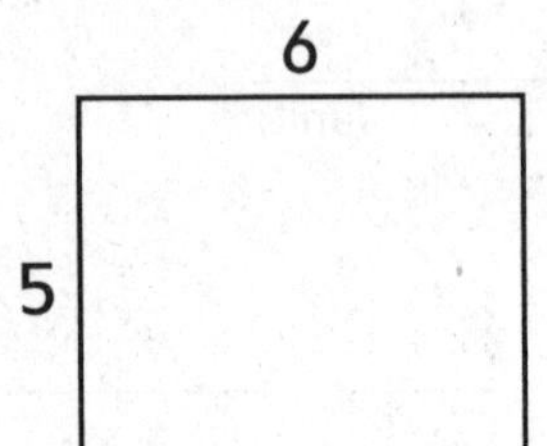

This is a 5-by-6 rectangle.

Area: _______ square units

Perimeter: _______ units

What strategies did you use? ____

____

____

Building a Rabbit Pen

Lesson 4-11

NAME DATE TIME

Miguel wants to build a rectangular pen for his rabbit. He has 24 feet of fence that he can use to make the pen. He plans to use all 24 feet of fence to make the best pen he can for his rabbit.

① Use the grid to draw **at least 2 different pens** that Miguel could build.

② **Find the area** of each pen and record it inside the pen.

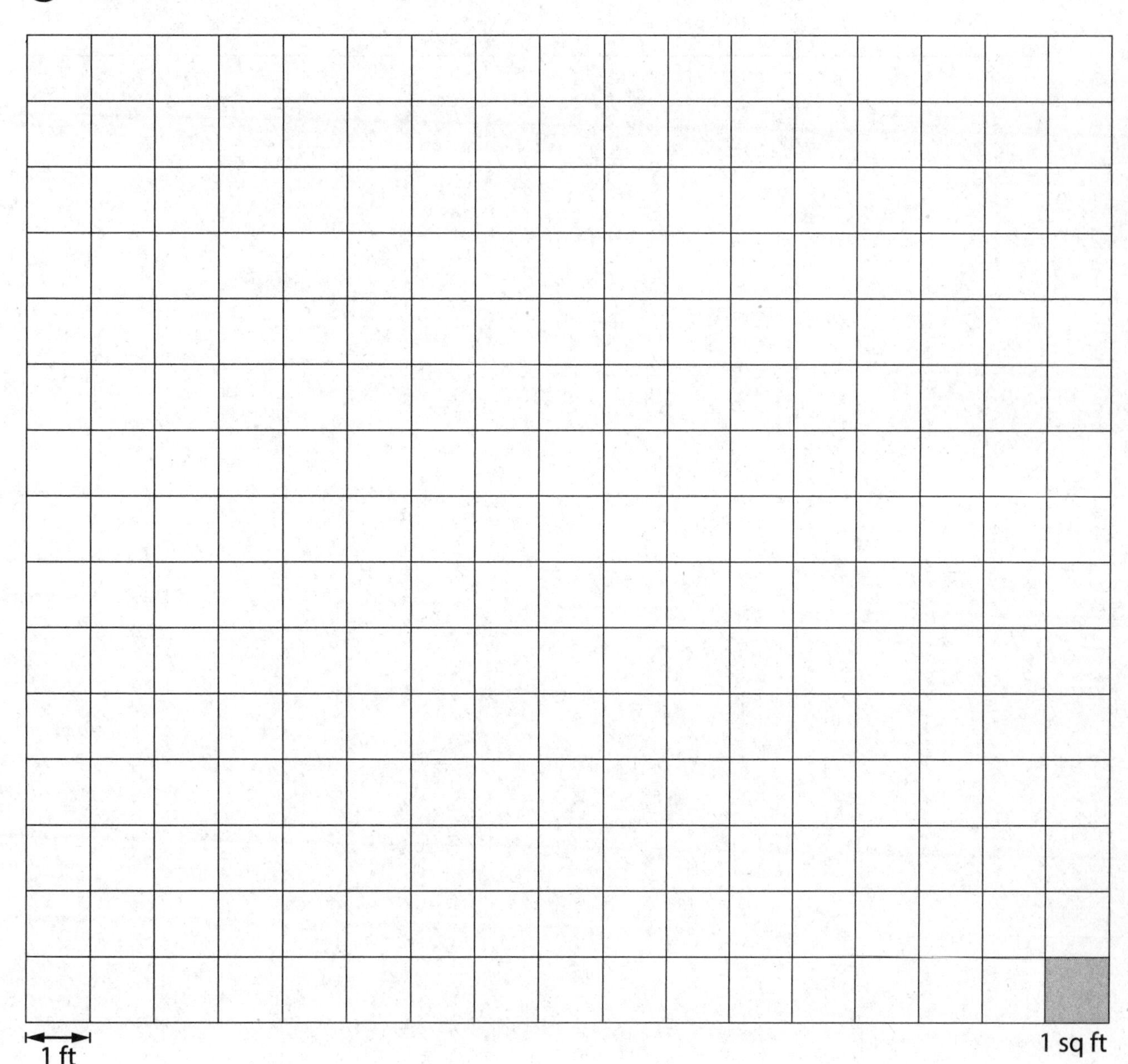

Building a Rabbit Pen (continued)

Lesson 4-11

NAME DATE TIME

(3) Which pen do you think would be the best for Miguel's rabbit?

____

(4) Use mathematical language to explain the reason for your choice.

____

____

Working with Perimeter and Area

Home Link 4-11

NAME DATE TIME

Family Note Today your child solved problems involving perimeter, the distance around a shape, and area, the amount of surface inside a shape. Ask your child to explain how area and perimeter are used in solving the two problems below.

Please return this Home Link to school tomorrow.

(1) All of the sides of the two figures below are 2 feet long. Find the perimeter of each figure. Remember to write the units with your answers.

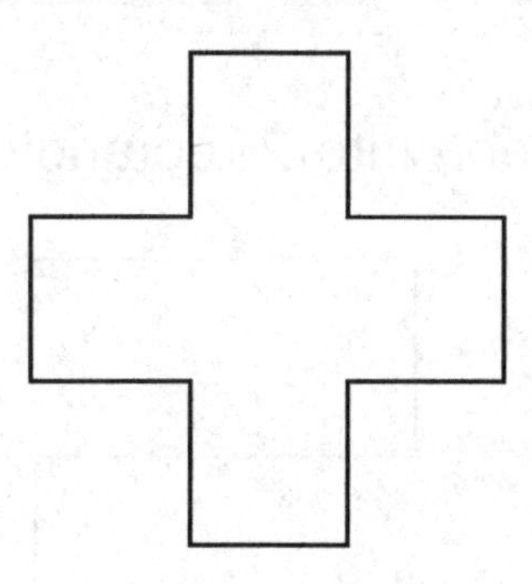

Perimeter = ______________ (unit)

Perimeter = ______________ (unit)

(2) Sue wants to paint the longest wall in her bedroom pink. She measured the wall and found that it is 10 feet long and 8 feet tall. When she went to the hardware store to buy paint, Sue learned that 1 quart of paint can cover 50 square feet.

Sue should buy ______________ (unit) of paint.

Show how you figured out how much paint Sue will need.

Dividing Polygons into Rectangles

Lesson 4-12

NAME DATE TIME

① Divide this polygon into 2 rectangles using 1 line segment.

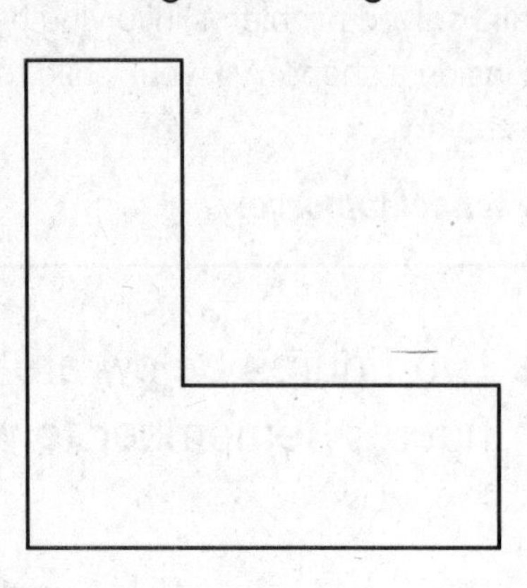

② Divide this polygon into 2 rectangles using 1 line segment.

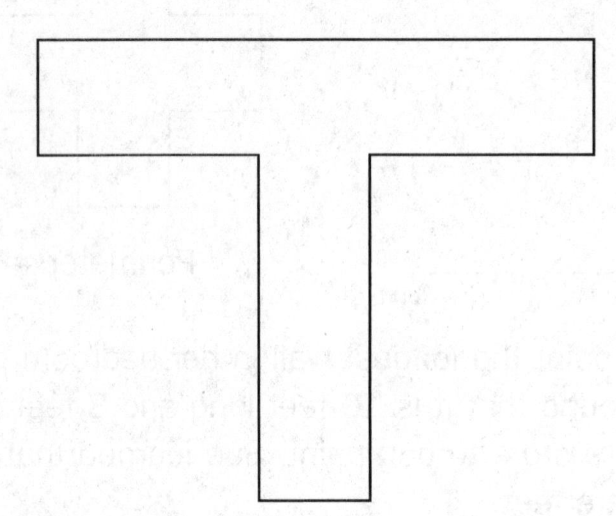

③ Divide this polygon into 3 rectangles using 2 line segments.

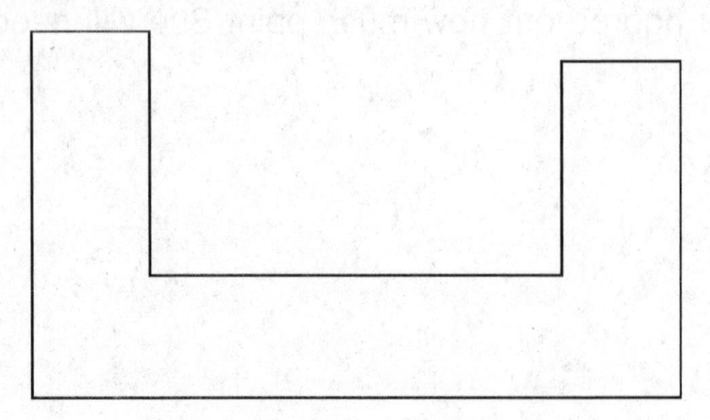

Decomposing Same-Size Rectilinear Figures

Lesson 4-12

NAME DATE TIME

Decompose each figure in a different way. Then use the smaller rectangles to find the area of each rectilinear figure. Show your work. Remember to write the units.

□ = 1 sq cm

① 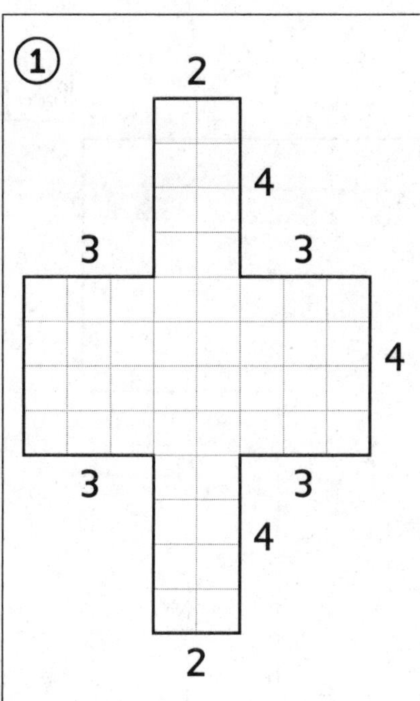

Area: ____________

② 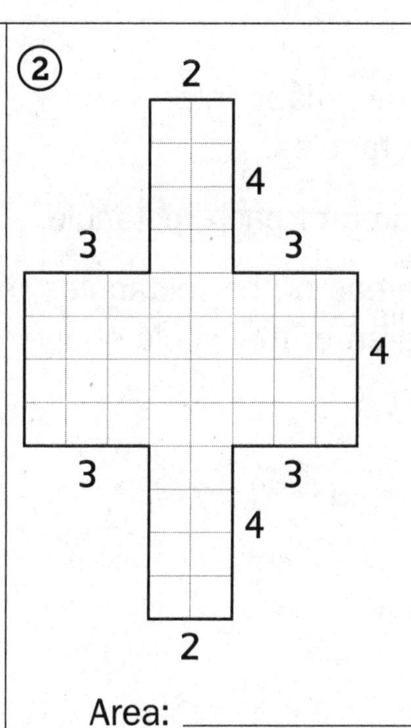

Area: ____________

③ 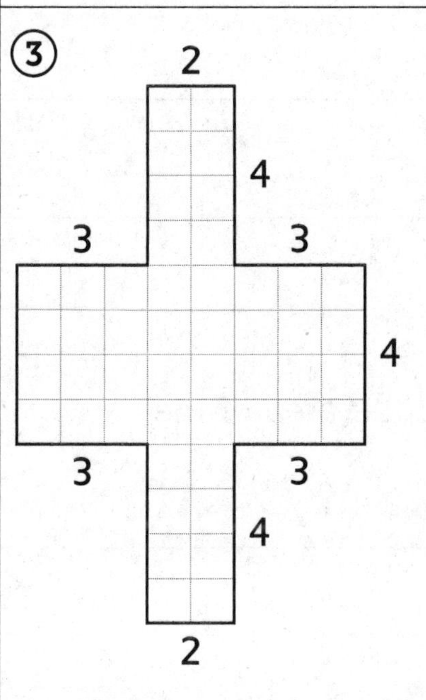

Area: ____________

④ On the back of this page, write what you notice about the areas of the rectilinear figures. Then explain your answer.

Finding the Area of Rectilinear Figures

Home Link 4-12

NAME DATE TIME

Family Note Today your child learned how to find the area of a rectilinear figure (a polygon whose sides all meet to make right angles) by decomposing, or separating, it into smaller rectangles. Help your child follow the steps to find the area of the rectilinear figure below.

Please return this Home Link to school tomorrow.

SRB 180-181

① Partition the shape into 2 or 3 rectangles.

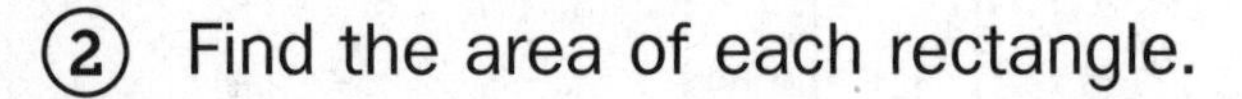

② Find the area of each rectangle.

③ Add the areas of the rectangles to find the area of the whole shape.

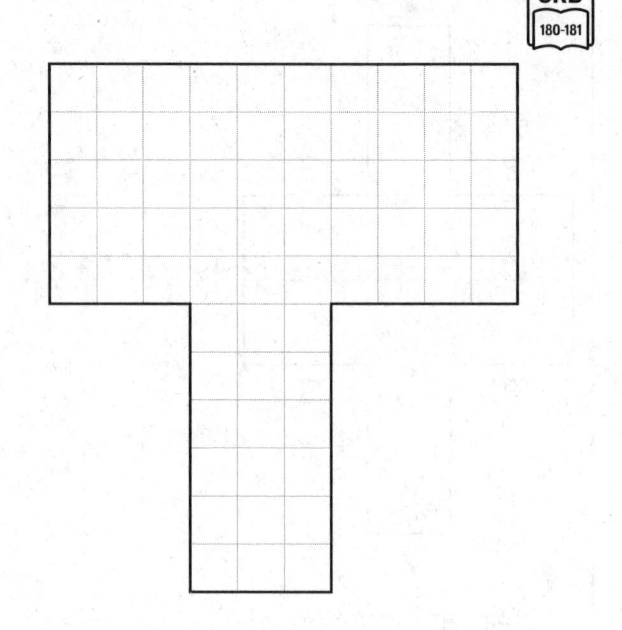

④ ______________________________

(number sentences for areas of rectangles)

____

(number sentence for area of whole shape)

Area of whole shape: ______ square units

⑤ How can the area of each rectangle help you find the area of the whole shape?

____

____

Unit 5: Family Letter

Home Link 4-13

NAME DATE TIME

Fractions and Multiplication Strategies

In this unit, your child will build on earlier experiences and continue to partition shapes and recognize fractions as equal parts of a whole. Children are formally introduced to standard notation for fractions $\left(\frac{1}{2}, \frac{3}{4}\right)$ and explore the relationship between the numerator and the denominator. They use fraction circle pieces to represent fractions of regions and to recognize equivalent fractions, such as $\frac{1}{2}$ and $\frac{2}{4}$.

Your child will also continue to develop multiplication strategies and work with properties of multiplication. Strategy application and discussion help children gain fluency and eventually automaticity with their multiplication facts.

In Unit 5, children will:

- Develop the understanding that the size of a fractional part changes with the size of the whole.
- Represent fractions using standard notation, words, numbers, and drawings.
- Recognize the importance of using the same whole when comparing fractions.
- Recognize equivalent fractions.
- Use known multiplication facts (helper facts) to solve unknown multiplication facts.
- Use doubling as a multiplication facts strategy.
- Play a game to find missing factors.
- Break apart a factor as a multiplication facts strategy.
- Identify and explain patterns in multiplication products.

−9	−8	−7	−6	−5	−4	−3	−2	−1	0
1	2	3	4	5	6	7	8	9	10
11	12	13	14	15	16	17	18	19	20
21	22	23	24	25	26	27	28	29	30
31	32	33	34	35	36	37	38	39	40
41	42	43	44	45	46	47	48	49	50
51	52	53	54	55	56	57	58	59	60
61	62	63	64	65	66	67	68	69	70
71	72	73	74	75	76	77	78	79	80
81	82	83	84	85	86	87	88	89	90
91	92	93	94	95	96	97	98	99	100
101	102	103	104	105	106	107	108	109	110
111	112	113	114	115	116	117	118	119	120

Vocabulary

Important lesson components and terms in Unit 5:

adding a group A multiplication strategy that involves thinking of a multiplication fact such as 6×4 as 6 groups of 4 and adding a group to a known fact to solve an unknown fact. *Example:* Knowing $5 \times 4 = 20$ can help solve $6 \times 4 = ?$. By adding a group of 4 to 20, you can solve $6 \times 4 = 24$.

break-apart strategy A multiplication strategy in which a factor is broken apart, or decomposed, into smaller numbers. To solve 7×8, children can break 7 into 5 and 2. Then they can solve. $5 \times 8 + 2 \times 8 = 40 + 16 = 56$, so $7 \times 8 = 56$.

denominator The number below the line in standard fraction notation, such as the 2 in $\frac{1}{2}$. The number of equal parts into which the whole has been divided.

numerator 3 ← number of parts shaded
denominator 4 ← number of equal parts

doubling A multiplication strategy in which the product of a known fact is doubled to solve an unknown fact. *Example:* Knowing $2 \times 7 = 14$ can help solve $4 \times 7 = ?$. By doubling 14, you can determine $14 + 14 = 4 \times 7 = 28$.

equivalent fractions Fractions that name the same value, such as $\frac{1}{2}$ and $\frac{4}{8}$.

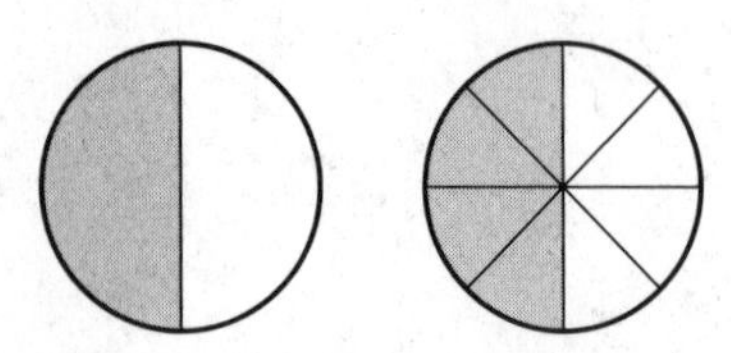

factor Any of the numbers that are multiplied to find a product. *Example:* In the problem $4 \times 7 = 28$, the factors are 4 and 7.

fraction A number in the form $\frac{a}{b}$. The numerator, *a*, can be any number. The denominator, *b*, cannot be 0. For example, $\frac{1}{4}$, $\frac{3}{8}$, and $\frac{5}{2}$ are fractions. A fraction may be used to name part of a whole, to compare two quantities, or to represent division.

helper facts A known fact that can be used to solve an unknown fact.

$6 \times 7 = 42$ $7 \times 7 = 49$ $8 \times 7 = 56$

$7 \times 7 = 49$ can be used as a helper fact to find 6×7 by subtracting a group or to find 8×7 by adding a group.

missing factor The unknown factor in a multiplication fact. *Example:* In $5 \times ? = 30$, the missing factor is 6.

near squares Facts that can be solved by adding or subtracting a group to square multiplication facts. *Example:* 3×4 is a near square because it is closely related to 4×4.

numerator The number above the line in standard fraction notation, such as the 1 in $\frac{1}{2}$. In a part-whole *fraction*, in which the *whole* is divided into a number of equal parts, the numerator is the number of equal parts being considered.

product The result of multiplying two factors. *Example:* In $4 \times 3 = 12$, the product is 12.

subtracting a group A multiplication strategy that involves subtracting a group from a known fact to solve an unknown fact.

unit fraction A *fraction* whose *numerator* is 1. For example, $\frac{1}{2}$, $\frac{1}{3}$, $\frac{1}{4}$, and $\frac{1}{8}$ are unit fractions.

Do-Anytime Activities

The following activities provide practice for concepts taught in this and previous units.

1. Help your child find fractions in the everyday world, in advertisements, on measuring tools, in recipes, and so on.
2. Discuss ways to cut a rectangular casserole, a circular pizza, or similar food to feed various numbers of people so that each person gets an equal portion. Draw pictures if you do not have the actual food item.
3. Continue to practice multiplication facts by playing games such as *Multiplication Draw* and *Salute!* (see *Building Skills through Games*) and by working with Fact Triangles.
4. Provide your child with problems that have missing factors for division and multiplication practice. *Example: 6 groups of how many pennies would equal 18 pennies?*
5. Discuss how various multiplication strategies, such as adding a group, subtracting a group, doubling, and breaking apart a factor, can help solve unknown facts.

21

×, ÷

3 **7**

Fact Triangle

$7 \times 3 = 21$

$3 \times 7 = 21$

$21 \div 7 = 3$

$21 \div 3 = 7$

Fact family for the numbers 3, 7, and 21

Building Skills through Games

In Unit 5 your child will practice multiplication facts by playing the following games. For detailed instructions, see the *Student Reference Book.*

Fraction Memory Players turn over two fraction cards to find equivalent fraction pairs. Pairs are collected, while other cards are turned back over for future turns.

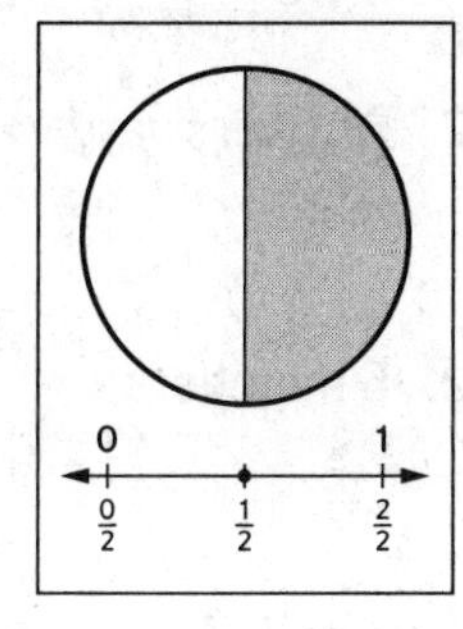

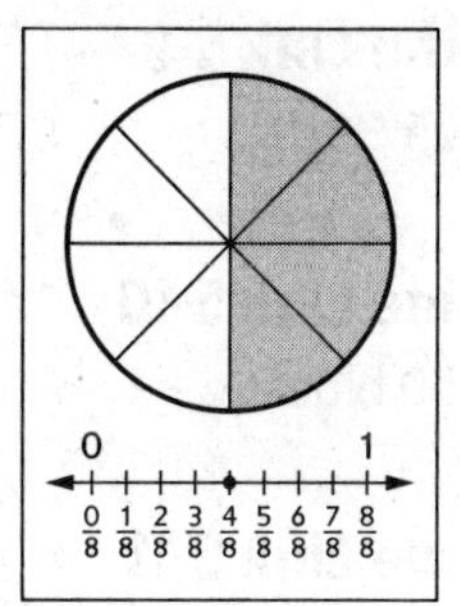

Multiplication Draw Players draw two number cards and multiply them. They add the products of five "draws" to try to get the largest sum.

Salute! The Dealer gives one card to each of two Players. Without looking at their cards, the Players place them on their foreheads facing out. The Dealer multiplies to find the product of the numbers on the cards and says it aloud. Each Player uses the product and the number on the opposing player's forehead to figure out the number (missing factor) on his or her own card.

As You Help Your Child with Homework

As your child brings home assignments, you may want to go over the instructions together, clarifying them as necessary. The answers listed below will guide you through this unit's Home Links.

Home Link 5-1

1. 1-half
2. 1-fourth
3. 1-half
4. 35
5. 165

Home Link 5-2

1. one-half or 1-half; $\frac{1}{2}$
2. $\frac{3}{8}$
3. five-sixths or 5-sixths; $\frac{5}{6}$
4. two-thirds or 2-thirds

Home Link 5-3

1.

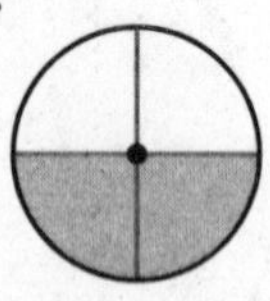

peach pie

2.

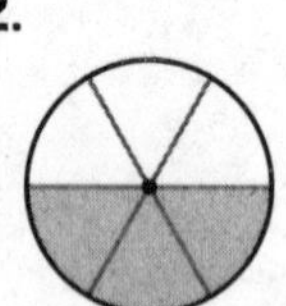

blueberry pie

3.

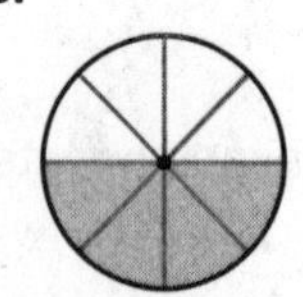

strawberry pie

Home Link 5-4

1. Sample helper fact: $2 \times 8 = 16$
   Sample answer: I start with 16 and add one group of 8 to get $16 + 8 = 24$.
   $3 \times 8 = 24$
2. Sample helper fact: $10 \times 7 = 70$
   Sample answer: I start with 70 and take away one group of 7 to get $70 - 7 = 63$.
   $9 \times 7 = 63$

Home Link 5-5

1.

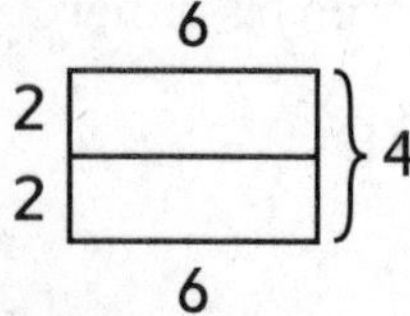

$2 \times 6 = 12$; $4 \times 6 = 24$

Sample answer: I started with the area of the first rectangle, which is 12. I doubled that by thinking $12 + 12 = 24$. So the area of the new rectangle is 24, which means $4 \times 6 = 24$.

Home Link 5-6

Sample answer: 8; $4 \times 6 = 24$, $24 + 24 = 48$; $8 \times 6 = 48$

Home Link 5-7

1–2.

1	2	3	4	5	6	7	8	9	10
11	12	13	14	15	16	17	18	19	20
21	22	23	24	25	26	27	28	29	30
31	32	33	34	35	36	37	38	39	40
41	42	43	44	45	46	47	48	49	50

3. Sample answer: With each 10s fact we add another 10, which is like moving down one row because the number grid has rows of 10. So they make a straight column.
4. Sample answer: With each 9s fact we add another 9. We can go down a row for the 10 but then go back a space because it is one less than 10. It makes a diagonal line.

Home Link 5-8

Round 1: Player 2: 5 Round 2: Player 1: 6
Round 3: Player 1: 4 Round 4: Player 2: 6
Sample answer: I thought: "What number do I have to multiply 5 by to get 30?" The answer is 6.

Home Link 5-9

1. 25; 36
2. Answers vary.

Home Link 5-10

1. 50 blocks
2. 6 members

Home Link 5-11

Sample answers: Factor: 9; Parts: 5; 4
Helper facts: $7 \times 5 = 35$, $7 \times 4 = 28$;
$7 \times 9 = 7 \times 5 + 7 \times 4$; $7 \times 9 = 63$

Exploring Fractions

Lesson 5-1

NAME DATE TIME

These show 1-fourth.	These do NOT show 1-fourth.

Explain how you can tell if a part of a shape shown above is 1-fourth.

Exploring Fractions

Lesson 5-1

NAME DATE TIME

These show 1-fourth.	These do NOT show 1-fourth.

Explain how you can tell if a part of a shape shown above is 1-fourth.

Completing the Whole

Lesson 5-1

NAME DATE TIME

① The first graders are building a small house with centimeter cubes. The drawing shows 1-half of the floor of the house. Use centimeter cubes to build the whole floor of the house. Then finish the picture.

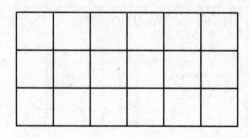

② This drawing shows 3-fourths of a line segment. Use your ruler to figure out how long this line segment is in inches. Then figure out how much longer the line segment should be to make it whole. Draw the rest of the line segment.

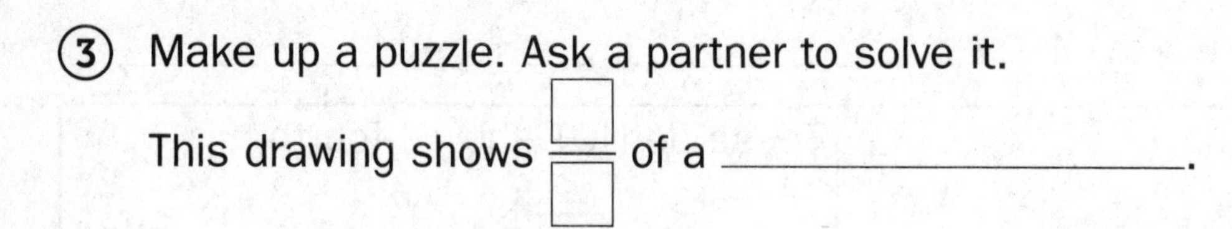

③ Make up a puzzle. Ask a partner to solve it.

This drawing shows $\frac{\square}{\square}$ of a ______________.

Draw the whole ______________.

Partitioning Halves of Different Wholes

Lesson 5-1

NAME DATE TIME

Partition each shape into halves. Label the fraction.

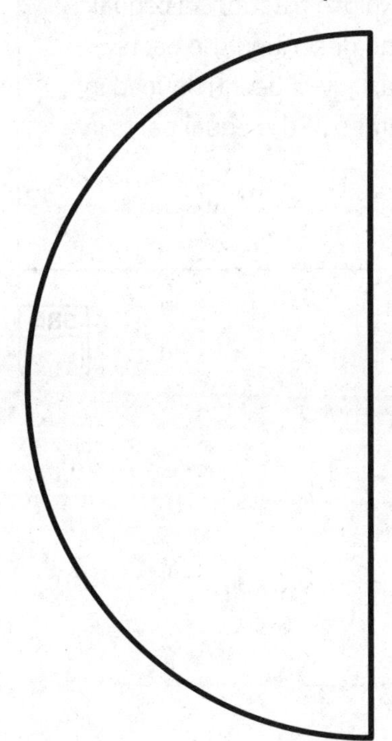

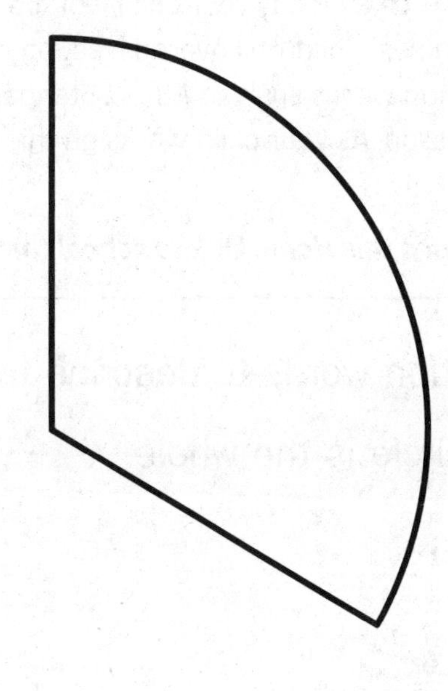

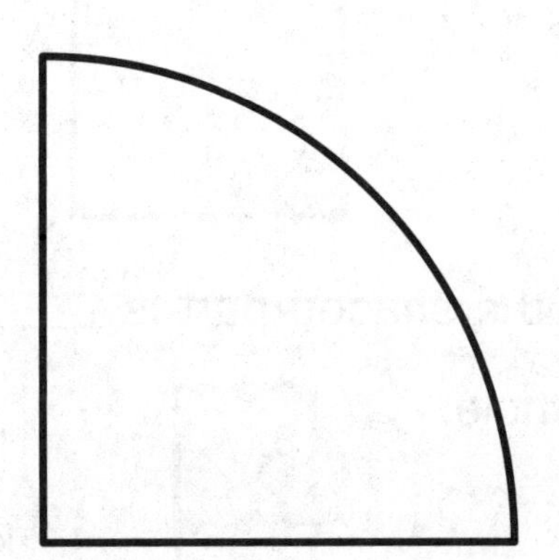

Try This

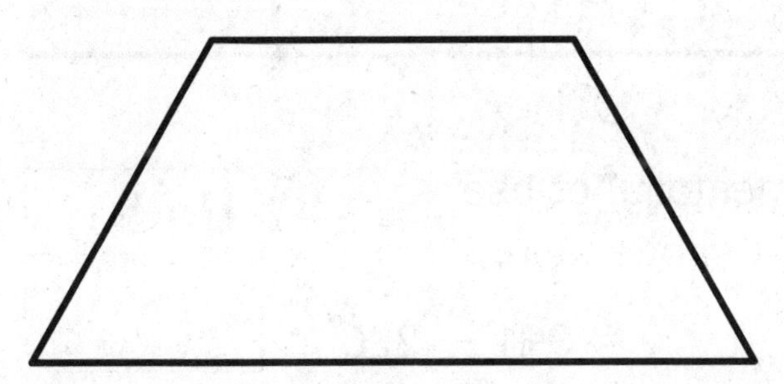

Fractions of a Whole

Home Link 5-1

NAME DATE TIME

Family Note Today your child worked with fraction circle pieces to explore fractions as equal parts of wholes. Children covered fraction circle pieces with equal parts and described the parts using fraction words such as *1-third.* Standard notation for fractions, such as $\frac{1}{3}$, will be introduced in the next lesson. As your child works on this activity, discuss the fraction names of the equal parts in each problem.

Please return this Home Link to school tomorrow.

Write fraction words to describe the shaded part.

① The circle is the whole.

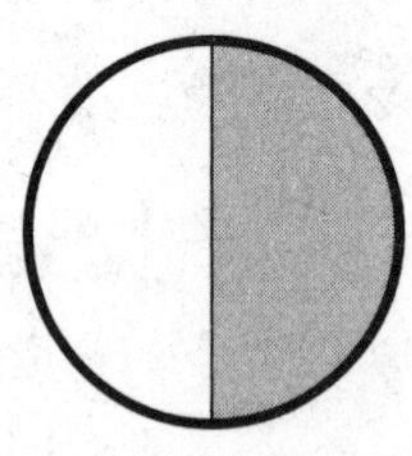

A fraction that names the shaded part is ______________.

② The square is the whole.

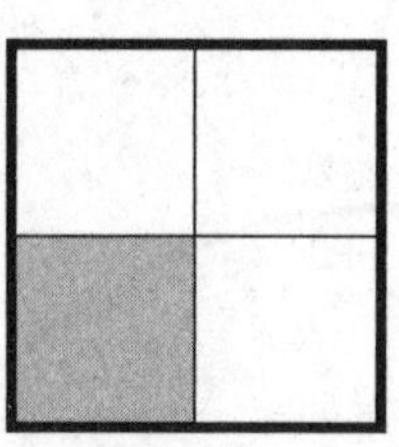

A fraction that names the shaded part is ______________.

③ The rectangle is the whole.

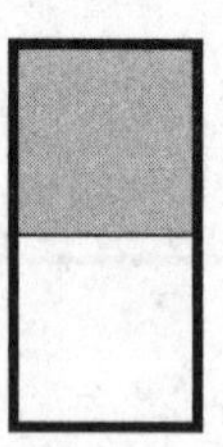

A fraction that names the shaded part is ______________.

Practice

Fill in the unit box. Solve these problems mentally, or use counting-up, expand-and-trade, or trade-first subtraction.

④ ________ = 326 − 291

⑤ ________ = 391 − 226

Comparing Fractional Amounts

Lesson 5-2

NAME DATE TIME

Quinn and Rob each brought $\frac{1}{4}$ of their money to spend at the dollar store. Quinn brought \$1 and Rob brought \$3.

① Explain how they could have different amounts even though they each brought $\frac{1}{4}$ of their money.

② The pictures below show how many dollar bills Quinn and Rob had when they went shopping. Draw dollar bills to show much money each had before they went shopping.

Quinn's money

\$1

Rob's money

\$1 \$1 \$1

Quinn had ___________.

Rob had ___________.

Exploring Numerators and Denominators

Lesson 5-2

NAME DATE TIME

Look at the shapes below.

① 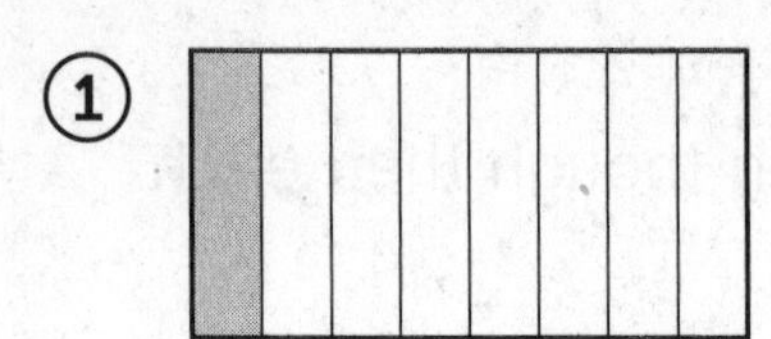

How many eighths are shaded? ____

How many eighths in all? ____

Fraction of shaded parts: $\frac{\square}{\square}$

②

Fraction of shaded parts: $\frac{\square}{\square}$

③ 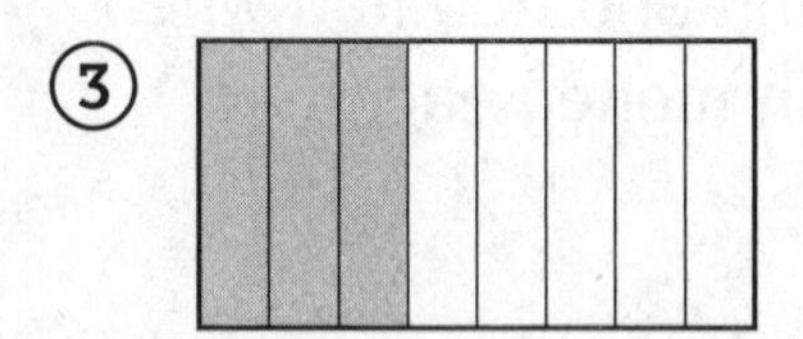

Fraction of shaded parts: $\frac{\square}{\square}$

④  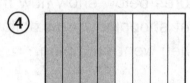

Fraction of shaded parts: $\frac{\square}{\square}$

⑤ 

Fraction of shaded parts: $\frac{\square}{\square}$

⑥ 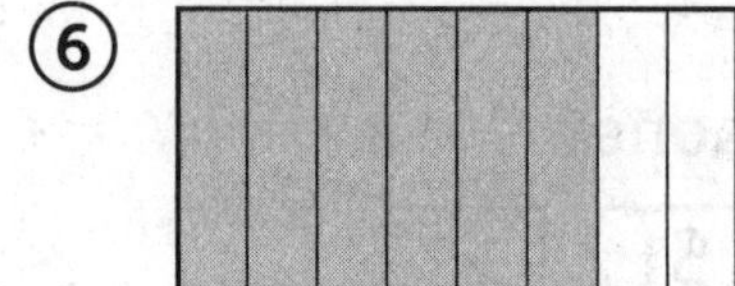

Fraction of shaded parts: $\frac{\square}{\square}$

⑦ 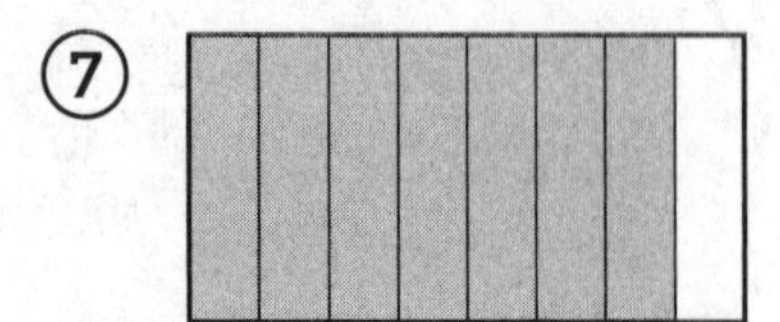

Fraction of shaded parts: $\frac{\square}{\square}$

⑧ 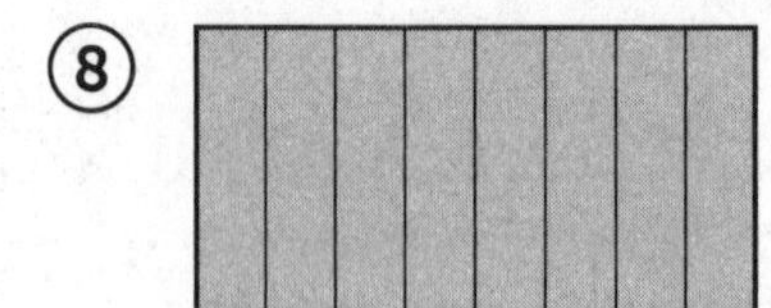

Fraction of shaded parts: $\frac{\square}{\square}$

⑨ What do you notice about the numerator and denominator in each shape?

____

____

Representing Fractions

Home Link 5-2

NAME DATE TIME

Family Note Today your child learned how to represent fractions in words, in standard notation ($\frac{1}{2}$, $\frac{3}{4}$, and so on), and with drawings. For each of the problems below, the whole is a different shape. Help your child relate each fraction representation to the number of equal-size parts each whole is divided into and the number of shaded equal parts.

Please return this Home Link to school tomorrow.

Complete the table.

SRB 134-136

Picture	Words	Number
Example:	three-fourths or 3-fourths	$\frac{3}{4}$
1.		
2.	three-eighths	
3.		
4.		$\frac{2}{3}$

Equivalent Fractions

Home Link 5-3

NAME DATE TIME

Family Note Today your child continued working with fractions by finding different fractions that name the same amount of the whole, or equivalent fractions. Children identify equivalent fractions by looking at the shaded area of a figure compared to the shaded area of another same-size figure.

Please return this Home Link to school tomorrow.

The pictures show three kinds of fruit pie. Use a straightedge to do the following:

1. Divide the peach pie into 4 equal pieces. Shade 2 of the pieces.
2. Divide the blueberry pie into 6 equal pieces. Shade 3 of the pieces.
3. Divide the strawberry pie into 8 equal pieces. Shade 4 of the pieces.

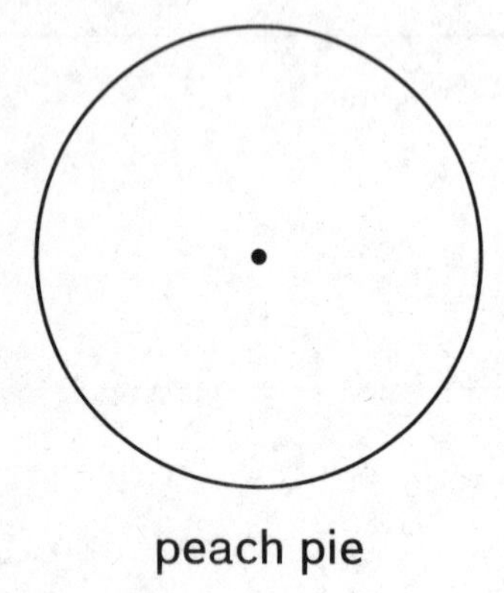

peach pie

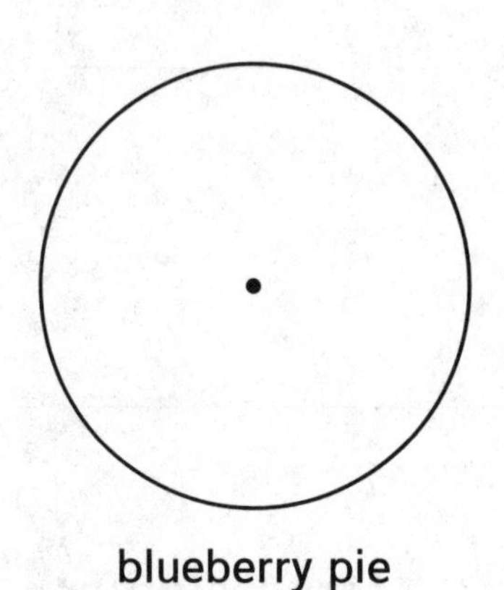

blueberry pie

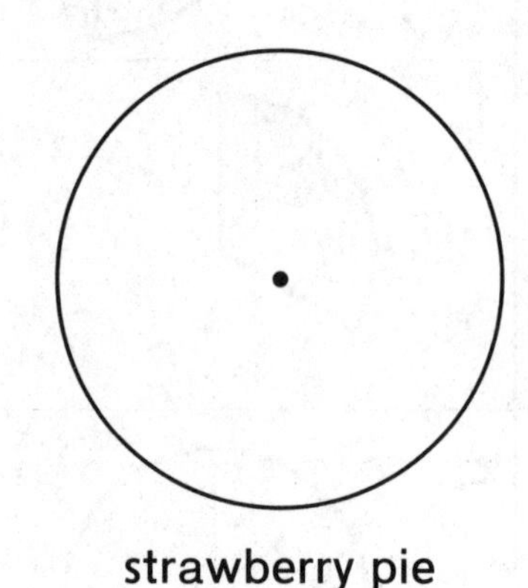

strawberry pie

What fraction of each pie did you shade?

4. I shaded ______ of the peach pie.

   Write another name for this fraction: ______

5. I shaded ______ of the blueberry pie.

   Write another name for this fraction: ______

6. I shaded ______ of the strawberry pie.

   Write another name for this fraction: ______

Explain to someone at home how you know that all of the fractions on this page are equivalent.

Calculating the Number of Mosaic Tiles

Lesson 5-4

NAME DATE TIME

Solve the following problem:

An artist made a square mosaic with 99 rows of tiles and 99 tiles in each row. How many tiles were used?

You can use this helper fact: $100 \times 100 = 10{,}000$

Do not use your calculator. Show your work. Explain what you did.

Identifying Helper Facts

Home Link 5-4

NAME DATE TIME

Family Note Today your child practiced identifying and using helper facts to solve unknown multiplication facts by adding or subtracting a group. For example, children added a group to "helper" 2s and 5s facts to solve 3s and 6s facts. They subtracted a group from "helper" 5s and 10s facts to solve 4s and 9s facts. Practice with efficient fact strategies such as these helps children become fluent with all multiplication facts.

Please return this Home Link to school tomorrow.

For each fact below:

- Think of a helper fact.
- Add or subtract a group from the product of the helper fact.
- Solve the fact.

Example: $6 \times 7 = ?$

Helper fact: $5 \times 7 = 35$

How I can use it: I can add one more group of 7 to 35 to get $35 + 7 = 42$.

$6 \times 7 =$ 42

① $3 \times 8 = ?$

Helper fact: ______

How I can use it: ______

______

$3 \times 8 =$ ______

② $9 \times 7 = ?$

Helper fact: ______

How I can use it: ______

______

$9 \times 7 =$ ______

Finding the Areas of Rectangles

Lesson 5-5

NAME DATE TIME

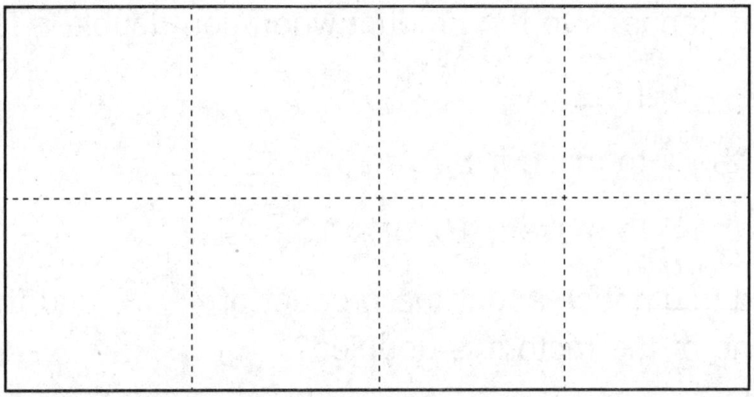

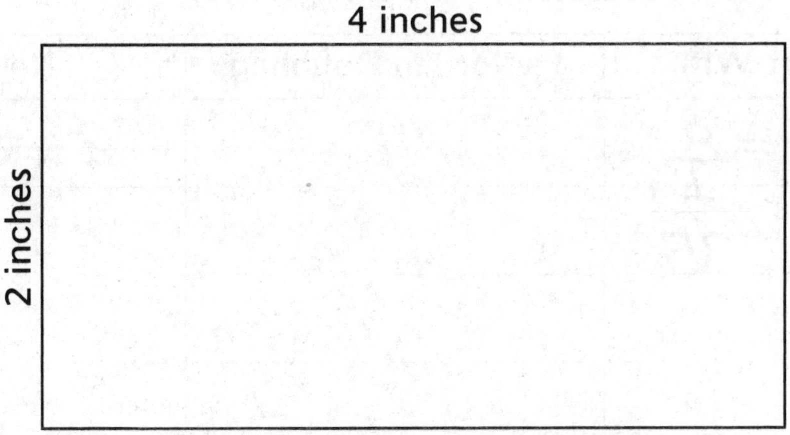
4 inches
2 inches

Exploring Factor Patterns

Lesson 5-5

NAME DATE TIME

① Explore what happens to the product when you double a factor.

a. Start. $4 \times 5 =$ ______

b. Double the 4 to make it 8. $8 \times 5 =$ ______

c. Double the 5 to make it 10. $4 \times 10 =$ ______

d. What do you notice about the product of 4×5 and the products when one of the factors is doubled?

____

② Double one of the factors in other multiplication facts. To start, record three facts. Show your work in the table below.

Fact I Start With	Factor I'm Doubling	New Fact
$2 \times 4 = 8$	4	$2 \times 8 = 16$

What do you notice about the product of your starting fact and the product of your new fact after you doubled a factor?

____

③ Think about your work with doubling factors. Now predict what will happen when you triple a factor.

____

Write an example of tripling a factor.

Fact I Start With	Factor I'm Tripling	New Fact

Doubling the Area of a Rectangle

Lesson 5-5

NAME DATE TIME

① Carefully cut out Rectangle A and Rectangle B on page 166.

Measure the side lengths of each rectangle to the nearest inch.

Then find the area. Write number sentences to show what you did.

a. Rectangle A

Number sentence:

____

Area: ______________________
(unit)

b. Rectangle B

Number sentence:

____

Area: ______________________
(unit)

② What do you notice about the areas of Rectangle A and Rectangle B?

____

Carefully tape the two rectangles together without gaps or overlaps to make a larger rectangle.

③ Predict the area of the larger rectangle. ______________________
(unit)

Explain how you made your prediction. ______________________

____

____

____

④ Measure the side lengths of the larger rectangle and find the area.

Area: ______________________
(unit)

Write a multiplication number sentence to show how you found the area of the larger rectangle.

Number sentence: ______________________

Doubling the Area of a Rectangle (continued)

NAME DATE TIME

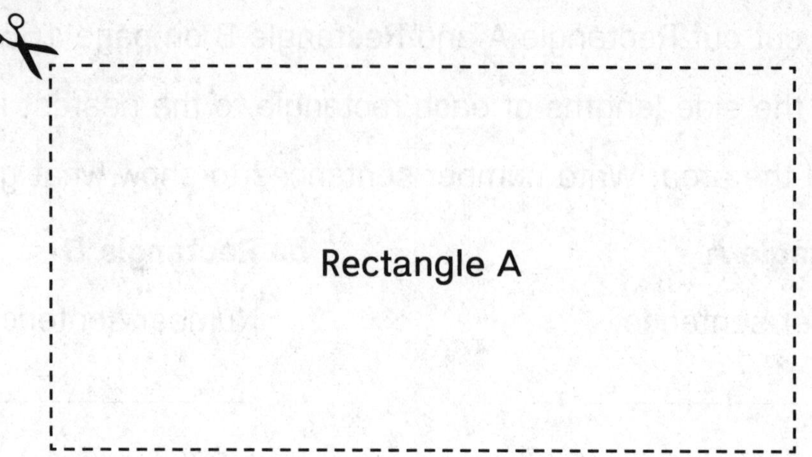

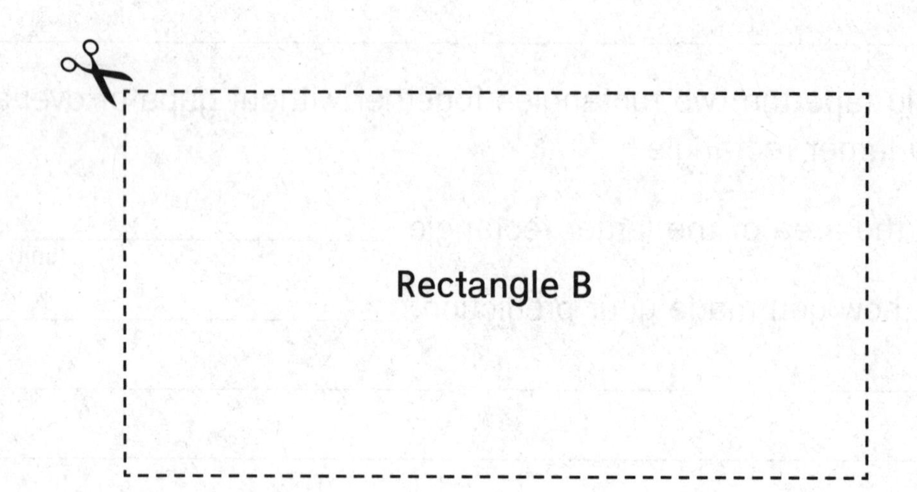

Doubling, Part 1

Home Link 5-5

NAME DATE TIME

Family Note Doubling (adding a number to itself or multiplying a number by 2) can be used as a strategy to solve facts with a "double" as a factor, such as the 4s, 6s, and 8s facts. For example, your child used doubling with the helper fact 2×7 to figure out 4×7 ($2 \times 7 = 14$ and $14 + 14 = 28$, so $4 \times 7 = 28$). Doubling the area of a rectangle can help children see and model the doubling fact strategy. Encourage your child to use drawings to solve the problem below.

Please return this Home Link to school tomorrow.

SRB 49

Maria wants to figure out $4 \times 6 = ?$.
She notices that 4 is the double of 2.
Help Maria use 2×6 to solve 4×6.
Maria starts by sketching a 2-by-6 rectangle. Add to Maria's picture to show how she could use doubling to find the answer to 4×6. Record your thinking below.

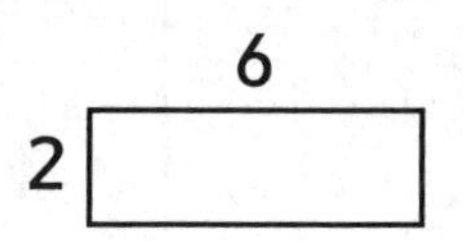

$2 \times 6 =$ ______

$4 \times 6 =$ ______

How I figured it out: ____

____

____

4-by-6 Rectangle

Lesson 5-6

NAME DATE TIME

6 inches

4 inches

Solving an Allowance Problem

Lesson 5-6

NAME DATE TIME

Sara is discussing a raise in allowance with her parents. They ask her to choose one of three plans.

Plan A Each week, Sara would get 1¢ on Monday, double Monday's amount on Tuesday, double Tuesday's amount on Wednesday, and so on. Her allowance would keep on doubling each day through Sunday. Then she would start with 1¢ again on Monday.

Plan B Sara would get 32¢ on Sunday, Monday, Wednesday, and Friday of each week. She would get nothing on Tuesday, Thursday, and Saturday.

Plan C Sara would get 16¢ on each day of each week.

Which plan should Sara choose to get the most money? ____________

Show your work on the back of this page. Explain how you found your answer. Use number sentences in your explanation.

____

____

____

____

____

Try This

For the plan you chose, how much money would Sara earn in a year?

____

____

____

8-by-6 Rectangle

Lesson 5-6

NAME DATE TIME

Cutting a Rectangle in Half to Find Area

Lesson 5-6

NAME DATE TIME

Follow the directions to find the area of the large rectangle.

1. Carefully cut out the rectangle on page 170.
2. Measure the lengths of the sides of the rectangle to the nearest inch.

   Lengths of sides: ______ inches and ______ inches
3. Cut your rectangle into 2 smaller but equal rectangles. You may fold it in half to help.
4. Measure the sides of the smaller rectangles. Use the measurements to find the areas of the smaller rectangles. Write number sentences to show what you did.

   Smaller rectangle 1

   Number sentence:

   ____

   Area: ______ square inches

   Smaller rectangle 2

   Number sentence:

   ____

   Area: ______ square inches
5. Add the areas of the 2 smaller rectangles to find the area of the original rectangle.

   Area: ______ square inches
6. Compare the side lengths of the original rectangle to the side lengths of the smaller rectangles. What do you notice?

   ____

   ____
7. Find someone whose smaller rectangles have different side lengths than yours. Is the area of each small rectangle the same or different? Explain.

   ____

   ____
8. Write a multiplication number sentence for the area of the large rectangle.

   Number sentence: ______________________________

Doubling, Part 2

Home Link 5-6

NAME DATE TIME

Family Note Today your child continued to work on the doubling facts strategy with facts that have even factors. Your child broke apart an even factor into two identical factors and used the resulting helper fact to derive the final product. For example, 6×7 can be broken into 3×7 and 3×7, making the total easier to find:

7 feet

6 feet

3	3×7	$3 \times 7 = 21$
3	3×7	$3 \times 7 = 21$

$21 + 21 = 42$
so $6 \times 7 = 42$

Please return this Home Link to school tomorrow.

SRB 49

Show how you can solve 8×6 using doubling.

Factor I will split in half: ______

Sketch:

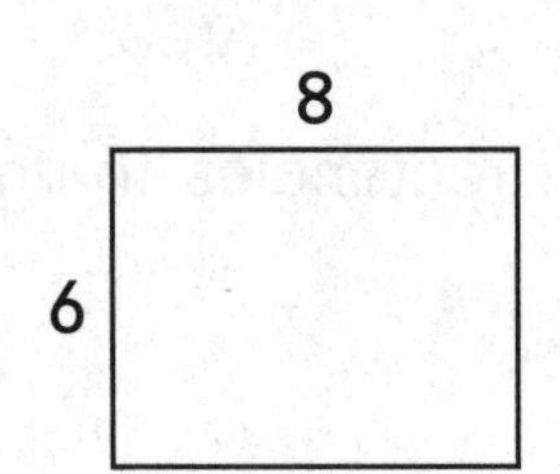

$8 \times 6 =$ ______

What helper fact did you double to solve 8×6?

____

Exploring a Pattern

Lesson 5-7

NAME DATE TIME

Find each product. Then look for patterns.

① $1 \times 2 =$ ______

② $2 \times 3 =$ ______

③ $3 \times 4 =$ ______

④ $4 \times 5 =$ ______

⑤ $5 \times 6 =$ ______

⑥ $6 \times 7 =$ ______

⑦ What happens when you subtract each product from the next larger product?

______

______

Read page 71 in your *Student Reference Book* to learn more about patterns in multiplication.

Patterns in Products

Home Link 5-7

NAME DATE TIME

Family Note Today your child used number grids and the Multiplication/Division Facts Table to explore patterns in multiples of 5s, 9s, and 10s, as well as in products of even and odd numbers. Recognizing and making sense of patterns is an important part of mathematics. Understanding the patterns in these products will also help your child become more fluent with multiplication facts.

Please return this Home Link to school tomorrow.

1	2	3	4	5	6	7	8	9	10
11	12	13	14	15	16	17	18	19	20
21	22	23	24	25	26	27	28	29	30
31	32	33	34	35	36	37	38	39	40
41	42	43	44	45	46	47	48	49	50

1. Circle the products of 10×1, 10×2, 10×3, 10×4, and 10×5.
2. Shade the products of 9×1, 9×2, 9×3, 9×4, and 9×5.
3. Explain why the pattern for the products of 10s facts is a straight column.

   ____

   ____

   ____

4. Explain why the pattern for the products of 9s facts is a diagonal.
   Hint: Think of how it compares to the products of the 10s facts pattern.

   ____

   ____

   ____

Extending Fact Families

Lesson 5-8

NAME DATE TIME

①

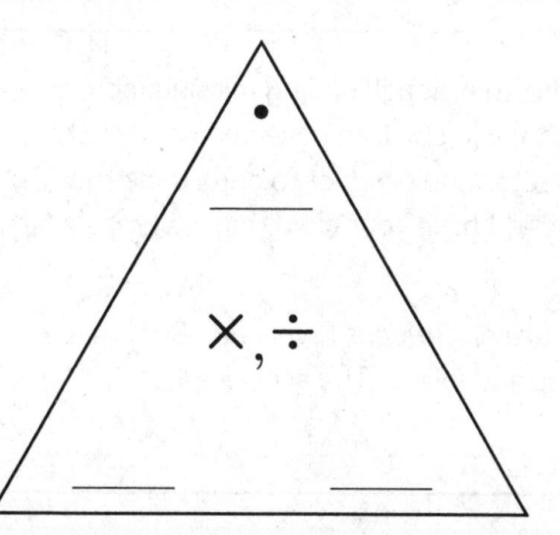

_____ × _____ = _____

_____ × _____ = _____

_____ ÷ _____ = _____

_____ ÷ _____ = _____

②

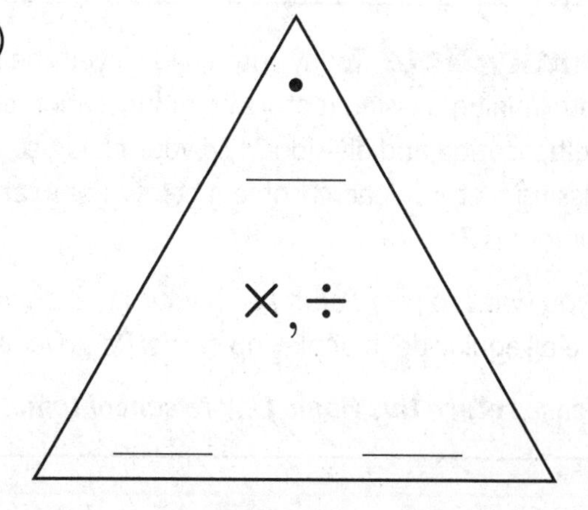

_____ × _____ = _____

_____ × _____ = _____

_____ ÷ _____ = _____

_____ ÷ _____ = _____

③

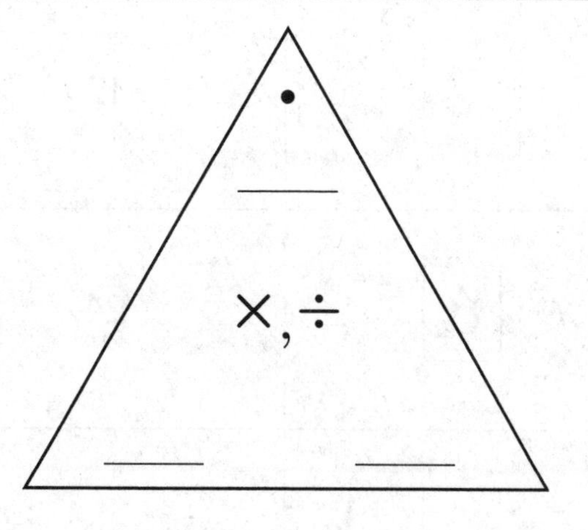

_____ × _____ = _____

_____ × _____ = _____

_____ ÷ _____ = _____

_____ ÷ _____ = _____

④

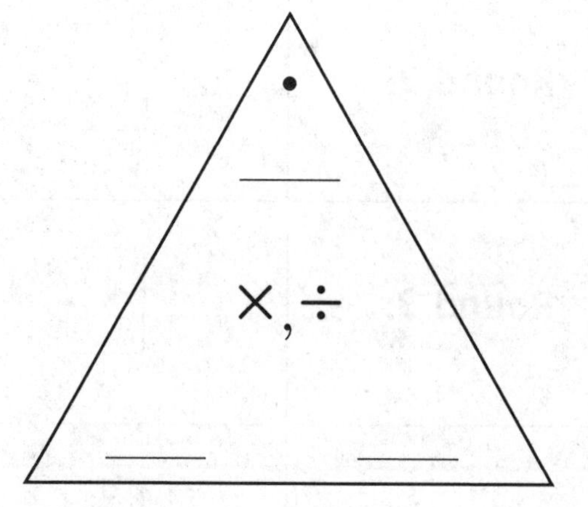

_____ × _____ = _____

_____ × _____ = _____

_____ ÷ _____ = _____

_____ ÷ _____ = _____

Finding Missing Factors

Home Link 5-8

NAME DATE TIME

Family Note Today your child played the game *Salute!* to practice finding missing factors. Determining missing factors in multiplication equations can help children develop fluency with multiplication and division. Help your child use the given factor and product to determine the missing factor in each problem below. For example, in Round 1 have your child think: *5 times what number is 25?*

If you want to play *Salute!* with your child, the directions are on *Student Reference Book*, page 255. Use a regular deck of playing cards. Remove all face cards and jokers. The aces are 1s.

Please return this Home Link to school tomorrow.

Write the missing factors for the rounds of *Salute!*

SRB 255

	Player 1	Player 2	Dealer says the product is:
Round 1:	5		25
Round 2:		2	12
Round 3:		6	24
Round 4:	5		30

Explain how you found the missing factor for one of the rounds.

Near Squares

Home Link 5-9

NAME DATE TIME

Family Note Today your child learned to use familiar multiplication squares, such as $3 \times 3 = 9$ and $8 \times 8 = 64$, to figure out near-squares facts by adding or subtracting groups. For example, the square $8 \times 8 = 64$ can be used as a helper fact for the near square 7×8. By subtracting one group of 8 from 64, children find that $7 \times 8 = 56$. Using squares as helper facts is one more strategy in your child's growing library of multiplication facts strategies.

Please return this Home Link to school tomorrow.

SRB 50

Example: $4 \times 3 = ?$

Square helper fact: $3 \times 3 = 9$

Near square: $4 \times 3 = 12$

X X X
X X X
X X X
O O O

How I solved it: *I added a group of 3 to find 4 × 3.*

① Solve the multiplication squares.

$5 \times 5 =$ _______ $6 \times 6 =$ _______

② Choose one of the squares facts from Problem 1. Write a near square and use your square to help solve the near square. Show your work.

Square helper fact: _______ × _______ = _______

Near square: _______ × _______ = _______

How I solved it: ______________________________

____

Button Dolls

Lesson 5-10

NAME DATE TIME

Read the problem below. Think about what you understand from the story.

Tonya is making button dolls for the school fair. On each doll, she uses 2 buttons for the eyes, 1 button for the nose, and 3 buttons for the clothes.

There are 8 buttons in each package.

Tonya needs to buy packages of buttons so that all of the buttons will be used without any left over.

How many packages could she buy? How many dolls will that make?

With your partner, show what you know about the problem using the pictures.

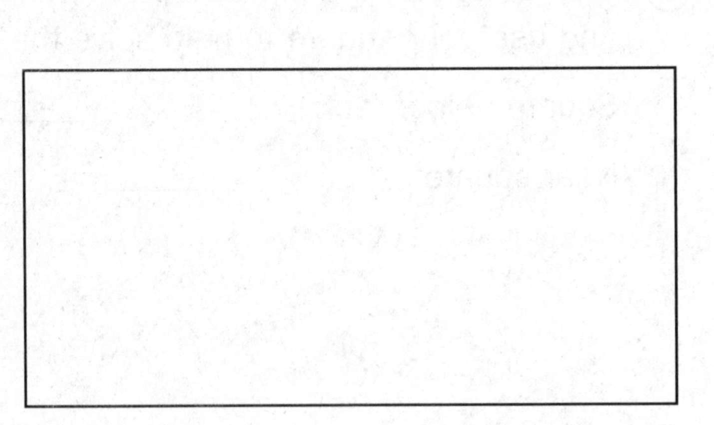

Take turns with your partner. Tell each other what you know about the problem and what you have to find out.

Button Dolls

(continued)

Lesson 5-10

NAME DATE TIME

Work with a partner to solve the problem. Use words or pictures to show how you solved it.

How many packages could Tonya buy? ____________________

How many dolls will that make? ____________________

Making Sense of a Problem

Home Link 5-10

NAME DATE TIME

Family Note Today your child solved a challenging number story. To solve it, your child had to make sense of the information in the problem. If your child has trouble getting started on the problems below, ask: *What does the problem tell you? What do you need to find out?* Encourage your child to draw a picture to show what he or she understands about each problem.

Please return this Home Link to school tomorrow.

① Danika lives 5 blocks from her school. If she walks to and from school 5 days each week, how many blocks does she walk in one week?

______________ (unit)

② It costs \$5 to join the art club. The club has collected \$70 from new members. The club's goal is to collect \$100. How many more members does the club need to meet its goal?

______________ (unit)

Extending the Break-Apart Strategy

Lesson 5-11

NAME DATE TIME

① Joey does not know the answer to 7 × 15.

Use the rectangle below and the break-apart strategy to help him solve the problem. Show your work.

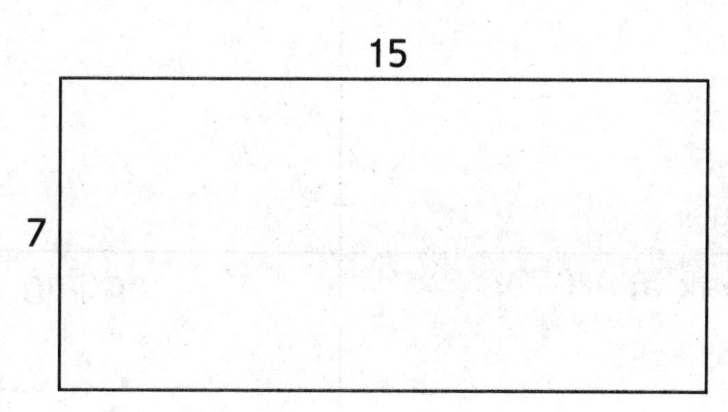

7 × 15 = ______

② Make up a multiplication problem like the one in Problem 1. Sketch a rectangle and use the break-apart strategy to help solve the problem.

Problem: ________________

Sketch:

Multiplication Strategies

Lesson 5-11

NAME DATE TIME

skip counting	turn-around rule
break apart	adding a group
repeated addition	subtracting a group
doubling	near squares

Matching Facts to Strategies

Lesson 5-11

NAME DATE TIME

5 × 8	2 × 7
5 × 1	7 × 8
5 × 5	4 × 4
8 × 3	4 × 7
9 × 4	3 × 7
9 × 3	9 × 6
6 × 4	8 × 6
6 × 7	8 × 9

The Break-Apart Strategy

Home Link 5-11

NAME DATE TIME

Family Note Today your child learned how to break apart one number in a multiplication fact in order to make two helper facts that are easier to solve. Using areas of rectangles helps to illustrate this, as in the example below.

- $7 \times 6 = ?$
- Break apart the 7 into 5 and 2.
- There are two helper facts: 5×6 and 2×6.
- So $7 \times 6 = 5 \times 6 + 2 \times 6$
  $7 \times 6 = 30 + 12$
  $7 \times 6 = 42$

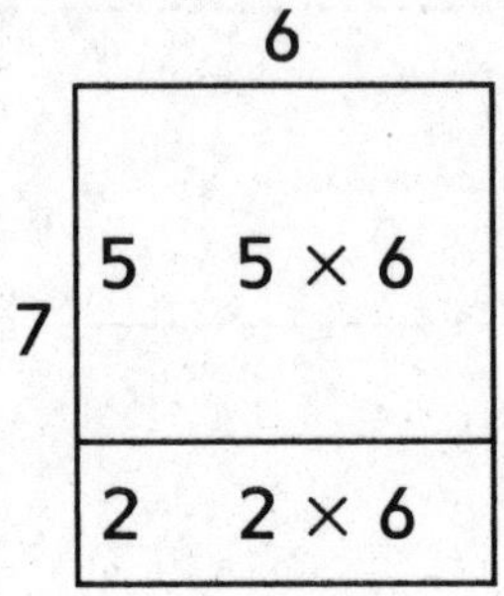

(Diagram: rectangle 7 by 6, split into 5 × 6 and 2 × 6)

6

7 | 5 | 5 × 6
  | 2 | 2 × 6

Please return this Home Link to school tomorrow.

Show one way you can solve $7 \times 9 = ?$.

I will break apart the factor ___ into ___ and ___.

Helper facts that match the areas of the smaller rectangles:

_______ × _______ = _______ and _______ × _______ = _______

Drawing:

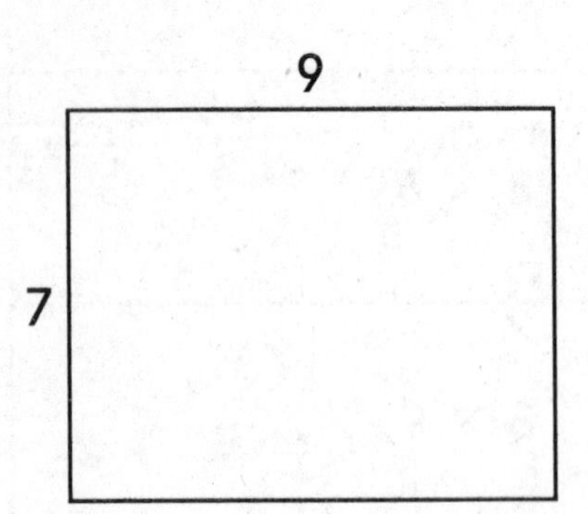

Write a number model using your helper facts:

$7 \times 9 =$ _______ × _______ + _______ × _______

$7 \times 9 =$ _______

Unit 6: Family Letter

Home Link 5-12

NAME DATE TIME

More Operations

In this unit, your child will apply multiplication facts strategies with a focus on using strategies that are efficient and appropriate for solving a given problem. Your child will also learn a new method for multidigit subtraction called trade-first subtraction. This method is highly efficient and relatively easy to learn, and it prepares children for learning the U.S. standard algorithm (introduced in *Fourth Grade Everyday Mathematics*). Children also build on their experiences with solving number stories. They represent number stories using single equations with multiple operations and letters for unknown quantities.

In Unit 6, children will:

- Use trade-first subtraction to solve subtraction problems.
- Identify and apply efficient and appropriate strategies for multiplication facts and problems with larger factors.
- Self-assess automaticity with multiplication facts.
- Use multiplication/division diagrams to represent an unknown quantity with a letter and make sense of multiplication and division number stories.
- Solve number sentences with parentheses.
- Apply the order of operations to solve multistep problems.
- Write number models to represent two-step number stories.
- Play multiplication games to build fact fluency.

$(2 \times 9) - 1 = 17$

One round of *Name That Number*

Vocabulary

Important lesson components and terms in Unit 6:

fact power In *Everyday Mathematics,* automaticity with basic arithmetic facts. Automatically knowing the facts is as important to arithmetic as knowing words by sight is to reading.

multiplication/division diagram A diagram used in *Everyday Mathematics* to model situations in which a total number is made up of equal-size groups. The diagram contains a number of groups, a number in each group, and a total number.

bowls	oranges per bowl	oranges in all
6	?	54

order of operations Rules that specify the order in which operations in a number sentence should be carried out. In *Third Grade Everyday Mathematics,* the order of operations is described as:

1. Do operations inside parentheses first. Follow rules 2 and 3 when computing inside parentheses.
2. Then multiply or divide, in order, from left to right.
3. Finally add or subtract, in order, from left to right.

parentheses ( ) Grouping symbols used to indicate which parts of a number sentence should be done first.

trade-first subtraction One method for solving subtraction problems in which all trades are made before subtracting.

$$\begin{array}{rrr} & 12 & \\ 1 & \cancel{13} & 11 \\ \cancel{2} & \cancel{3} & \cancel{1} \\ -\ 1 & 7 & 4 \\ \hline 0 & 5 & 7 \end{array}$$

Trade-first subtraction

Do-Anytime Activities

The following activities provide practice for concepts taught in this and previous units.

1. As your child subtracts multidigit numbers, talk about the value of each digit and the trades your child makes.
2. Challenge your child to mentally solve unknown multiplication facts faster than you can solve them on a calculator. Have your child explain the strategy he or she used.

3. Ask questions that involve equal sharing and equal grouping. Encourage your child to act out, draw pictures or diagrams, or use numbers to show his or her thinking.

   Example: *8 children each have 3 books. How many books do they have in all? 24 books*

children	books per child	books in all
8	3	?

$$8 \times 3 = ?$$

   Example: *Each box has 8 crayons. There are 24 crayons in all. How many boxes are there? 3 boxes*

boxes	crayons per box	crayons in all
?	8	24

$$24 \div 8 = ? \text{ or } ? \times 8 = 24$$

4. Pose simple number stories that your child can solve with two calculations. Help your child make sense of them by asking questions such as: *What do you know from the story? What do you need to figure out? What can you do first? Next?* Examples:
   - *I have 50 cents and want to buy peanuts for 25 cents and popcorn for 40 cents. Do I have enough money?*
   - *We want to make 6 party bags with 2 glitter pencils and 1 mechanical pencil in each. How many pencils do we need in all?*

Building Skills through Games

In Unit 6 your child will practice multiplication facts and strategies by playing the following games. For detailed instructions, see the *Student Reference Book*.

Baseball Multiplication Players use multiplication facts to score runs. Team members take turns generating two factors (1–10) by rolling 10-sided dice. Then players on the batting team take turns multiplying the two factors and moving around the bases.

Beat the Calculator Three children each take a different role in this fact game. The "Caller" calls out a multiplication fact; the "Brain" tries to multiply the two numbers mentally to beat the "Calculator," who multiplies the two numbers with a calculator.

Multiplication Top-It Players play with number cards 1–10 (four of each). They turn two cards and find the product of the numbers. The player with the larger product wins the round and takes both cards.

Name That Number Players try to name a target number by adding, subtracting, multiplying, or dividing the numbers on 2 or more of 5 cards.

As You Help Your Child with Homework

As your child brings home assignments, you may want to go over the instructions together, clarifying them as necessary. The answers listed below will guide you through this unit's Home Links. Answers to Home Links 6-4 and 6-7 are not shown.

Home Link 6-1

1. Sample answer: $400 - 300 = 100$; 79
2. Sample answer: $120 - 90 = 30$; 28
3. Sample answer: $500 - 350 = 150$; 135

Home Link 6-2

Hidden message: Math is fun!

Home Link 6-3

1. Strategy: Sample answer: subtracting a group

   How I solved: Sample answer: $10 \times 6 = 60$, $60 - 6 = 54$, so $9 \times 6 = 54$.
2. Strategy: Sample answer: near squares

   How I solved: Sample answer: $8 \times 8 = 64$, $64 + 8 = 72$, so $8 \times 9 = 72$.
3. Strategy: Sample answer: adding a group

   How I solved: Sample answer: $5 \times 8 = 40$, $40 + 8 = 48$, so $6 \times 8 = 48$.
4. Strategy: Sample answer: doubling

   How I solved: Sample answer: $2 \times 7 = 14$, 4 is the double of 2 and $14 + 14 = 28$, so $4 \times 7 = 28$.

Home Link 6-5

1. Sample answers:
2. Sample answers:
3. Sample answer:
4. 184 **5.** 90 **6.** 134

Home Link 6-6

1. Sample answer: *B* for balls

tennis courts	balls per court	balls in all
6	*B*	42

Number model: Sample answers: $42 \div 6 = B$; $6 \times B = 42$; Answer: 7 tennis balls

Home Link 6-8

1. 10; 4 **2.** 1; 13 **3.** 30; 38
4. Sample answers: $4 \times (8 - 6) = 8$; $(4 \times 8) - 6 = 26$
6. The parentheses are placed incorrectly. The number model should be $(8 \times 4) - 2 = 30$.

Home Link 6-9

1. 4 pears; Explanations vary.

Home Link 6-10

1. $8 \div (\underline{7 - 3}) = 2$ **2.** $26 = \underline{2 \times 10} + 6$
3. $10 + (\underline{6 \times 2}) = 22$ **4.** $15 - \underline{3 \times 2} = 9$
5. Sample answer: If we didn't have rules for the order of operations, we could get different answers for the same problem.

Home Link 6-11

1. Sample answer: *B* for banana muffins; $(6 \times 4) - 18 = B$; $(6 \times 4) = 18 + B$

   6 banana muffins; Sample answer: $(6 \times 4) - 18 = 6$; $(6 \times 4) = 18 + 6$
2. 50 **3.** 55 **4.** 80 **5.** 88

Solving Subtraction Problems

Home Link 6-1

NAME DATE TIME

Family Note Today your child learned trade-first subtraction, a method for solving subtraction problems that involves making all of the necessary trades before subtracting. Trade-first subtraction builds on children's understanding of place value and helps them solve subtraction problems more efficiently. The example below shows the trade-first method.

Please return this Home Link to school tomorrow.

Fill in the unit box. Then solve each problem. Choose a strategy that works best for you. You may use your estimates to check your work.

Example: Estimate: 510 - 250 = 260 10 4 ~~11~~ 13 ~~5~~ ~~1~~ ~~3~~ − 2 4 7 266	① Estimate: ______ 3 7 5 − 2 9 6
② Estimate: ______ 1 1 5 − 8 7	③ Estimate: ______ 5 0 3 − 3 6 8

Check: Do your answers make sense? How do you know?

____

____

Practicing Multiplication Facts with Arrays

Lesson 6-2

NAME DATE TIME

List the times ______ facts. If you are not sure of a fact, draw an array.

① 2 × ______ = ______

② 3 × ______ = ______

③ 4 × ______ = ______

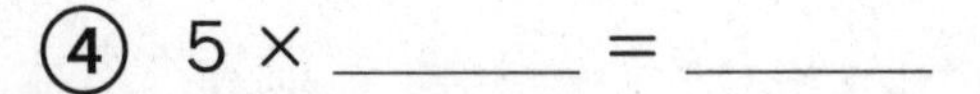

④ 5 × ______ = ______

⑤ 6 × ______ = ______

⑥ 7 × ______ = ______

⑦ 8 × ______ = ______

⑧ 9 × ______ = ______

⑨ 10 × ______ = ______

Talk to a partner about the patterns you find in your list.

Solving *Baseball Multiplication* Number Stories

Lesson 6-2

NAME DATE TIME

Each problem is about *Baseball Multiplication.* When you explain the strategies, write the product for each pitch.

1. Madison's partner pitches a 4 and a 9. Explain at least one way Madison could find the product of 4×9.

___

___

___

2. Taurean's partner pitches a 6 and an 8. Explain two ways Taurean could find the product.

___

___

___

3. Liam pitches a 7 and a 4 to Eli. Eli says that he can use a near-squares strategy to solve the problem. Liam says a near-squares strategy is not efficient for this problem. Explain why Liam said this.

___

___

What is an efficient strategy that Eli could use to solve 7×4?

___

___

___

___

Multiplication Hidden Message

NAME DATE TIME

Family Note Today your child worked toward automaticity with multiplication facts by learning to play *Baseball Multiplication.* Fact games and ×/÷ Fact Triangles provide opportunities for multiplication facts practice at home. Continue to work with your child on multiplication facts practice for brief periods of time (no more than 5 to 10 minutes) on a daily basis.

Please return this Home Link to school tomorrow.

Find the hidden message. Solve the facts below. You do not have to write the products. Use the key to decide whether to shade the shapes.

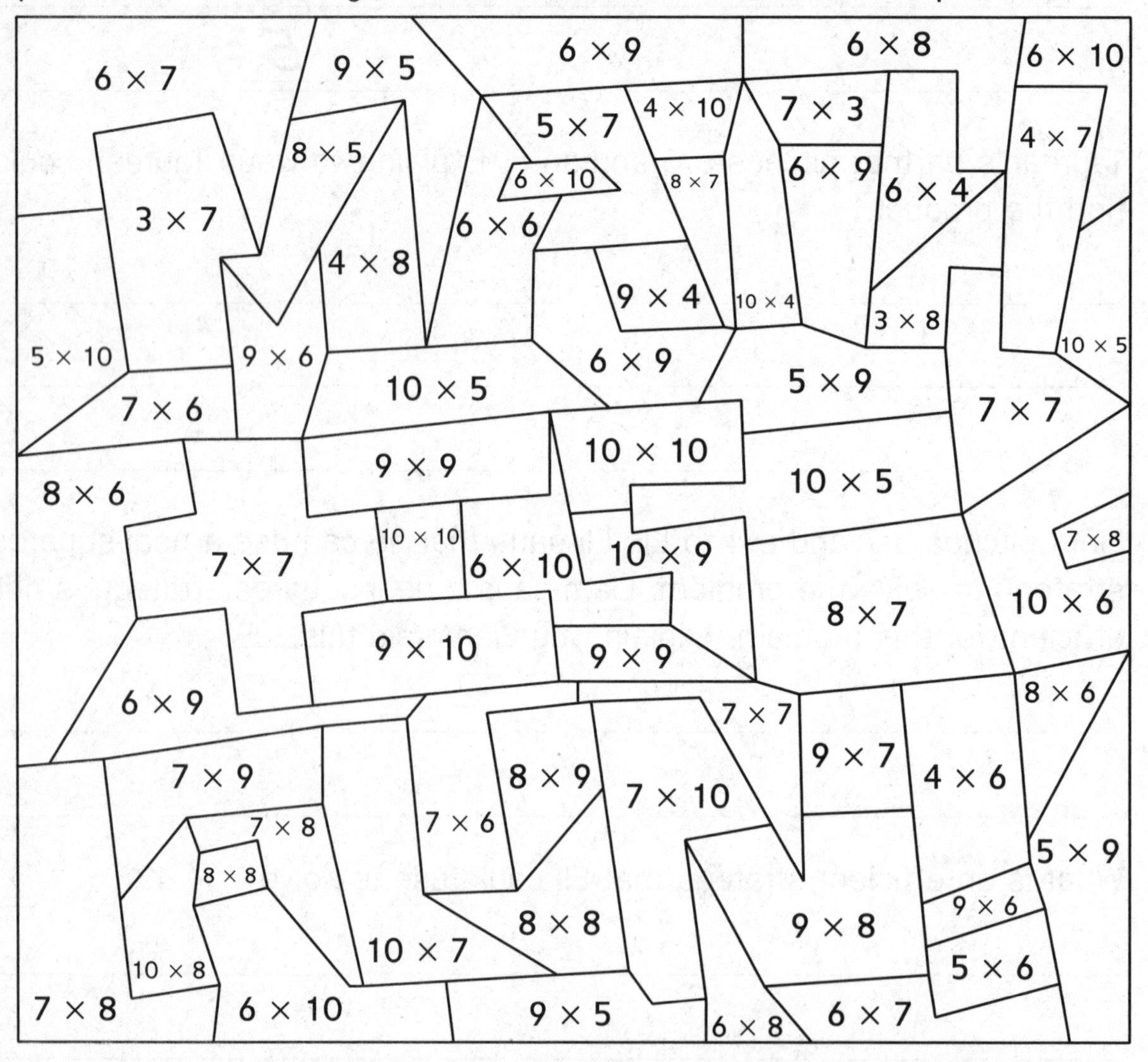

Key			
0–20	Do not shade	41–60	Do not shade
21–40	Shade	61–100	Shade

Applying Strategies to Multiplying by 11

Lesson 6-3

NAME DATE TIME

① Record and solve four more multiplication number sentences with 11 as one factor and a single-digit number as the other. Then explain how you solved the number sentence.

11s Facts	How I Solved It
Example: $11 \times 5 =$ 55	I know $10 \times 5 = 50$. I add one more group of 5 to find $11 \times 5 = 55$.

② What pattern do you notice in the products?

___

___

___

③ Write a rule to describe a strategy that you can use to multiply 11 by a single-digit number.

___

___

Multiplication Facts Strategies

Home Link 6-3

NAME DATE TIME

Family Note Today your child practiced applying appropriate and efficient strategies to solve less-familiar multiplication facts. Talk with your child about why he or she chose to use a particular strategy to solve the facts below.
Have your child cut apart and practice the Fact Triangles on the next page. Watch as your child sorts the Fact Triangles into 2 piles—those that are known and those that are unknown. Help your child identify strategies to help solve the unknown facts.

Please return this Home Link to school tomorrow.

For each fact below:

- Choose one of the strategies from the box.
- Solve the fact using that strategy.
- Explain how you solved the fact.

doubling	subtracting a group	adding a group
near squares	breaking apart	

① 9×6	② 8×9
Strategy: ______ How I solved:	Strategy: ______ How I solved:
③ 6×8 Strategy: ______ How I solved:	④ 4×7 Strategy: ______ How I solved:

⑤ Explain the strategy you used to solve Problem 2 to someone at home.

×, ÷ Fact Triangles: Remaining Facts

Home Link 6-3

NAME DATE TIME

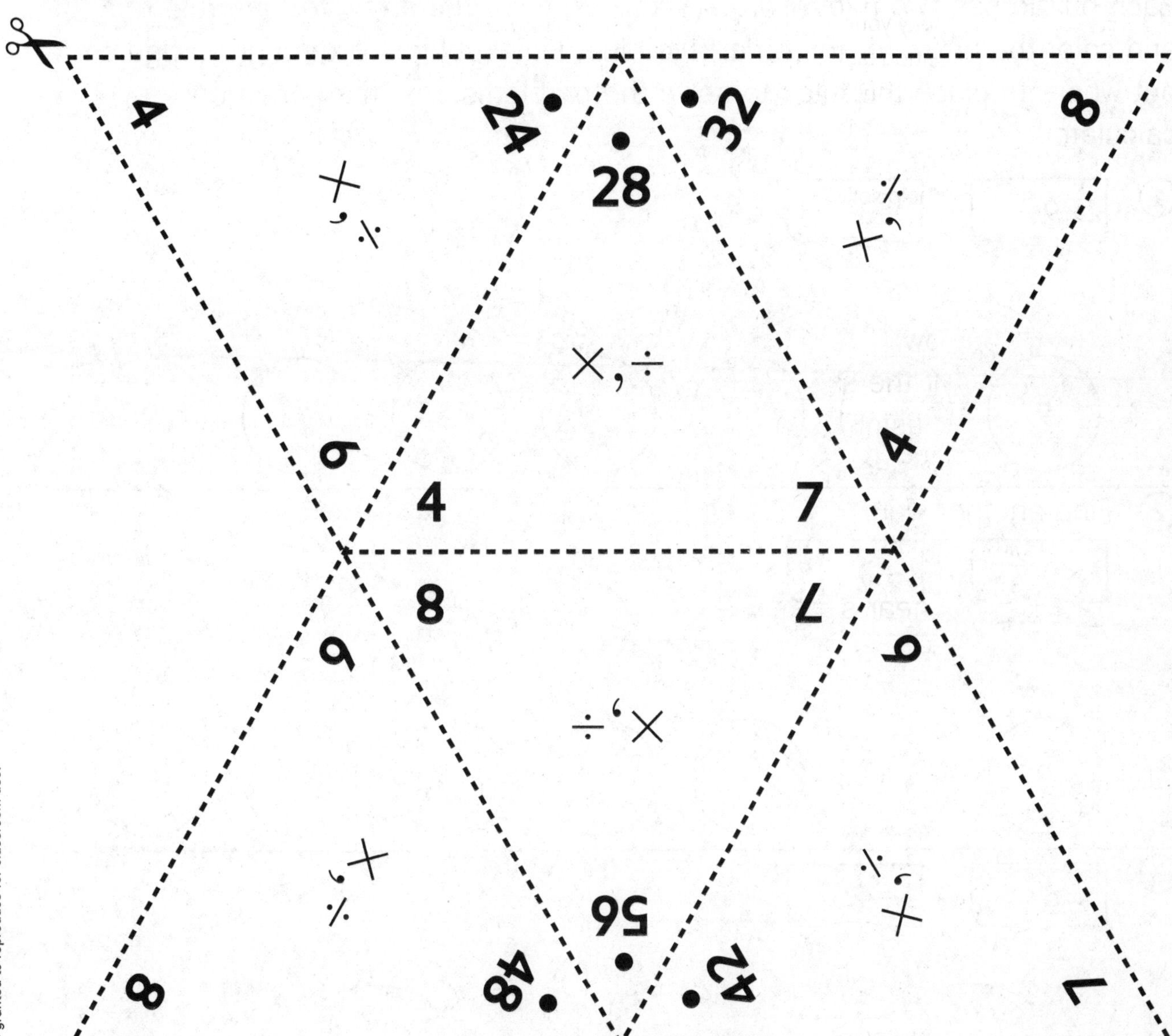

Frames-and-Arrows Puzzles

Lesson 6-4

NAME DATE TIME

Each puzzle has two rules. For each puzzle, color the arrow for one rule red and color the arrow for the other rule blue. For Problems 1 through 3, figure out where to place the rules to solve the problems. You may use your calculator.

①

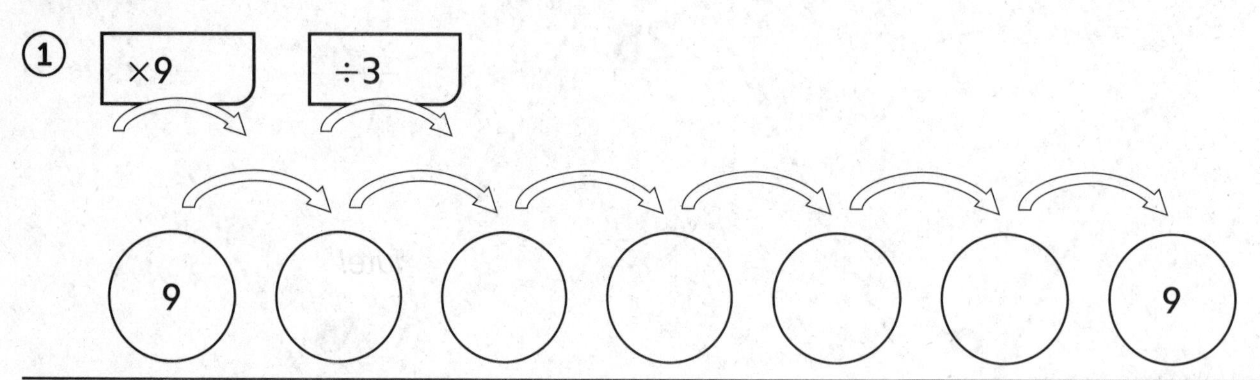

② Find another way.

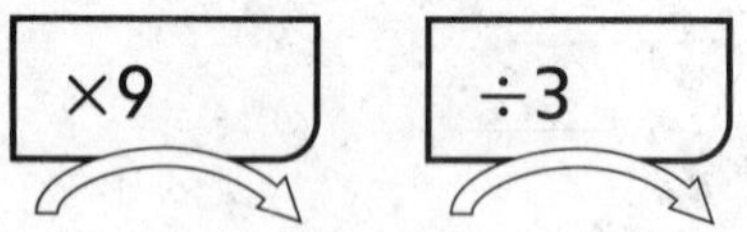

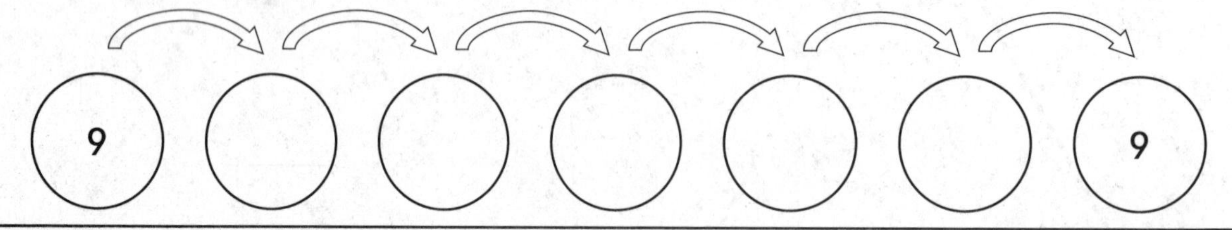

③

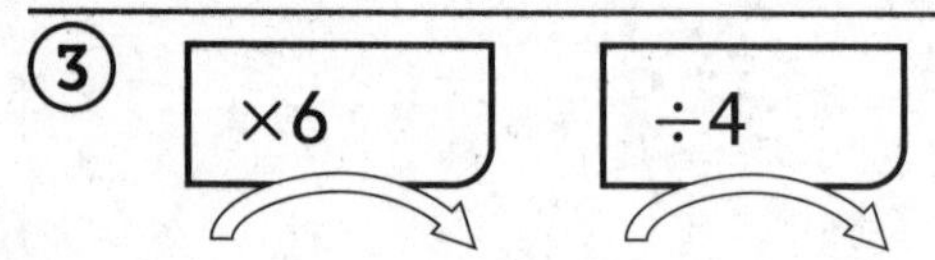

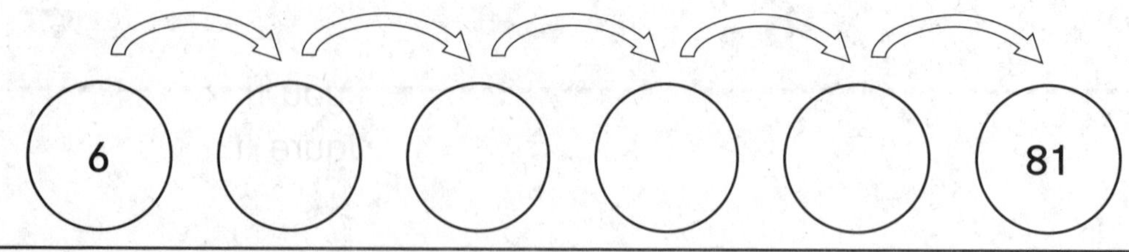

④ Make up a puzzle. Ask a partner to solve it.

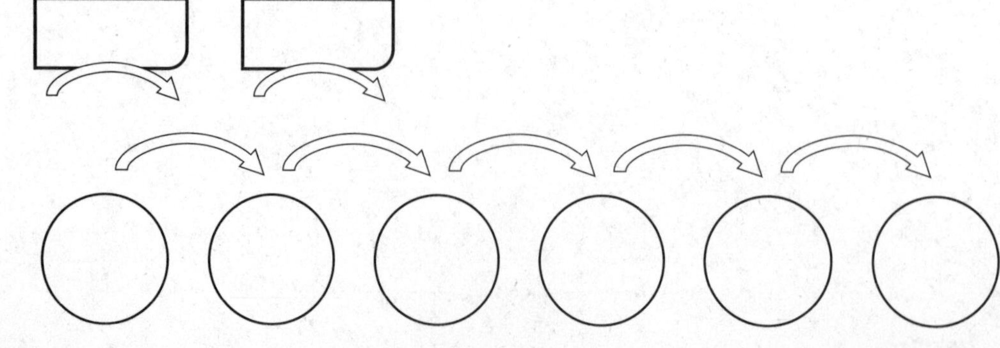

Fact Power

Home Link 6-4

NAME DATE TIME

Family Note Today your child learned about "fact power," or the ability to solve multiplication facts quickly and easily. Children practiced developing fact power by playing a multiplication game called *Beat the Calculator*. You can help your child develop fact power by playing multiplication fact games and practicing with Fact Triangles at home.

Please return this Home Link to school tomorrow.

1. Choose a way to practice multiplication facts from the list of activities below:

   Beat the Calculator Fact Triangle practice *Salute!*

2. Use the activity you chose to practice your multiplication facts with someone at home.

3. In the boxes below, record six facts for which you have "fact power."

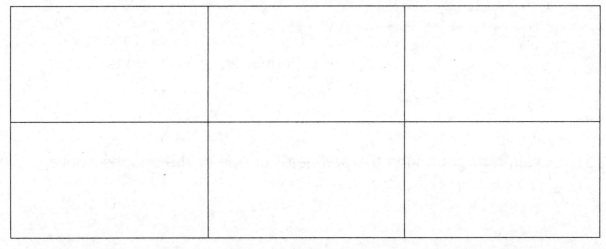

4. Record one fact that you are still practicing or that you think might be challenging for someone else. Show how you can figure it out efficiently. Explain your strategy to someone at home.

Finding Perimeters of Rectilinear Figures

Lesson 6-5

NAME DATE TIME

Find the perimeter of each figure.

① 5, 4, 2, 2, 3

Perimeter: _______ units

② 12, 4, 2, 4, 4

Perimeter: _______ units

③ Explain how you found the perimeter of one of the shapes above.

Solving Geometry Problems

Home Link 6-5

NAME DATE TIME

Family Note Today your child used straws and twist ties to create different quadrilaterals, or shapes with four sides. Help your child draw quadrilaterals that match each of the descriptions below.

Please return this Home Link to school tomorrow.

Draw one or more shapes to match each description.

① A parallelogram that is not a square:

② A quadrilateral that is not a rhombus:

③ A quadrilateral that is not a square, a parallelogram, or a rhombus:

Practice

Fill in the unit box. Then solve.

Unit

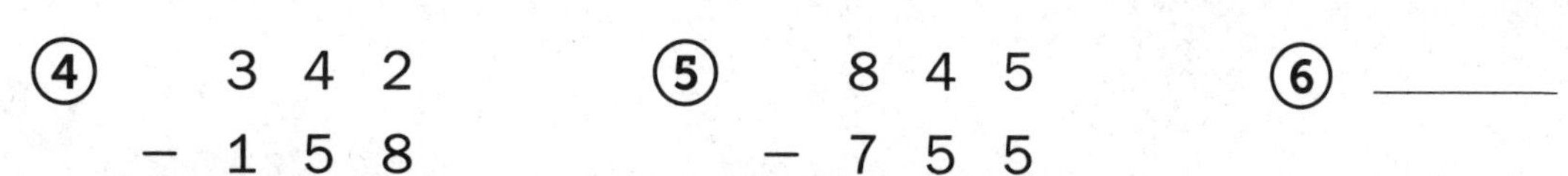

④ $\begin{array}{r} 342 \\ -\ 158 \\ \hline \end{array}$

⑤ $\begin{array}{r} 845 \\ -\ 755 \\ \hline \end{array}$

⑥ ______ = 231 − 97

Practicing with Number Stories

Lesson 6-6

NAME DATE TIME

For each number story:

- Complete the diagram. Use a letter to represent the unknown amount.
- Write a number model.
- Solve the number story. You may draw a picture to help.
- Write your answer with the unit. Does your answer make sense?

① Ty has 5 packages of markers. He has 40 markers in all. How many markers are in each package?

Letter and what it represents: ______________________

packages	markers per package	markers in all

____
(number model with letter)

Answer: ______________
(unit)

② Allie has 32 batteries. There are 8 batteries in each pack. How many packs of batteries does she have?

Letter and what it represents: ______________________

packs	batteries per pack	batteries in all

____
(number model with letter)

Answer: ______________
(unit)

Multiplication/ Division Diagrams

Home Link 6-6

NAME DATE TIME

Family Note Today your child learned to organize number story information in a multiplication/ division diagram. A properly filled out diagram can help children write an equation with a letter representing the unknown quantity. Help your child choose a letter that has something to do with the unknown quantity in the story. For example, in Problem 1, because children need to find the number of balls, *B* can represent the unknown quantity.

Please return this Home Link to school tomorrow.

- Complete the diagram. Use a letter to represent the unknown amount.
- Write a number model.
- Solve the number story. You may draw a picture to help.
- Write your answer with a unit. Does your answer make sense?

① You have 42 tennis balls to share among 6 tennis courts. How many tennis balls will you place on each court?

Letter and what it represents: ______________________________

tennis courts	balls per court	balls in all

____
(number model with letter)

Answer: ______________________
(unit)

② Explain to someone at home how you know your answer makes sense.

Applying Strategies to Multiplying by 12

Lesson 6-7

NAME DATE TIME

① Record four facts with the factor 12 multiplied by a single-digit number, for example, 12×5.

12s Facts	Strategy I Can Use

② Write a rule to describe a strategy that you can use to solve any 12s fact.

____

____

____

Applying Strategies to *Multiplication Top-It*

Lesson 6-7

NAME DATE TIME

① Lei and Simon are playing *Multiplication Top-It.* Lei draws a 6 and a 7 and says $6 \times 7 = 42$. Simon disagrees and says the product is 36.

Who is correct? ______________

Use a multiplication strategy to explain your answer.

____

____

____

② Justin and Ernesto are also playing *Multiplication Top-It.* Justin draws an 8 and a 7 and says $8 \times 7 = 48$. Ernesto disagrees and says the product is 56.

Who is correct? ______________

Use a multiplication strategy to explain your answer.

____

____

____

Multiplication Top-It

Home Link 6-7

NAME DATE TIME

Family Note Today your child learned a new game for practicing multiplication facts, *Multiplication Top-It*. Follow the directions below, and play at least one round of *Multiplication Top-It* with your child. You can make a deck of number cards by labeling index cards or slips of paper 1–10, or you may alter a regular deck of playing cards by removing the face cards and making each ace a 1.

Please return this Home Link to school tomorrow.

Directions for *Multiplication Top-It*

1. Shuffle the cards. Place the deck number-side down on the table.
2. Each player turns over 2 cards and calls out the product of the numbers.
3. The player with the larger product wins the round and takes all the cards.
4. In case of a tie for the largest product, each tied player turns over 2 more cards and calls out the product of the numbers. The player with the largest product then takes all the cards from both plays.
5. The game ends when there are not enough cards left for each player to have another turn.
6. The player with the most cards wins.

Record two of your rounds. Explain how a multiplication fact strategy could help someone who didn't know the fact.

My cards: ______ and ______ Strategy that could be used: Fact: ___ × ___ = ___	My cards: ______ and ______ Strategy that could be used: Fact: ___ × ___ = ___

Dot Patterns with Number Sentences

Lesson 6-8

NAME DATE TIME

The number of dots in this dot array can be found by using patterns.

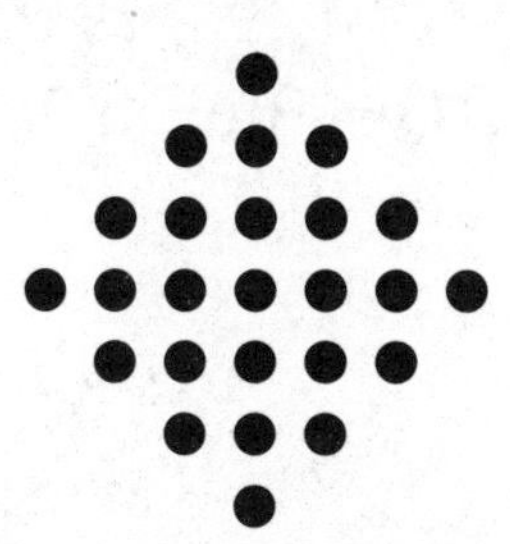

Here is one way to find the number of dots:

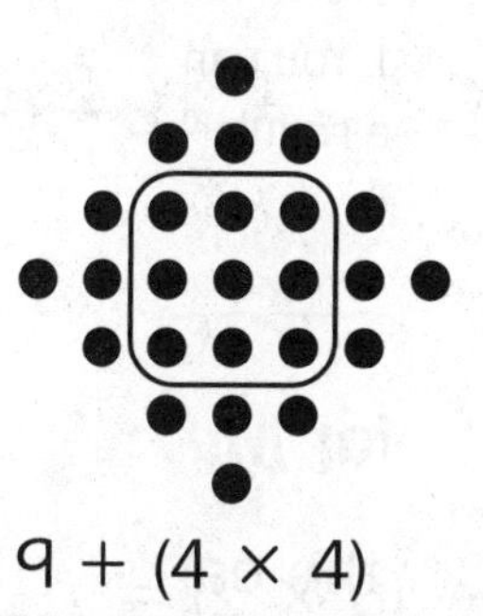

$9 + (4 \times 4)$

Find as many ways as you can to use patterns to find the number of dots. Show each pattern on the dot array, and write a number sentence to describe the pattern. Use parentheses in your number sentence if you can.

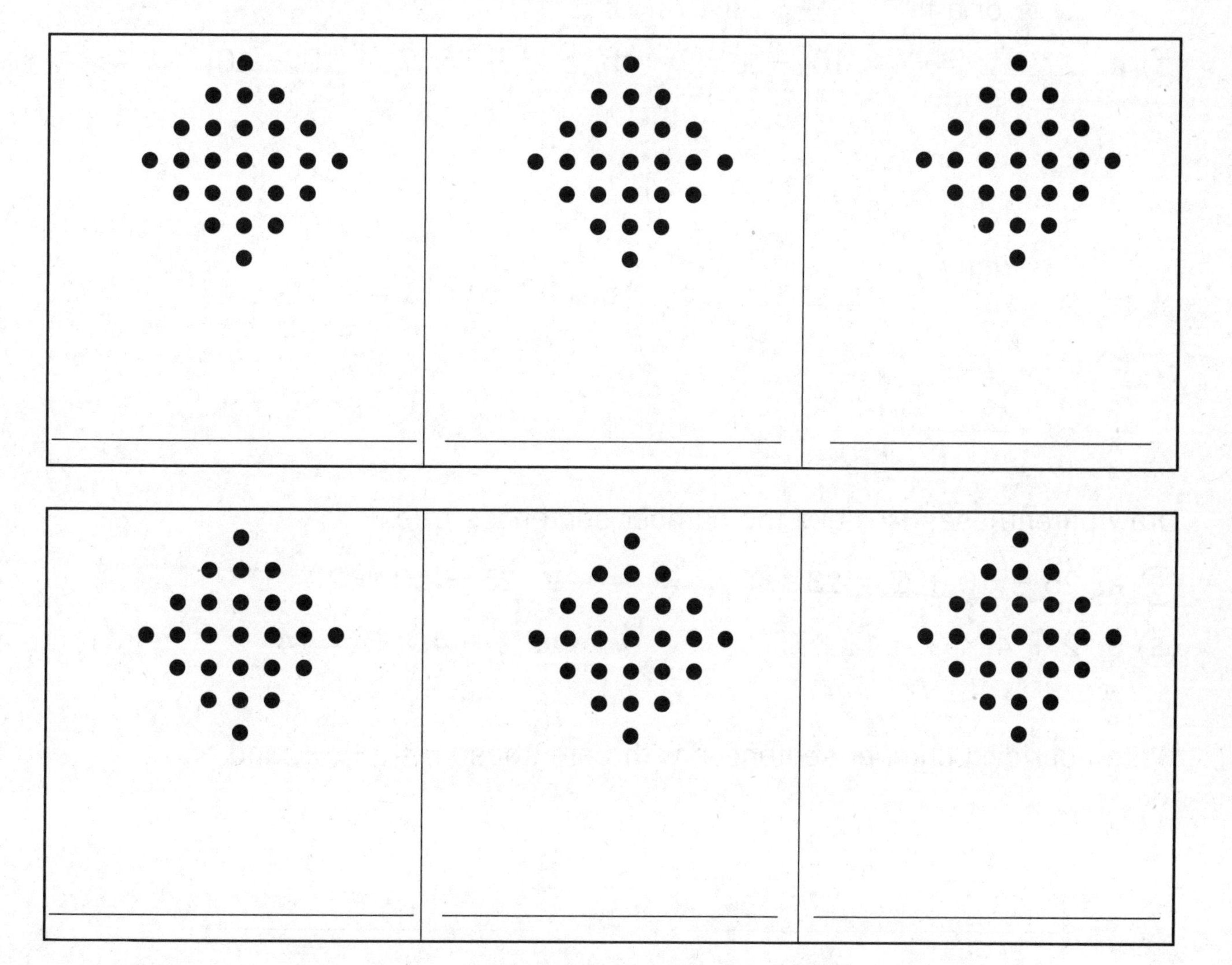

Practicing with Parentheses

Lesson 6-8

NAME DATE TIME

Complete these number sentences.

① a. ______ $= 21 - (9 + 6)$ b. ______ $= (21 - 9) + 6$

② a. $4 + (6 \times 2) =$ ______ b. $(4 + 6) \times 2 =$ ______

③ a. ______ $= (7 \times 10) - 10$ b. ______ $= 7 \times (10 - 10)$

④ a. $4 + (5 \times 3) =$ ______ b. $(4 + 5) \times 3 =$ ______

Draw parentheses to make the number sentences true.

⑤ a. $25 - 10 + 2 = 13$ b. $25 - 10 + 2 = 17$

⑥ a. $2 + 4 \times 3 = 18$ b. $2 + 4 \times 3 = 14$

Write your own number sentences with parentheses, =, +, −, and ×.

⑦ a. ______________________ b. ______________________

Parentheses Puzzles

Home Link 6-8

NAME DATE TIME

Family Note Today your child learned that parentheses are grouping symbols. Parentheses are used in number sentences to indicate which calculations to perform first.

Please return this Home Link to school tomorrow.

Show someone at home how to complete the number sentences below. Remember that the parentheses are used to show what you do first.

1. $(17 - 10) + 3 =$ ______ $17 - (10 + 3) =$ ______

2. ______ $= (24 - 17) - 6$ ______ $= 24 - (17 - 6)$

3. $5 \times (8 - 2) =$ ______ $(5 \times 8) - 2 =$ ______

Make up another parentheses puzzle and write it below.

4. ______ ______

Try This

5. There are 8 fish tanks at the pet store. Each tank has 4 fish. Dalia buys 2 fish. How many fish are left at the store?

   You may draw a picture to help.

6. Walter wrote this number model to fit the number story in Problem 5:
   $8 \times (4 - 2) = 16$

   Explain Walter's mistake. ______

   ______

Writing a Number Story to Fit a Number Sentence

Lesson 6-9

NAME DATE TIME

Write a two-step number story to fit the number sentence below.

$12 - (4 \times 2) = 4$

Number Stories and Number Sentences

Home Link 6-9

NAME DATE TIME

Family Note Today your child worked on writing a number story to fit a number sentence that includes parentheses. In Problem 1, the parentheses indicate that the 7 + 4 must be done first before subtracting 11 from 15. Ask your child to match each number and operation in the number sentence to a part of the number story. For Problem 2, ask your child to explain how his or her number story fits the number sentence. For both problems ask your child, "What do the parentheses mean?"

Please return this Home Link to school tomorrow.

① Shawn bought 15 pears at the farmers' market to share with his friends. He gave 7 pears away on Monday and 4 on Tuesday. How many pears does Shawn have left to share?

Use this number model to solve the problem.

$15 - (7 + 4) =$ ______ pears

Explain how the number model fits the number story.

② Write a number story to fit this number sentence.

$20 - (3 \times 6) = 2$

______

______

______

______

______

Solving Problems with Parentheses

Lesson 6-10

NAME DATE TIME

Write T if the number sentence is true. Write F if the number sentence is false.

① $16 - (9 + 3) = 10$ ______

② $15 - (10 \div 2) = 10$ ______

③ $15 = 23 - (16 - 8)$ ______

④ $24 = 2 \times (6 + 12)$ ______

Insert parentheses to complete the number sentences. Remember that parentheses show what you do first.

⑤ $20 - 10 + 6 = 16$

⑥ $16 = 3 \times 4 + 4$

⑦ $6 = 3 \times 5 - 9$

⑧ $24 = 17 - 9 \times 3$

Write your own number sentences with parentheses.

⑨ ______________________________

⑩ ______________________________

More Practice with Order of Operations

Lesson 6-10

NAME DATE TIME

Underline the part of each number sentence that should be completed first and then solve. See *Student Reference Book,* pages 69–70 for information about the order of operations. Show your work.

① $10 - 2 \times 4 =$ ______	② ______ $= 11 - 6 \div 3$
③ $24 - 5 + 4 =$ ______	④ ______ $= 10 \div 5 \times 8$
⑤ ______ $= 10 \times 5 - 8$	⑥ $23 - 9 \div 3 =$ ______

Order of Operations

Home Link 6-10

NAME DATE TIME

Family Note Today your child learned how to solve problems using the order of operations, a list of rules mathematicians follow when solving multistep problems.

Please return this Home Link to school tomorrow.

Use the order of operations to solve each number sentence below. Underline the part of each number sentence that should be completed first and then solve. Show your work.

Rules for the Order of Operations

1. Do operations inside parentheses first. Follow rules 2 and 3 when computing inside parentheses.

2. Then multiply or divide, in order, from left to right.

3. Finally add or subtract, in order, from left to right.

① $8 \div (7 - 3) =$ ______	② ______ $= 2 \times 10 + 6$
③ $10 + (6 \times 2) =$ ______	④ $15 - 3 \times 2 =$ ______

⑤ Tell someone at home why it is important to have rules for the order of operations.

Solving Number Stories

Lesson 6-11

NAME DATE TIME

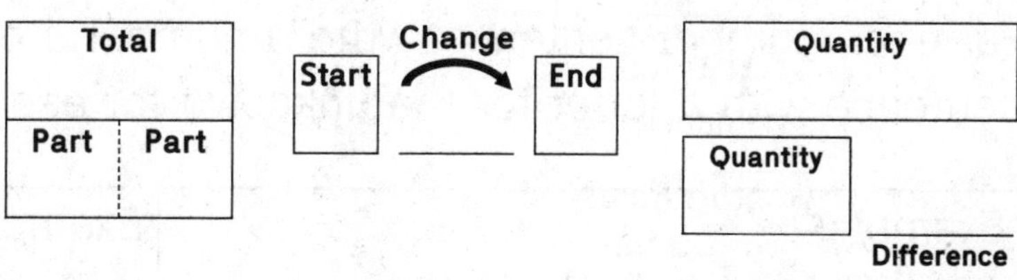

For each number story:

- Write a number model. Use a letter for the unknown. You may draw a diagram.
- Solve the problem. Show your work.
- Check that the answer makes your number model true. Write a summary number model.

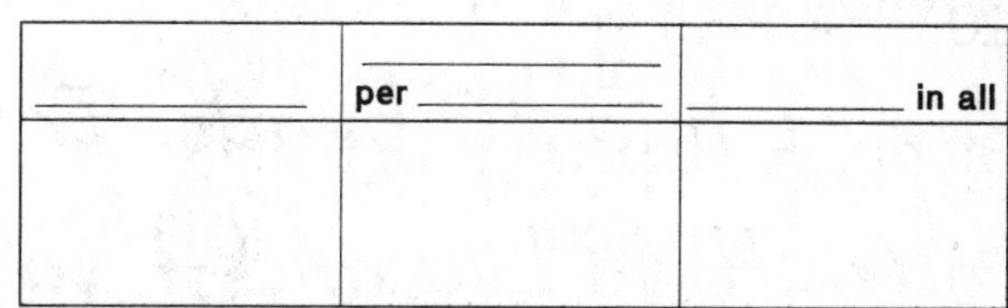

① Hiroko had 9 pencils. Her friend gave her some more pencils. Now she has 12 pencils. How many pencils did her friend give her?

Letter and what it represents: ______ for ____________

____
(number model with letter)

Answer: ____________
(unit)

____
(number model with answer)

② Elijah started with 15 grapes. He ate some. Now he has 5 grapes. How many grapes did he eat?

Letter and what it represents: ______ for ____________

____
(number model with letter)

Answer: ____________
(unit)

____
(number model with answer)

Writing Two-Step Number Stories

Lesson 6-11

NAME DATE TIME

For each number sentence, write a number story to fit. Then write a number sentence with a letter for the unknown for each story and solve your story.

Example: $10 - (4 \times 2) = ?$ Story: I had 10 beads. I gave 4 to Lila and 4 to Molly. How many beads did I have left over? The letter *B* stands for the number of beads left over. $10 - (4 \times 2) = B$ (number model with letter) Answer: 2 beads (unit)	*Example:* $(10 - 4) \times 2 = ?$ Story: I had 10 beads but lost 4. Gina gave me double the number I had left. How many beads do I have now? The letter *B* stands for the number of beads left over. $(10 - 4) \times 2 = B$ (number model with letter) Answer: 12 beads (unit)
$(2 \times 3) + 5 = ?$ Story: ______ The letter ___ stands for ______. ______ (number model with letter) Answer: ______ (unit)	$2 \times (3 + 5) = ?$ Story: ______ The letter ___ stands for ______. ______ (number model with letter) Answer: ______ (unit)

Solving Two-Step Number Stories

Lesson 6-11

NAME DATE TIME

For each number story:

- Write a number model. Use a letter for the unknown. You may draw a diagram to help.
- Solve. Write your answer.
- Check that the answer makes your number model true. Write a summary number model.

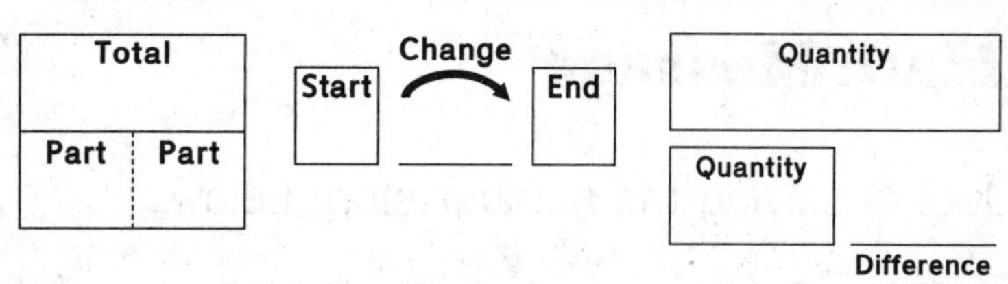

______	per ______	______ in all

① Andrew divided 12 dinner rolls among 4 baskets. Then he added 6 more rolls to each basket. What was the total number of rolls in each basket?

Letter and what it represents: ______ for ____________

Number model: ____________________________
(number model with letter)

Answer: ______ rolls

Summary number model: ____________________________
(number model with answer)

② Shoshanna has $10. She wants to buy 3 bags of giant balloons for $3 each. How much money will she have left?

Letter for the unknown: ______ for ____________

____
(number model with letter)

Answer: $______

Summary number model: ____________________________
(number model with answer)

Representing a Number Story

NAME DATE TIME

Math Message

Judi is solving the number story below.

> Mrs. Burns buys 4 packs of brilliant colored markers.
> There are 5 markers in each pack.
> She gives some markers away and ends up with 8 markers for herself.
> How many markers did she give away?

To help, Judi drew the situation diagram shown below.

She wrote the letter M to represent the number of *markers* given away in the story.

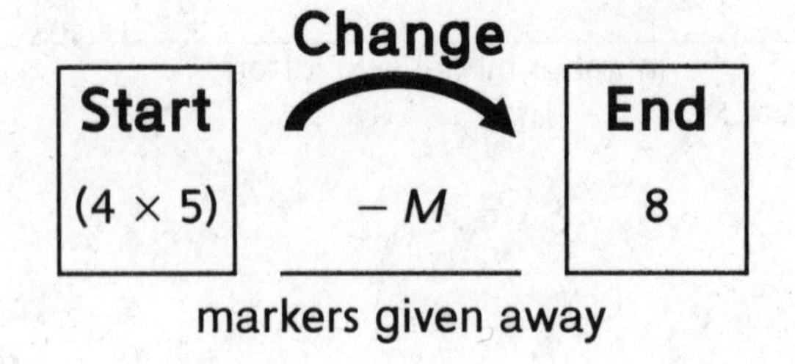

Solving a Number Story

Home Link 6-11

NAME DATE TIME

> **Family Note** Today your child used diagrams to organize information in number stories. These diagrams can be used to help write single-number models for multistep problems.
>
> Please send clean, empty containers to school for an upcoming lesson.
>
> ***Please return this Home Link to school tomorrow.***

Write a number model. Use a letter for the unknown.
You may draw a diagram to help.

Total
Part | Part

Start — Change → End

Quantity
Quantity
Difference

______	per ______	______ in all

Solve the story and check to make sure that your answer makes sense. Then write the number model with the answer.

① Andrea made 4 pans of muffins. Each pan holds 6 muffins. She made 18 pumpkin muffins. The rest were banana muffins. How many banana muffins did Andrea make?
Letter and what it represents: ______ for ____________

____
(number model with letter)

Answer: ______ banana muffins

____
(number model with answer)

Practice

Solve.

② $10 \times 5 =$ ______

③ $11 \times 5 =$ ______

④ ______ $= 10 \times 8$

⑤ ______ $= 11 \times 8$

Unit 7: Family Letter

Home Link 6-12

NAME DATE TIME

Fractions

In this unit, children explore liquid volume and develop a deeper understanding of fractions.

Through a variety of hands-on activities, children will estimate, measure, and solve problems with liquid volume. They learn that 1 liter = 1,000 milliliters. They will solve number stories to review and extend previous work with measures of length and mass. To prepare for activities involving liquid volume, please have your child bring in a clean, empty container that can hold liquid, such as a jar, bowl, water bottle, plastic food container, milk jug, or plastic cup. These items can be returned after the conclusion of the unit.

Children will use a variety of models and tools to build on fraction work from previous units. They will continue to use fraction circle pieces to represent and compare equal parts of a whole. They also learn about fractions as distances on number lines and begin to understand fractions as numbers between whole numbers. They use these visual models to recognize and generate equivalent fractions and make comparisons.

In Unit 7, your child will:

- Measure and estimate liquid volumes using liters and milliliters.
- Solve number stories involving mass, volume, and length.
- Partition fraction strips and use them to name and compare fractions.

$\frac{1}{2}$	$\frac{1}{2}$

$\frac{1}{4}$	$\frac{1}{4}$	$\frac{1}{4}$	$\frac{1}{4}$

- Develop an understanding of fractions as distances on a number line.
- Represent whole numbers as fractions.
- Recognize and generate equivalent fractions using fraction circle pieces, fraction strips, and number lines.
- Identify and locate fractions greater than, less than, and equal to 1 on a number line.
- Use <, >, and = to compare fractions.
- Solve number stories involving fractions.
- Share collections equally and represent the resulting groups with fractions.

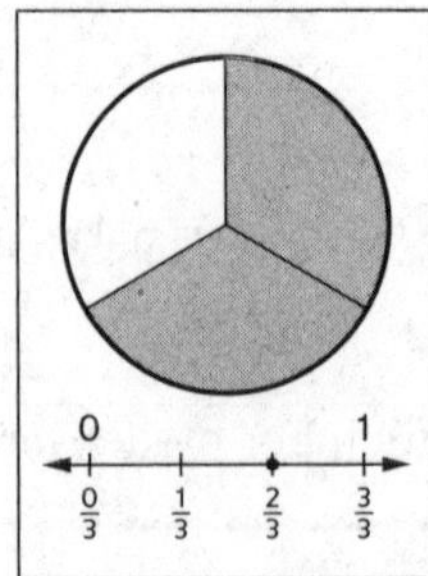

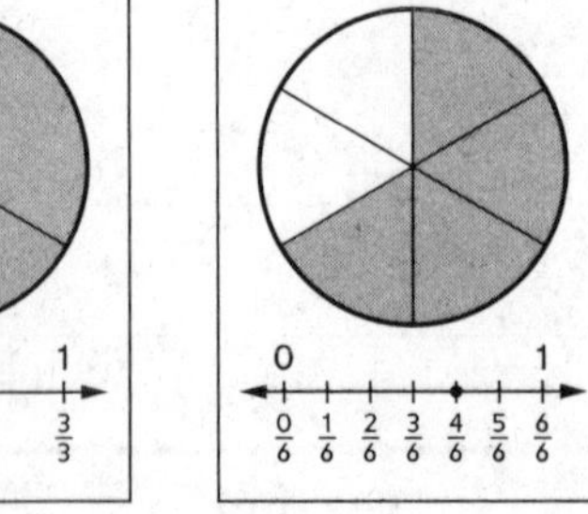

Please keep this Family Letter for reference as your child works through Unit 7.

Vocabulary

Important terms in Unit 7:

denominator The number below the line in standard fraction notation, such as the 2 in $\frac{1}{2}$; the number of equal parts into which the *whole* has been divided.

equivalent fractions Fractions that name the same value, such as $\frac{1}{2}$ and $\frac{4}{8}$.

fraction A number in the form $\frac{a}{b}$. The *numerator, a,* can be any whole number. The *denominator, b,* can be any whole number except 0. For example, $\frac{1}{4}$, $\frac{3}{8}$, and $\frac{5}{2}$ are fractions. A fraction may be used to name part of a *whole,* to compare two quantities, or to represent division.

liquid volume How much liquid a container holds.

numerator The number above the line in standard fraction notation, such as the 1 in $\frac{1}{2}$. In a part-whole *fraction* in which the *whole* is divided into a number of equal parts, the numerator is the number of equal parts being considered.

unit fraction A *fraction* whose *numerator* is 1. For example, $\frac{1}{2}$, $\frac{1}{3}$, $\frac{1}{4}$, and $\frac{1}{8}$ are unit fractions.

volume A measure of how much 3-dimensional space something occupies. Volume is often measured in liquid units such as liters (L) or milliliters (mL).

whole An entire object, collection of objects, or quantity being considered in a problem situation. To accurately compare fractions of a whole, the whole must be the same size. For example, $\frac{1}{4}$ of a whole sandwich cannot be compared to $\frac{1}{4}$ of a whole pizza.

Do-Anytime Activities

Help your child find fractions in the everyday world, such as in advertisements, on measuring tools, in recipes, and so on.

1. Find containers that hold up to about 1 liter and $\frac{1}{2}$ liter of liquid. Provide 1-liter and $\frac{1}{2}$-liter bottles for your child to use as liquid-volume benchmarks. Your child can pour water into various containers to help estimate their volumes.
2. Compare two fractions and tell which is larger. For example:
   - Which would give you more of a pizza, $\frac{1}{8}$ or $\frac{1}{4}$?
   - Sam has $\frac{2}{3}$ of a granola bar left and Rose has $\frac{1}{3}$ of the same kind of granola bar left. Who has the larger amount left?
   - Ty drinks $\frac{1}{2}$ of a bottle of water. Dion drinks $\frac{1}{4}$ of a same-size bottle of water. Who has more water left in his bottle?
3. Pose different fraction stories to solve. Encourage your child to use real objects to act out the stories and justify his or her answers. For example:
   - $\frac{3}{4}$ of the napkins are white. What fraction of the napkins are not white?
   - $\frac{4}{8}$ of the fish have stripes. What fraction of the fish do not have stripes?
4. Read stories about fractions, such as *Give Me Half!* by Stuart Murphy (Great Source, 1996).

Building Skills through Games

In Unit 7 your child will play the following games. For detailed instructions, see the *Student Reference Book.*

Fraction Memory Players turn over two fraction cards to find equivalent fraction pairs. Pairs are collected, while other cards are turned back over for future turns.

Fraction Top-It Players turn over two fraction cards and compare the fractions. The player with the larger fraction keeps all the cards. The player with more cards at the end wins.

As You Help Your Child with Homework

As your child brings home assignments, you may want to go over the instructions together, clarifying them as necessary. The answers listed below will guide you through this unit's Home Links.

Home Link 7-1

4. Answers vary.

Home Link 7-2

1. Sample answer: I disagree. I can see that more of the circle is shaded on $\frac{5}{6}$ than on $\frac{6}{8}$, so they are not equivalent.

2. 54 **3.** 72 **4.** 56

Home Link 7-3

1. 450 mL **2.** 25 minutes **3.** $7\frac{1}{2}$ liters

Home Link 7-4

1.

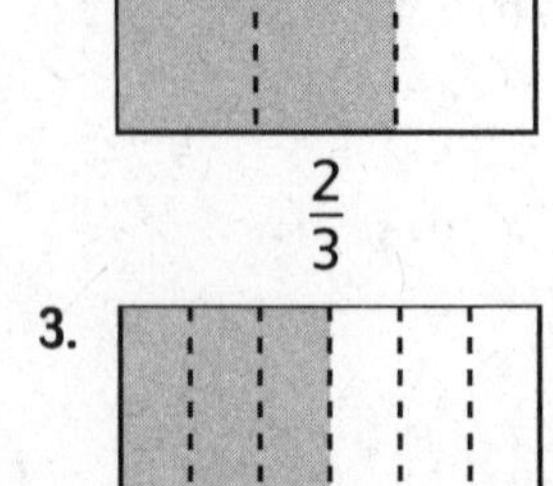

$\frac{2}{3}$

2.

$\frac{3}{8}$

3.

$\frac{3}{6}$

4.

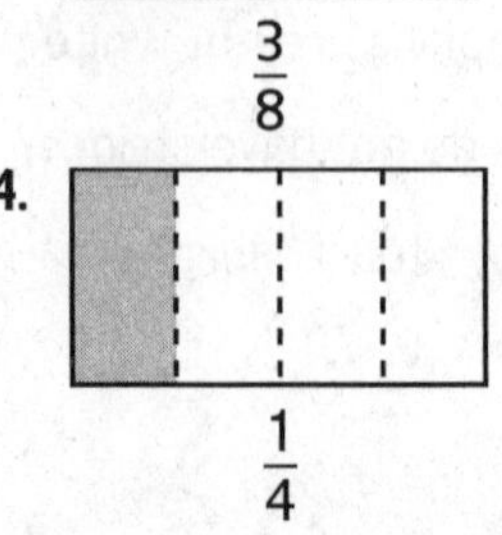

$\frac{1}{4}$

5. < **6.** >

7. 415; Sample estimate: 950 − 550 = 400

8. 710; Sample estimate: 400 + 300 = 700

Home Link 7-5

1. $\frac{1}{2}, \frac{3}{8}$ **2.** $\frac{1}{3}, \frac{2}{3}, \frac{0}{4}, \frac{1}{4}, \frac{4}{4}$

3. 825 **4.** 210

Home Link 7-6

1. $\frac{1}{3}, \frac{2}{3}, \frac{2}{3}$ **2.** $\frac{1}{3}, \frac{3}{3}, \frac{5}{3}, \frac{3}{3}$

3. $\frac{1}{6}, \frac{2}{6}, \frac{3}{6}, \frac{4}{6}, \frac{5}{6}, \frac{7}{6}, \frac{8}{6}, \frac{9}{6}, \frac{10}{6}, \frac{11}{6}, \frac{9}{6}$

4. $\frac{9}{6}$ **5.** 260 **6.** 1,000

Home Link 7-7

1. 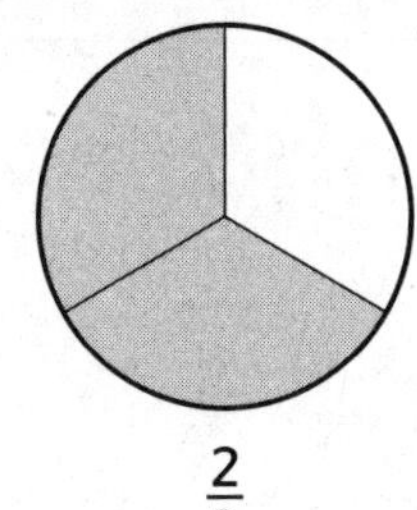

$\frac{2}{3}$

2. 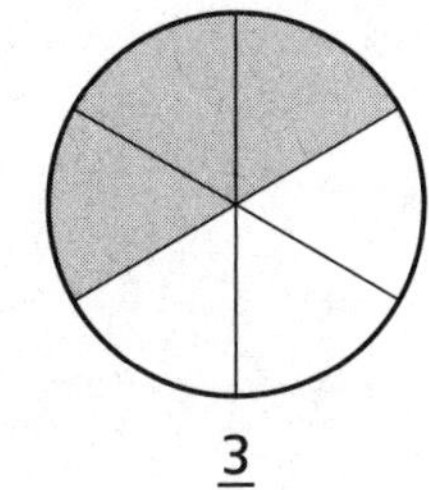

$\frac{3}{6}$

3. 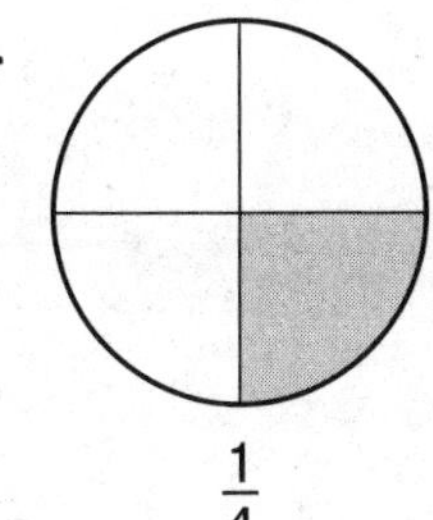

$\frac{1}{4}$

4. $\frac{1}{4}$ 5. $\frac{3}{6}$ 6. $\frac{2}{3}$ 7. $>$

Home Link 7-8

1. Less Than 1: $\frac{1}{2}, \frac{2}{3}, \frac{7}{8}, \frac{6}{8}$; Greater Than 1: $\frac{6}{4}, \frac{3}{2}, \frac{5}{3}, \frac{7}{6}$
2. Sample answer: In fractions that are less than 1, the numerator is less than the denominator. In fractions that are greater than 1, the numerator is greater than the denominator.

Home Link 7-9

1. 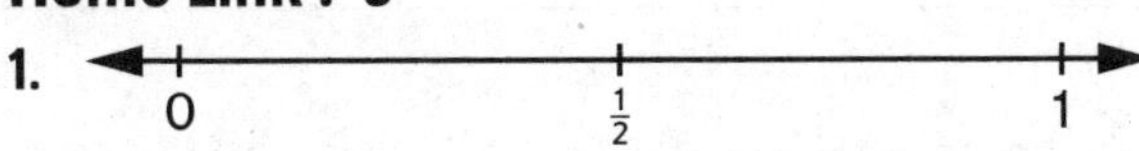

2. 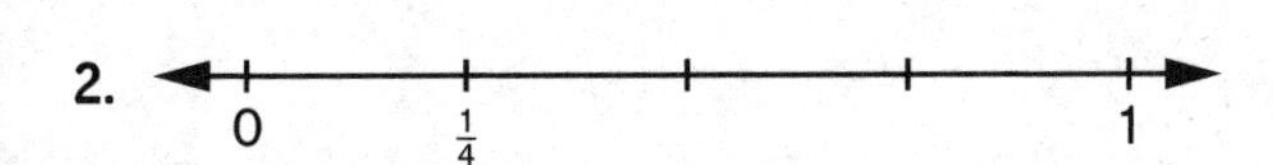

3. 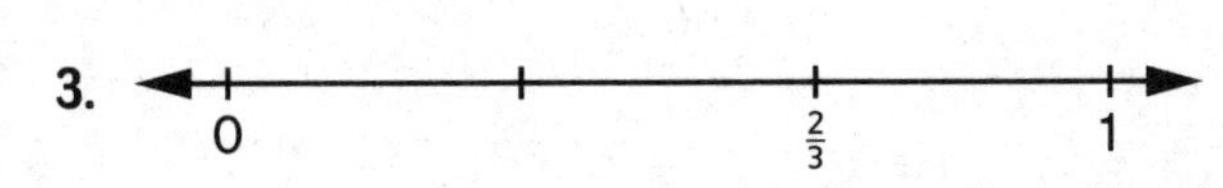

4. 0 $\frac{5}{6}$ 1

5. 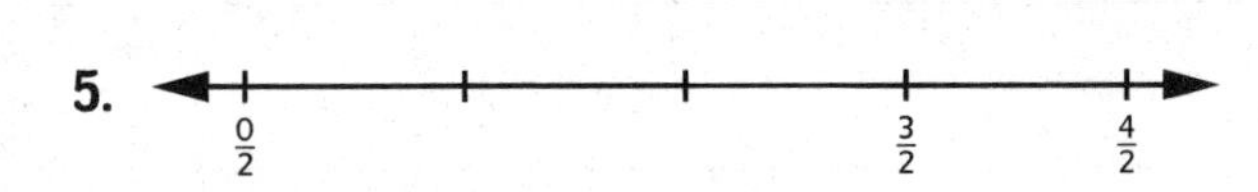

Home Link 7-10

1. $\frac{1}{2} > \frac{1}{8}$;

$\frac{1}{2}$

2. $\frac{2}{3} = \frac{4}{6}$;

0 1

0 1

3. $\frac{3}{8} < \frac{3}{6}$;

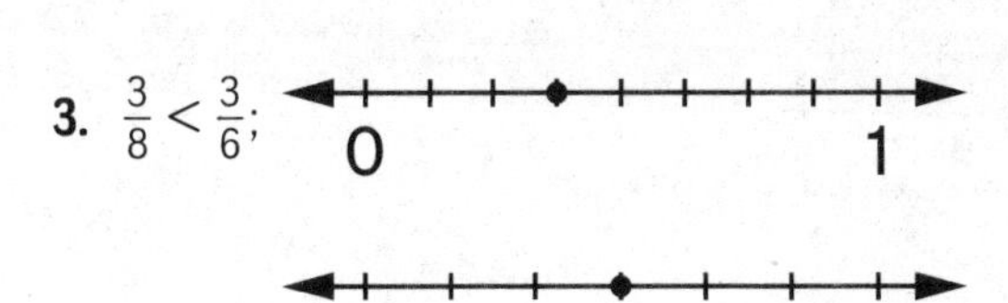

4. $\frac{2}{6} < \frac{5}{6}$;

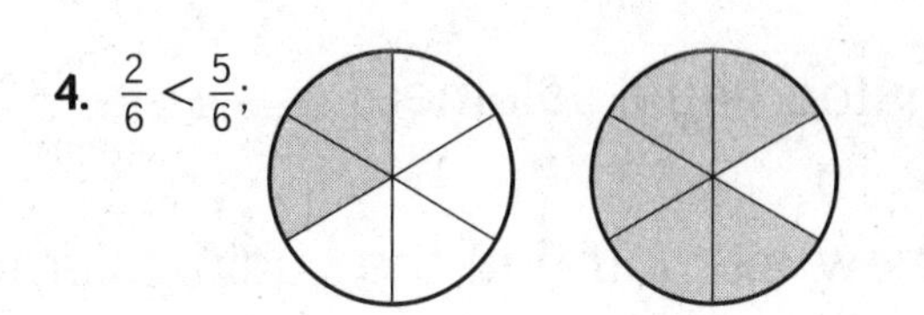

Home Link 7-11

1. $\frac{7}{8}$
2. $\frac{1}{4} + \frac{1}{4} = \frac{2}{4}; \frac{1}{2}$
3. They both rode the same distance. $\frac{2}{2} = 1$ whole block, $\frac{4}{4} = 1$ whole block

Home Link 7-12

1. $\frac{2}{12}$ 2. $\frac{4}{7}$ 3. $\frac{12}{16}$
4. Carlie; Sample answer:

Lisa

$\frac{2}{6}$ yellow

Carlie

$\frac{4}{6}$ yellow

5. 265 6. 886

Estimating and Measuring Liquid Volumes

Lesson 7-1

NAME DATE TIME

Follow the directions on Activity Card 77.

Draw your container below.

Estimated liquid volume: ________ (unit) Measured liquid volume: ________ (unit)

① How did you find the liquid volume of your container?

____

____

____

② Which benchmark beaker has a volume closest to your container's volume?

____

How do you know?

____

____

____

Estimating Liquid Volume

Lesson 7-1

NAME DATE TIME

For each question, circle the best answer. You may look at the class benchmark beakers and containers to help you decide.

① About how much hot chocolate could a mug hold?

400 milliliters or 400 liters

② About how much water does a bathtub hold?

200 milliliters or 200 liters

③ Which of these containers is most likely to hold about 1 liter?

cup   water bottle   bathtub

④ Which of these containers is most likely to hold about 5 liters?

milk carton   swimming pool   bucket

⑤ Which of these containers is most likely to hold about 250 milliliters?

kitchen sink   juice box   fish tank

⑥ For Problem 5, how can benchmark measurements help you make an estimate?

____

____

____

Liquid Volume Hunt

Home Link 7-1

NAME DATE TIME

Family Note Today your child used a set of benchmark beakers to estimate and measure liquid volume in liters (L) and milliliters (mL). Liquid volume is a measure of how much liquid a container can hold. Help your child look at labels to find containers of liquids that are measured in milliliters and liters. Have your child record the items in the table below.

Please send clean, empty, unbreakable containers to school for our next lesson.

Please return this Home Link to school tomorrow.

① Examine labels on items for liquid volume measured in liters or milliliters. Record your findings in the table below.

Item	Liquid Volume Units
flavored water bottle	530 mL

② Circle an item that you can use as a benchmark for 1 liter.

③ Put a star next to an item that you can use as a benchmark for 500 milliliters.

Try This

④ Estimate the liquid volume of a clean dinner plate: about __________ mL

If you have a measuring tool marked with milliliters, find the liquid volume of your dinner plate by measuring how much water it holds before spilling over the edges.

about __________ mL

On the back of this page, explain how you found the liquid volume of the dinner plate.

Estimating the Number of Seats in an Auditorium

Lesson 7-2

NAME DATE TIME

The picture below shows the seating chart for Starlight Auditorium.

Starlight Auditorium Seating Chart

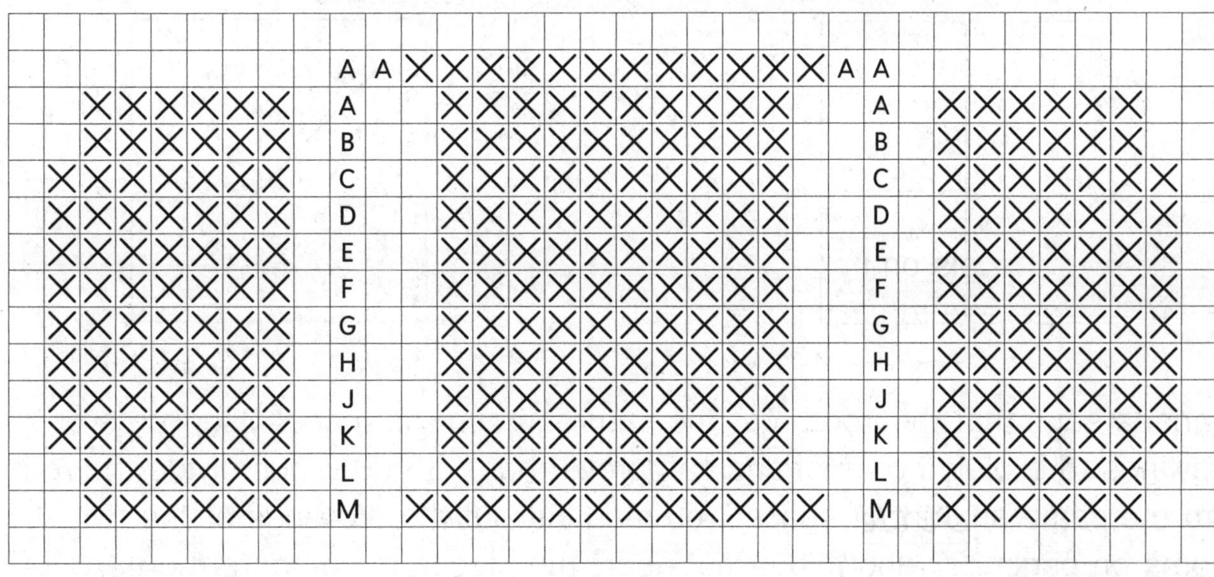

Left Center Right

Use what you know about arrays and estimation to help you make a reasonable estimate of how many people can be seated in the auditorium. Show your thinking.

About how many people can be seated in the auditorium?

About ______________________ (unit)

Explain how you made your estimate. ______________________

Justifying Equal Parts

Lesson 7-2

NAME DATE TIME

Mariana and Abby are sisters who share a bedroom.

Mariana divided the room this way:

Abby divided the room this way:

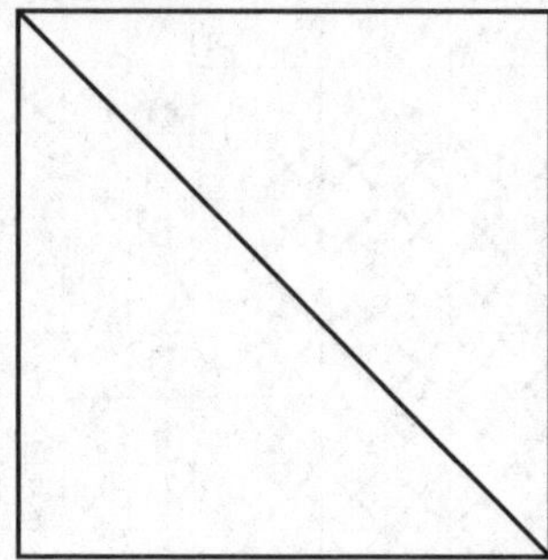

Mariana says that they both get the same amount of space. Use words or drawings to explain how Mariana can prove that both drawings divide the room into the same amount of space. You may cut out the drawings of the two rooms on page 227 and fold or cut apart the pieces to compare the parts of the room.

Justifying Equal Parts (continued)

Lesson 7-2

NAME DATE TIME

Our Room

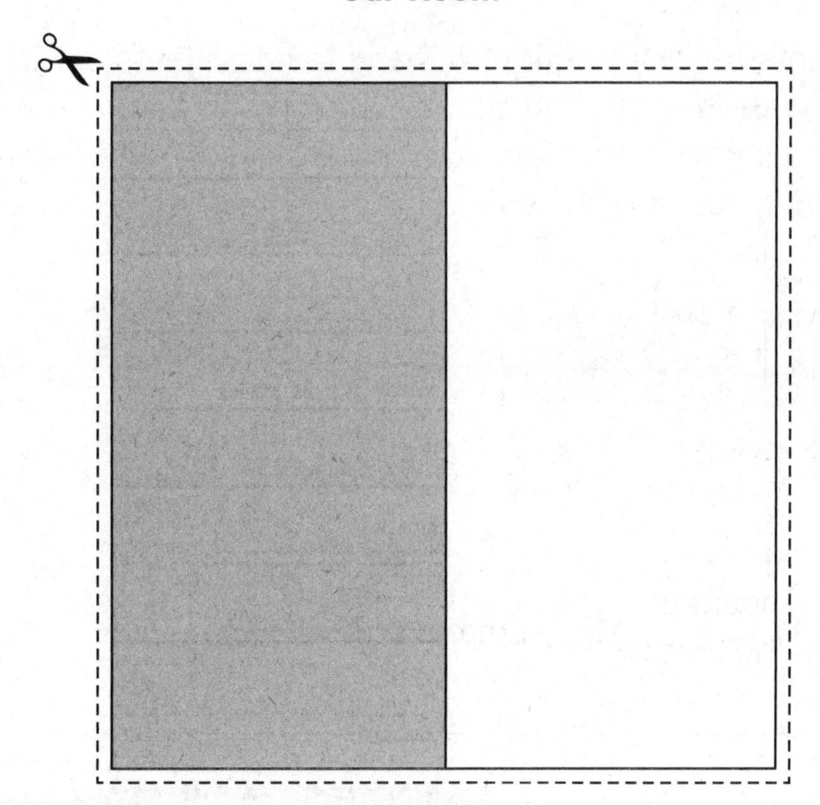

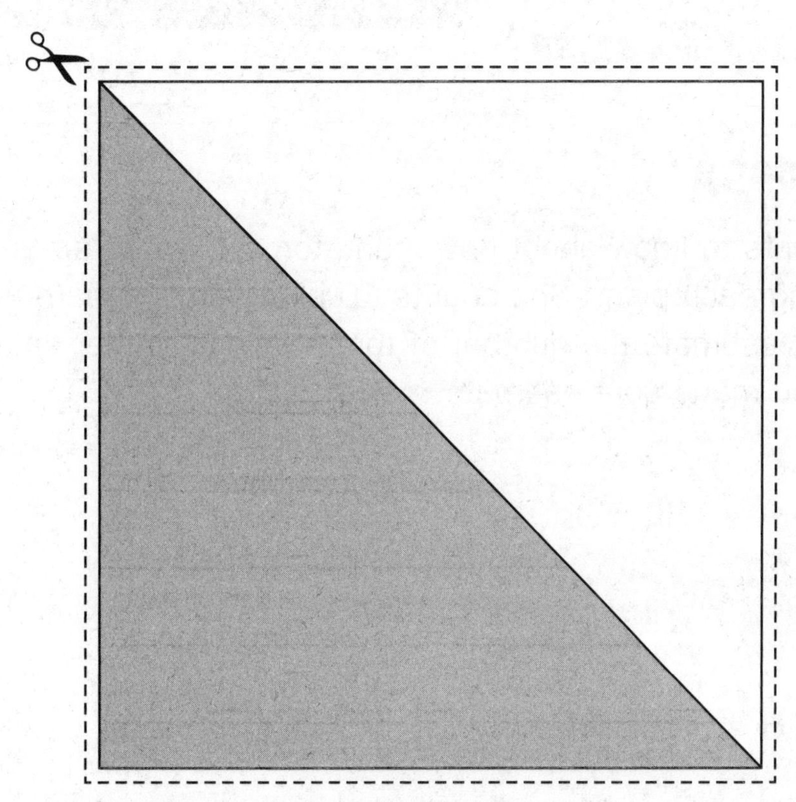

How Many Plants?

Lesson 7-2

NAME DATE TIME

Math Message

Mrs. Reyes wants to know about how many tomato plants are in her garden without counting each plant. She counts 10 plants in 1 square yard. Using the diagram below, estimate the number of tomato plants in her entire garden. Explain how you made your estimate.

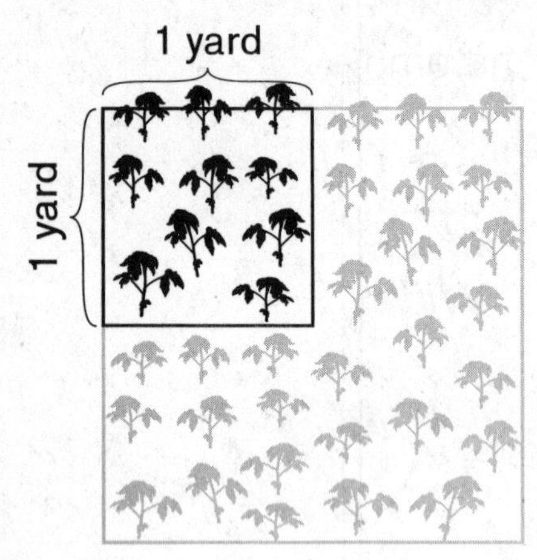

My estimate: about ______________________
(unit)

How Many Plants?

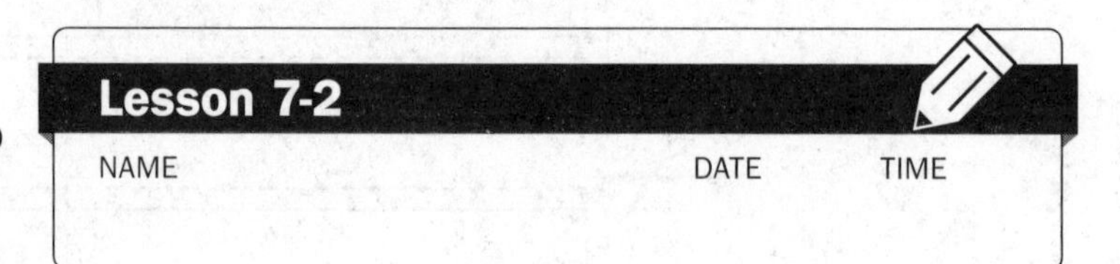

Math Message

Mrs. Reyes wants to know about how many tomato plants are in her garden without counting each plant. She counts 10 plants in 1 square yard. Using the diagram below, estimate the number of tomato plants in her entire garden. Explain how you made your estimate.

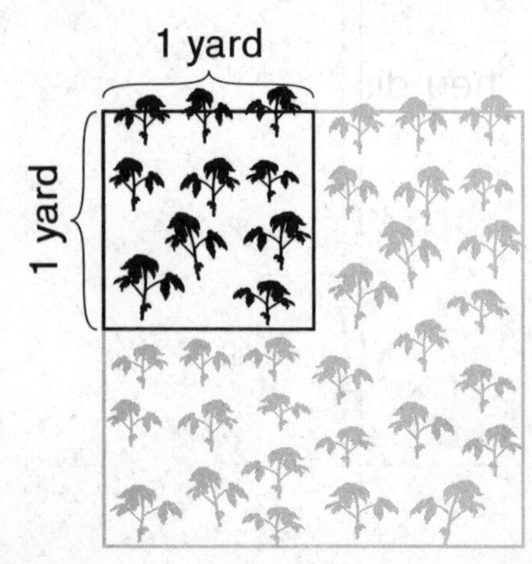

My estimate: about ______________________
(unit)

Copyright © McGraw-Hill Education. Permission is granted to reproduce for classroom use.

Measuring Liquid Volume

Lesson 7-2

NAME DATE TIME

① My predictions:

I think ________________ will displace the most amount of water.

I think ________________ will displace the least amount of water.

② Follow the directions on Activity Card 78 to measure the amount of water displaced by four objects. Record the object and the amount of displaced water in the table below.

Name of Object	**Liquid Volume of Displaced Water** (remember to include units)

③ Write the objects in order by the amount of water they displaced.

______________ ______________ ______________ ______________
(most water displaced) (least water displaced)

Shares of Cornbread

NAME DATE TIME

Exploration C

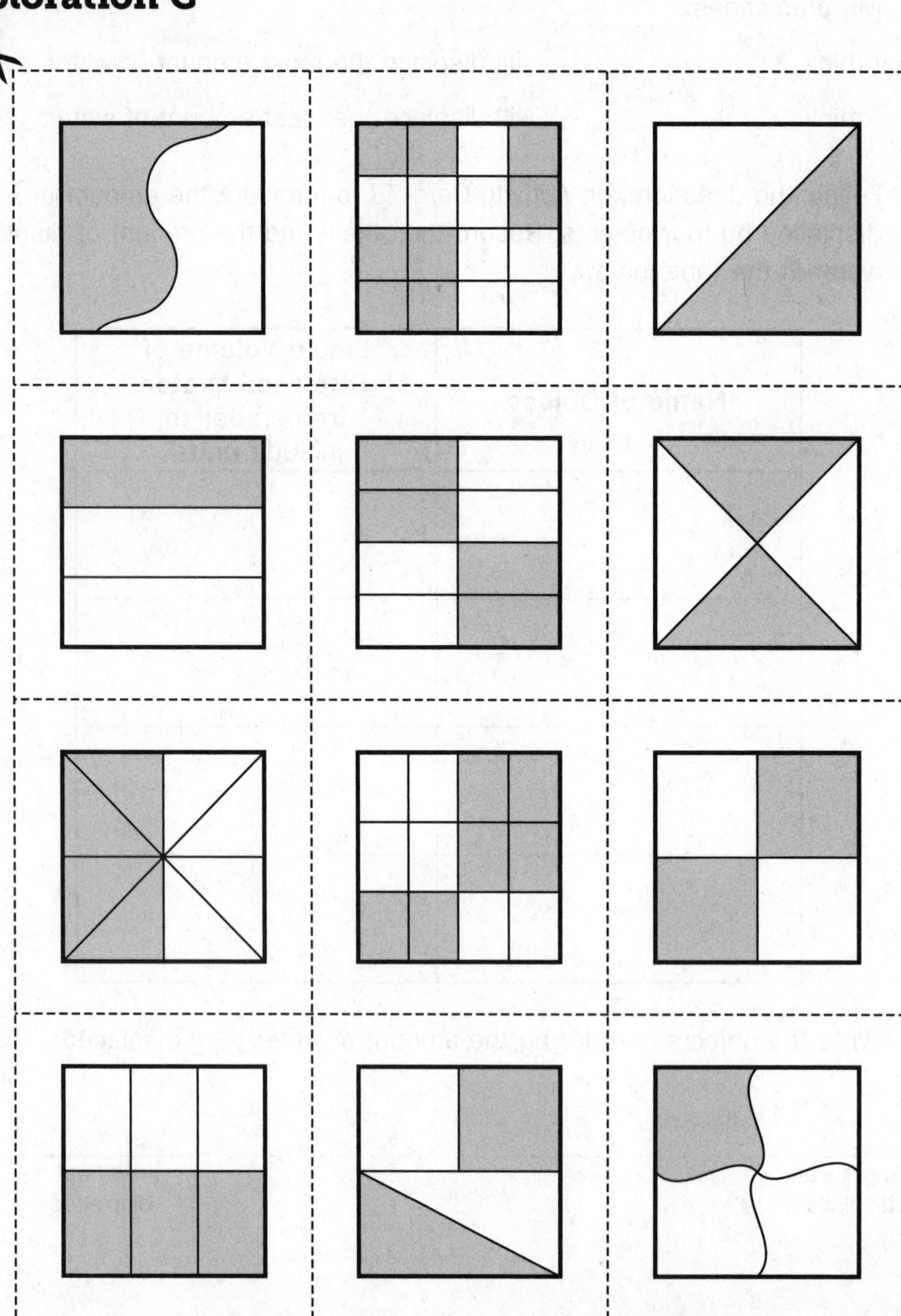

Exploring Equivalent Fractions

Home Link 7-2

NAME DATE TIME

Family Note Today your child explored different representations of equal shares and played *Fraction Memory*. Help your child make sense of the *Fraction Memory* round below.

Please return this Home Link to school tomorrow.

① Nash chose these two cards in a round of *Fraction Memory:*

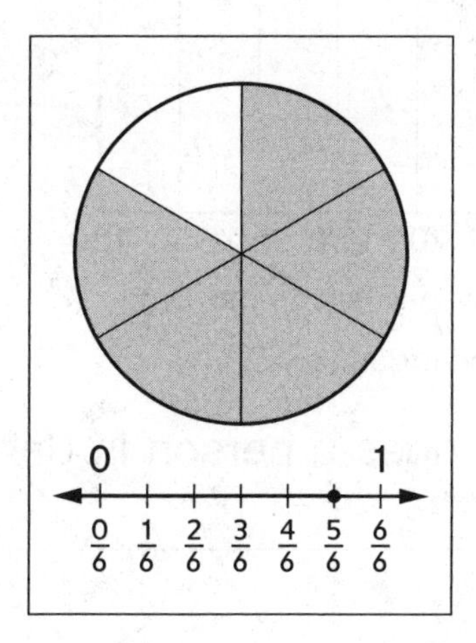

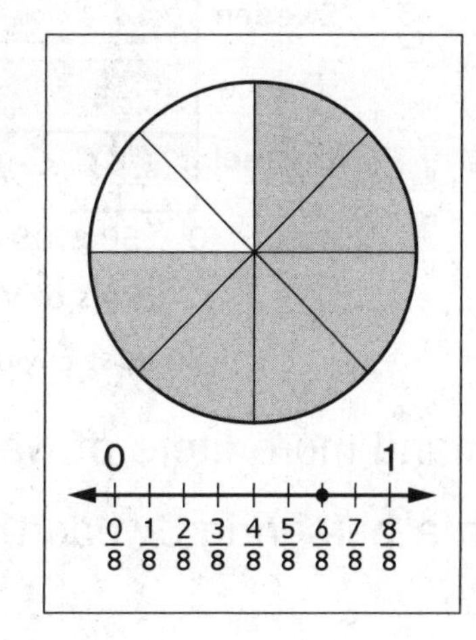

Nash says that these cards show equivalent fractions. Do you agree or disagree? Explain.

____

____

____

Practice

Solve.

② $6 \times 9 =$ ______ ③ $9 \times 8 =$ ______ ④ ______ $= 7 \times 8$

Solving Problems Using a Bar Graph

Lesson 7-3

NAME DATE TIME

Use the information in the graph to solve and write number stories.

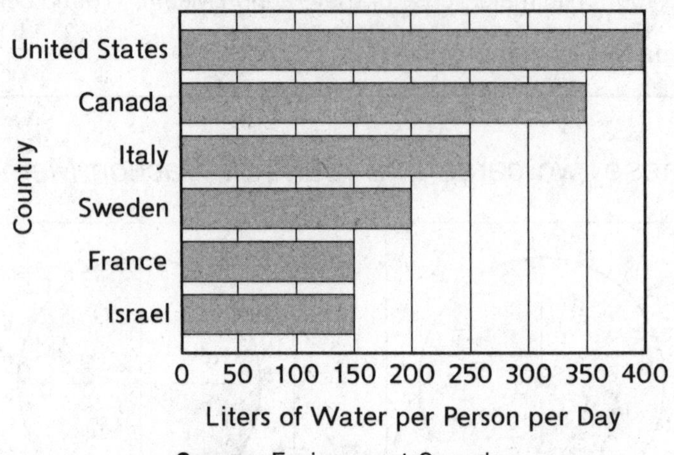

① About how many more liters of water does a person in Canada use per day than a person in Sweden? ______________

② About how many liters of water in all do a person in France, a person in Italy, and a person in Sweden use per day?

____

③ Is France and Israel's total per person daily water use more or less than Canada's? ______________

How many liters more or less? ______________

④ Write your own number story using information from the graph.

____

____

____

____

Number Stories with Measures

Home Link 7-3

NAME DATE TIME

Family Note Today your child solved number stories involving time, volume, mass, and length. Help your child make sense of the stories below. Problems 1 and 2 are similar to those we solved in class. For the Try This problem, you may wish to remind your child that 2 halves make 1 whole.

Please return this Home Link to school tomorrow.

Solve. Use drawings or number models to show your work.

① The liquid volume of 1 juice box is about 150 mL.

What is the liquid volume of 3 juice boxes?

Answer: about ______________
(unit)

② Art club ends at 3:30 P.M. Your mom arrives to pick you up at 3:10 P.M. If the teacher lets you out 5 minutes late, about how long does your mom have to wait?

Answer: about ______________
(unit)

Try This

③ Anastasia's water bottle has a liquid volume of about 1 liter. She drinks about $1\frac{1}{2}$ bottles of water every day.

About how many liters of water does Anastasia drink in 5 days? You may draw a picture.

Answer: about ______________
(unit)

Equal Parts

Lesson 7-4

NAME DATE TIME

Use a straightedge or your Pattern-Block Template.

① Divide the shape in half. 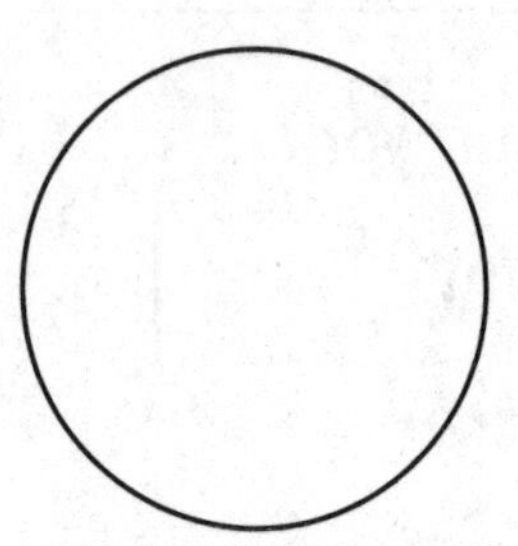 What fraction names each part? __________	② Divide the shape into 4 equal parts. 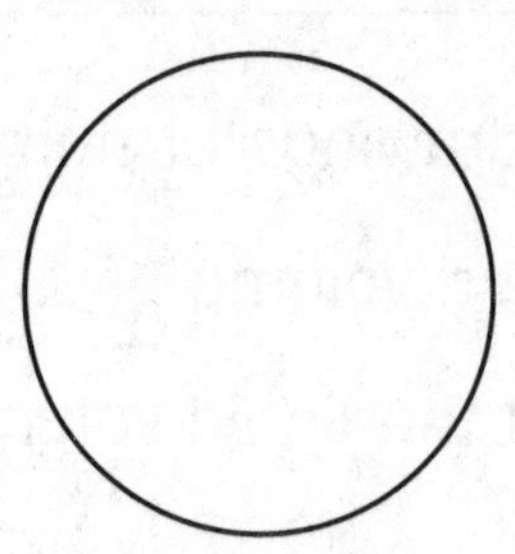 What fraction names each part? __________
③ Divide the shape in half. 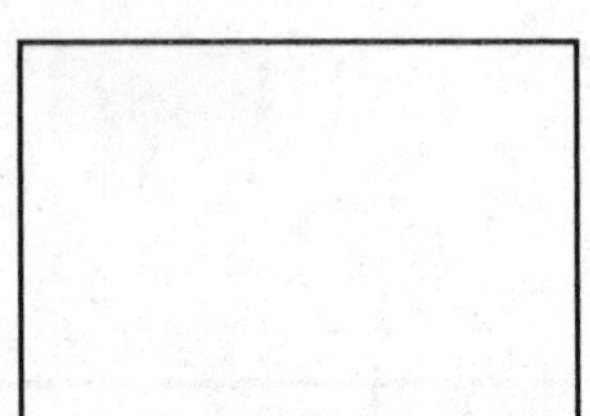 What fraction names each part? __________	④ Divide the shape into 4 equal parts. 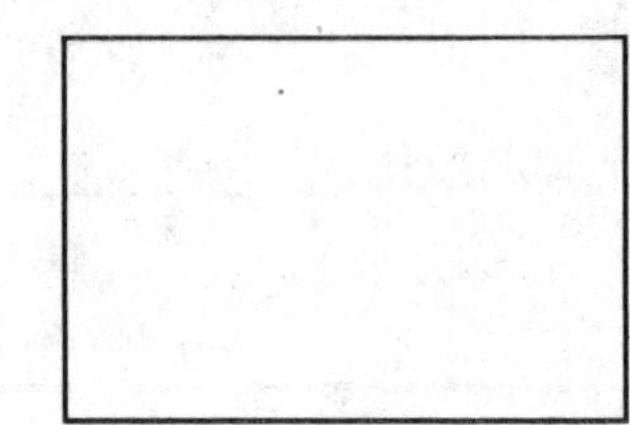 What fraction names each part? __________

Try This

⑤ Use your Pattern-Block Template triangle.
Divide the shape into 3 equal parts.

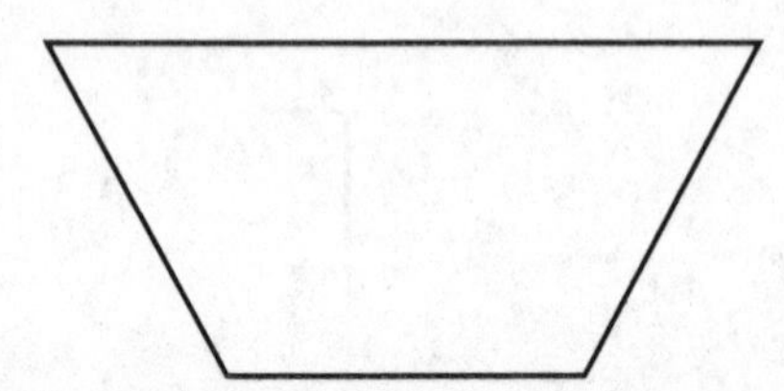

What fraction names each part? __________

Fifths, Tenths, and Twelfths

Lesson 7-4

NAME DATE TIME

Fifths, Tenths, and Twelfths

Lesson 7-4

NAME DATE TIME

Copyright © McGraw-Hill Education. Permission is granted to reproduce for classroom use.

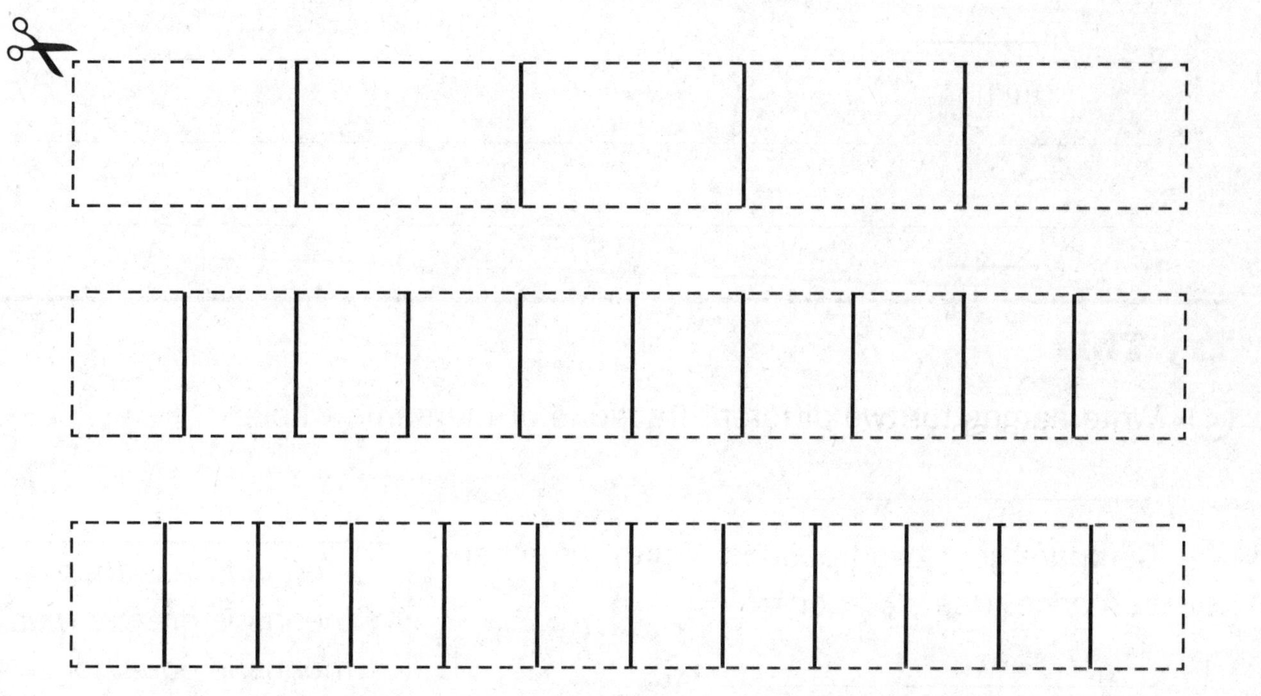

Comparing Fractions with Fraction Strips

Lesson 7-4

NAME DATE TIME

Use your fraction strips to help solve these problems.

① Savita has $\frac{1}{2}$ of a granola bar and Jeremiah has $\frac{1}{3}$ of the same-size granola bar. Who has the larger amount? Explain your answer.

____

____

____

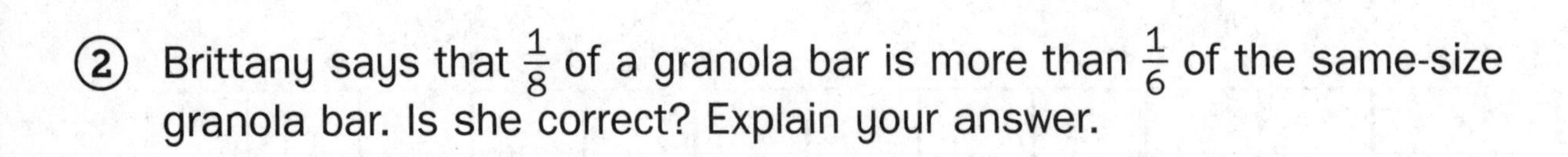

② Brittany says that $\frac{1}{8}$ of a granola bar is more than $\frac{1}{6}$ of the same-size granola bar. Is she correct? Explain your answer.

____

____

____

③ Luis says $\frac{1}{2}$ of a granola bar is the same as $\frac{2}{4}$ of the same-size granola bar. Is he correct? Explain your answer.

____

____

____

Try This

④ Write names for two different fractions of the same whole.

__________ __________

Compare your two fractions. Write a number sentence. Use <, >, or =.

____

< means *is less than*
> means *is greater than*
= means *is equal to*

Fraction Strips

Home Link 7-4

NAME DATE TIME

Family Note Today your child made a set of fraction strips. Fraction strips are equal-length strips folded into equal parts. Each equal part is labeled with the appropriate unit fraction, such as $\frac{1}{2}$ and $\frac{1}{4}$. The strips can be used to compare fractions.

Help your child shade rectangles to show each fraction and write fractions that match the shaded parts.

Please return this Home Link to school tomorrow.

SRB 139

Shade each rectangle to match the fraction below it.

Example: $\frac{2}{4}$ 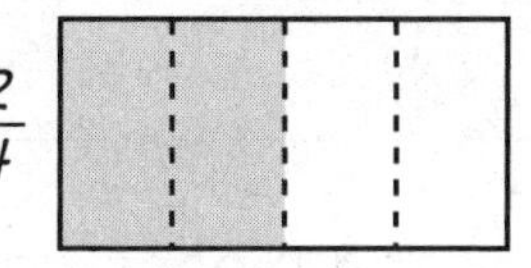

1. 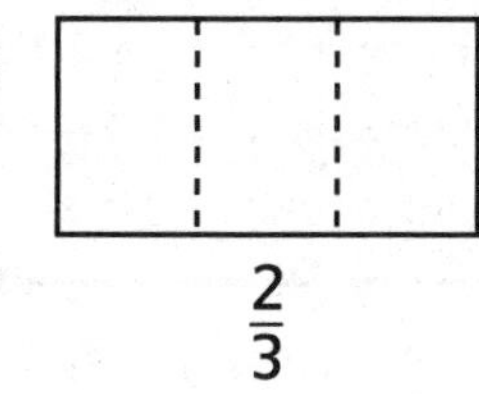 $\frac{2}{3}$

2. 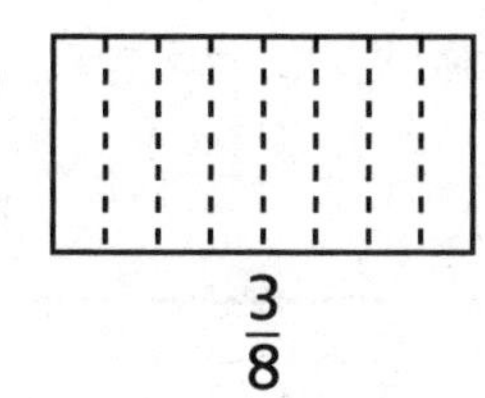 $\frac{3}{8}$

3. 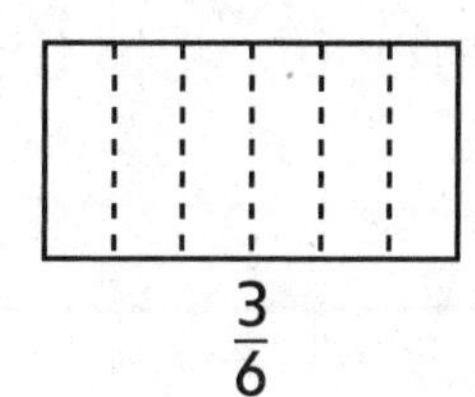 $\frac{3}{6}$

4. 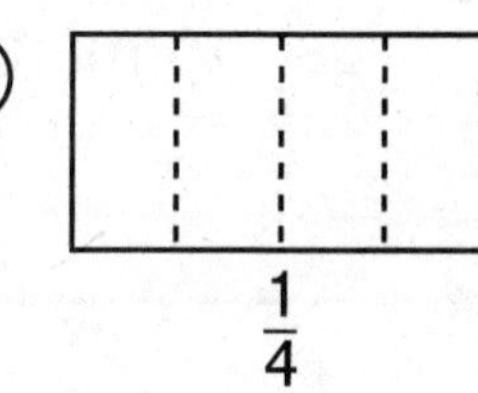 $\frac{1}{4}$

Compare the shaded parts of the fraction strips. Write >, <, or = to make the number sentence true.

< means *is less than*
> means *is greater than*
= means *is equal to*

5.

$\frac{1}{4}$	$\frac{1}{4}$	$\frac{1}{4}$	$\frac{1}{4}$

$\frac{1}{3}$	$\frac{1}{3}$	$\frac{1}{3}$

$\frac{1}{4}$ ____ $\frac{1}{3}$

6. 

$\frac{1}{6}$	$\frac{1}{6}$	$\frac{1}{6}$	$\frac{1}{6}$	$\frac{1}{6}$	$\frac{1}{6}$

$\frac{1}{6}$	$\frac{1}{6}$	$\frac{1}{6}$	$\frac{1}{6}$	$\frac{1}{6}$	$\frac{1}{6}$

$\frac{4}{6}$ ____ $\frac{2}{6}$

Practice

Unit

Make an estimate. Then show how you solve each problem on the back of this page. Explain to someone how you can use your estimate to check whether your answer makes sense.

7. 963 − 548 = ________

   My estimate: ________________

8. 412 + 298 = ________

   My estimate: ________________

Solving Fraction-Strip Problems

Lesson 7-5

NAME DATE TIME

Use your fraction strips to solve the problems on this page. You can fold each strip to different lengths to model each problem. Then shade and label both fraction-strip pieces inside the blank strip below the number line.

Example: Without using eighths, which 2 different fraction-strip pieces could you use to make a fraction strip that is as long as $\frac{6}{8}$?

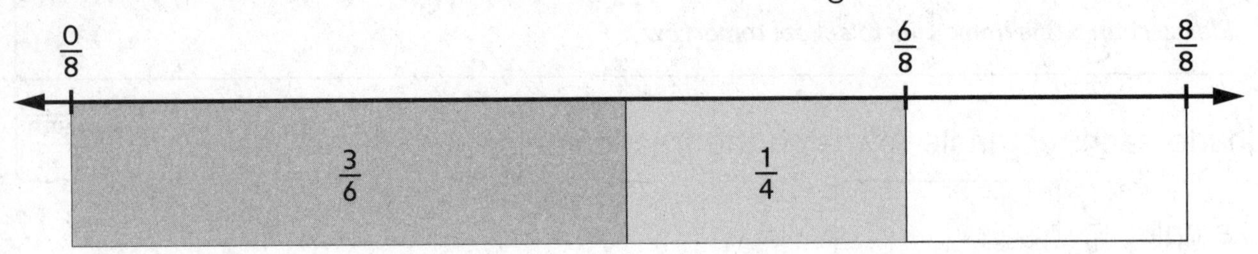

① Without using fourths, which 2 different fraction-strip pieces could you use to make a fraction strip that is as long as $\frac{3}{4}$?

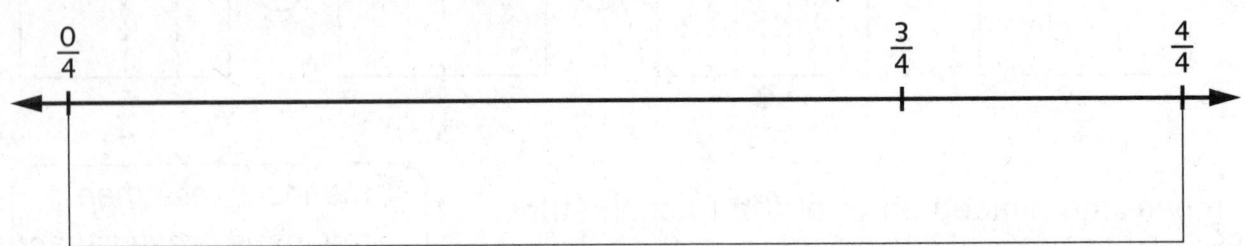

② Without using thirds, which 2 different fraction-strip pieces could you use to make a fraction strip that is as long as $\frac{2}{3}$?

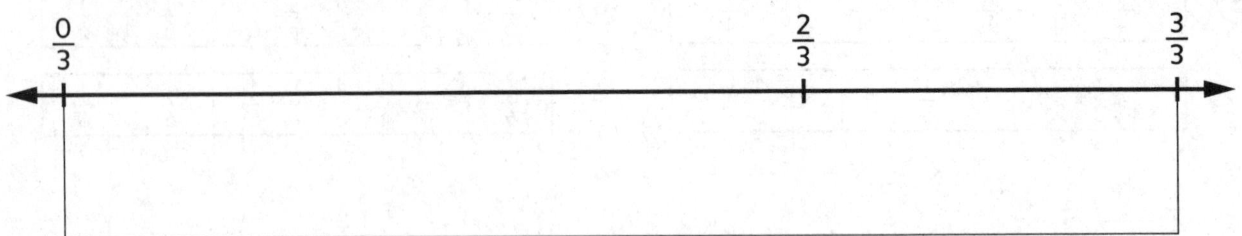

③ Without using sixths, which 2 different fraction-strip pieces could you use to make a fraction strip that is as long as $\frac{5}{6}$?

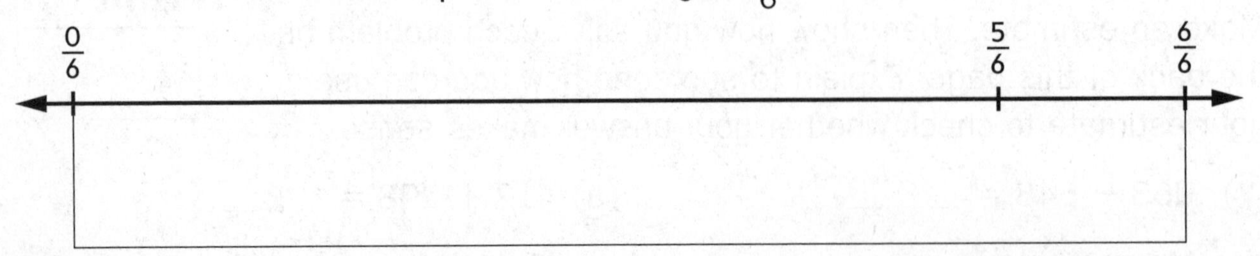

④ On the back of this page, make up your own fraction-strip problem.

Fractions on Number Lines

Home Link 7-5

NAME DATE TIME

Family Note Today your child learned about fractions as numbers on a number line. Children made their own Fraction Number-Line Posters by dividing number lines from 0 to 1 into equal-size parts, or *distances*. They labeled the tick marks with the appropriate fractions. Support your child in locating fractions on the number lines below.

Please return this Home Link to school tomorrow.

① Write the fraction that represents the distance the triangle moved.

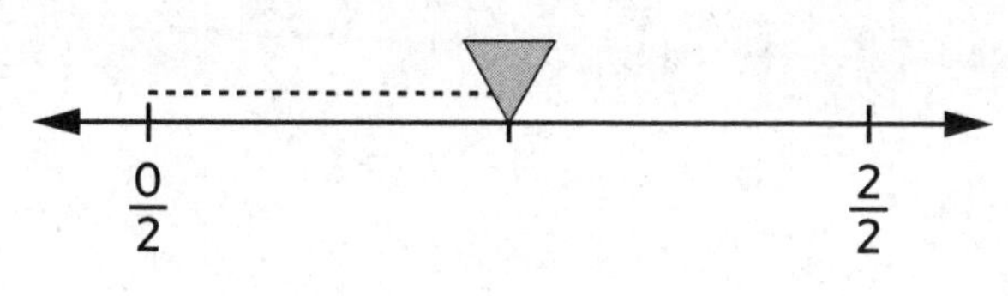

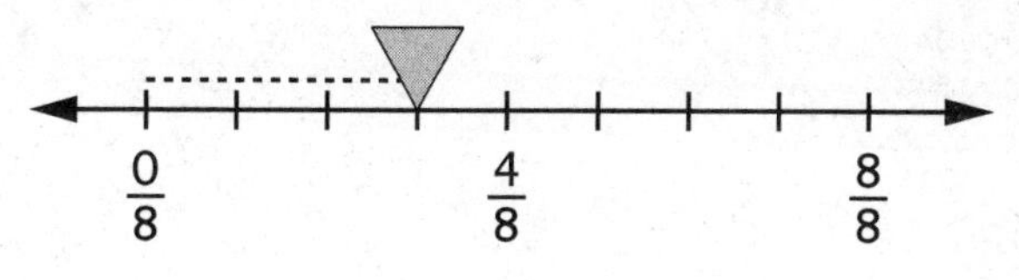

______ ______

② Write the missing fractions.

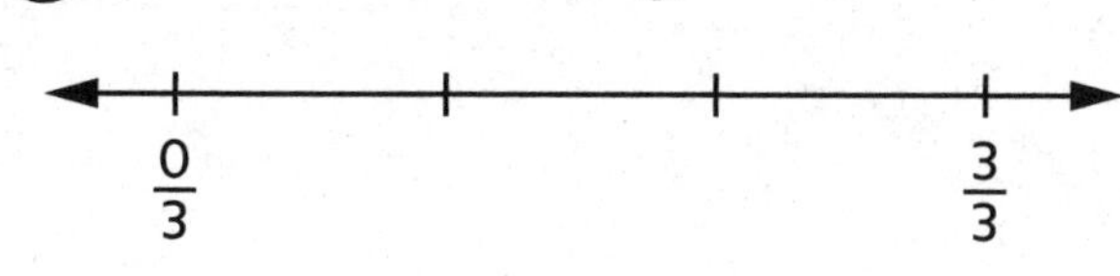

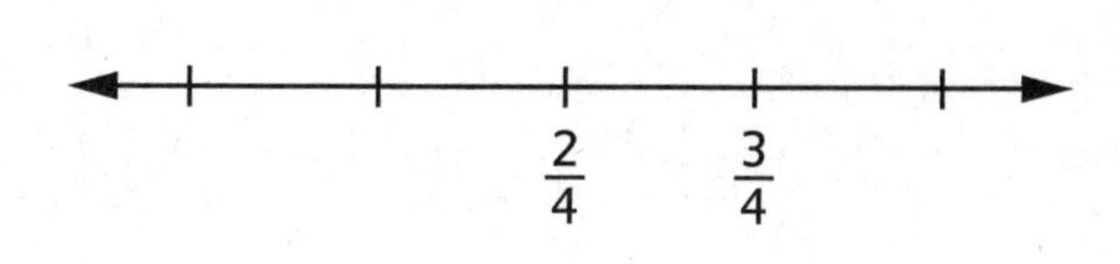

______ ______ ______ ______ ______

Practice

Fill in the unit. Solve. Show your work.

Unit

③ 333 + 492 = ______

④ ______ = 888 − 678

Identifying Missing Fractions on Number Lines

Lesson 7-6

NAME DATE TIME

For each problem, write the missing fractions.

①

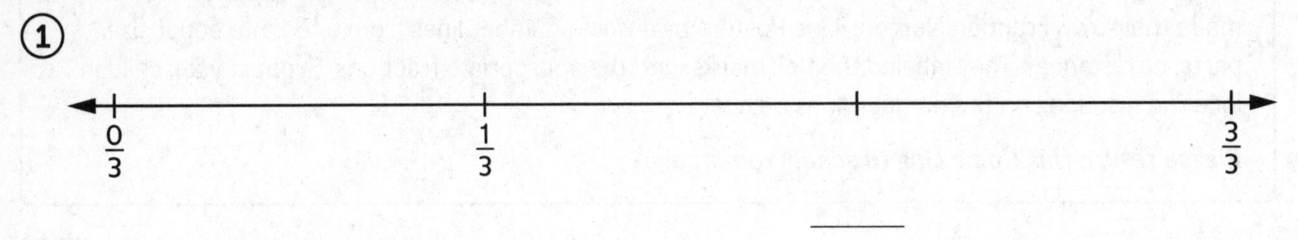

②

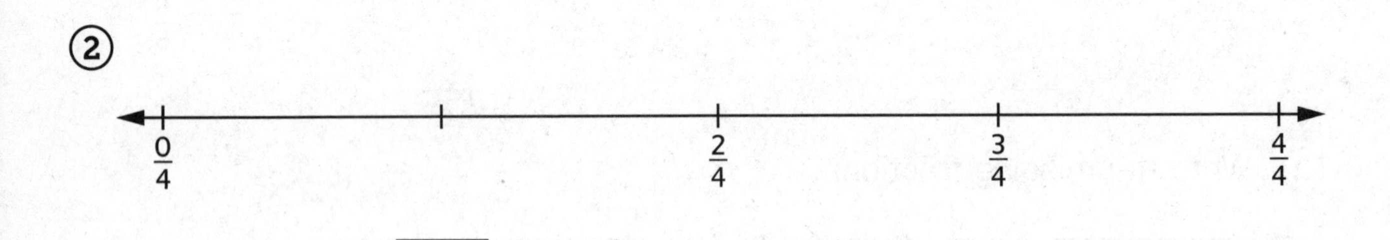

③

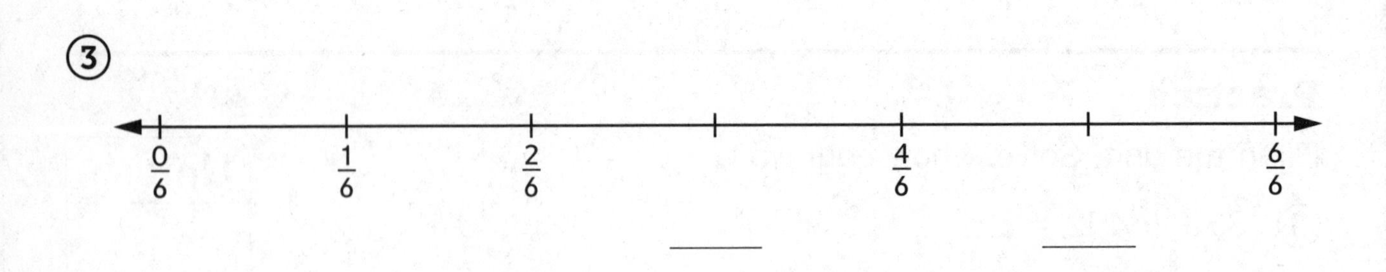

④

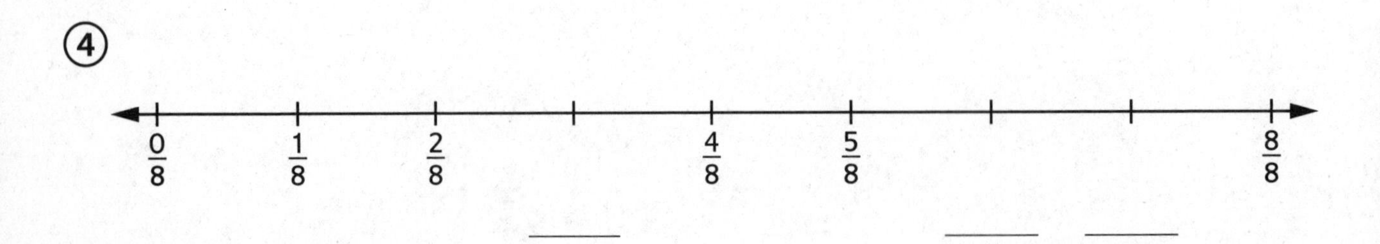

Identifying Fractions on Number Lines

Lesson 7-6

NAME DATE TIME

For each problem:

- Think of fractions that are between the whole numbers on the number line.
- Decide where each fraction is located on the number line.
- Make points on the number line to show where the fractions belong.
- Label the points with the fractions.

① Show at least three fractions that are between 0 and 2.

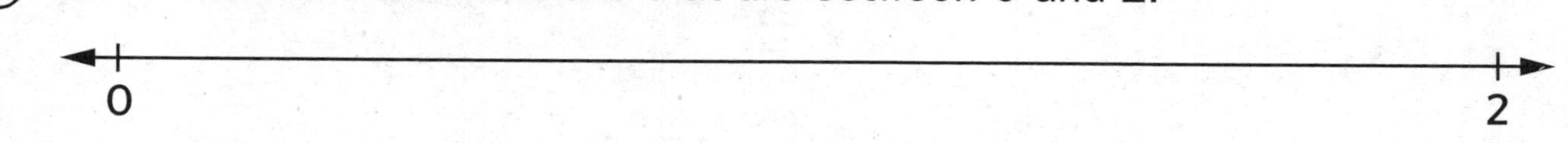

② Show at least three fractions that are between 2 and 4.

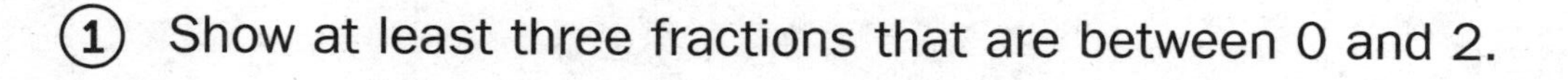

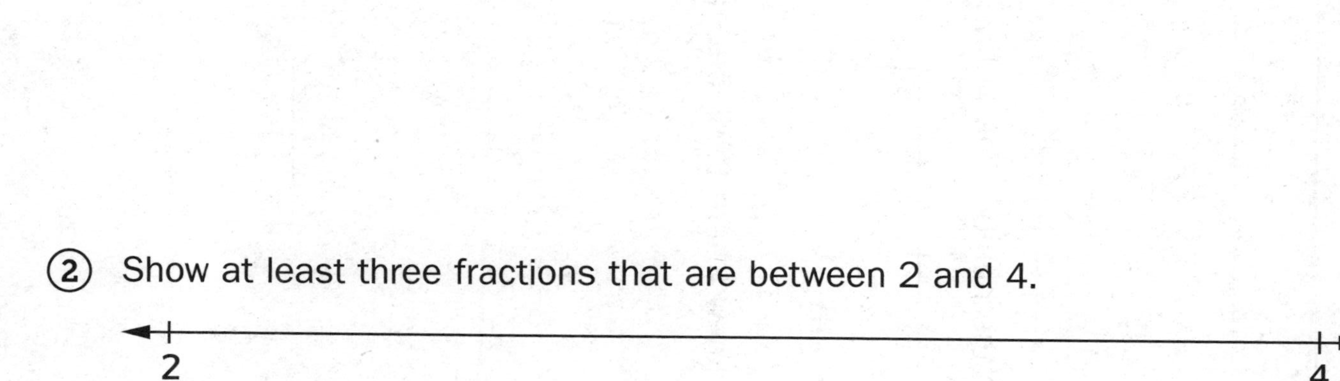

Recognizing Fractions Greater Than 1

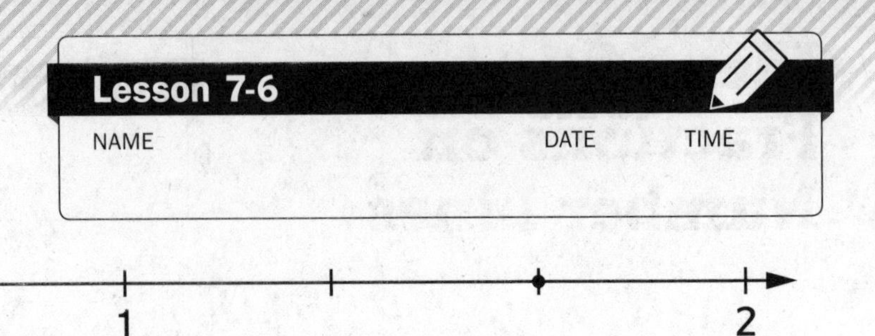

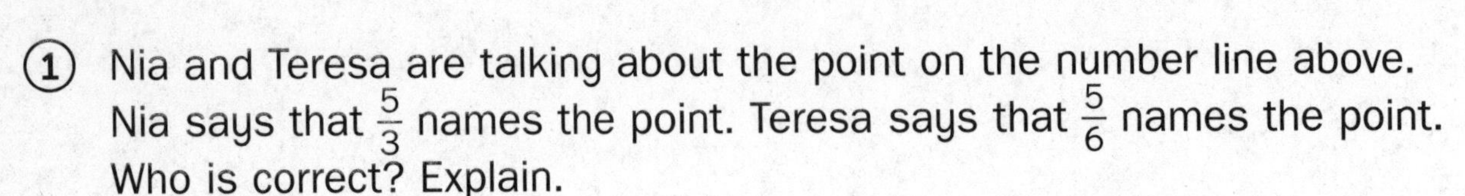

① Nia and Teresa are talking about the point on the number line above. Nia says that $\frac{5}{3}$ names the point. Teresa says that $\frac{5}{6}$ names the point. Who is correct? Explain.

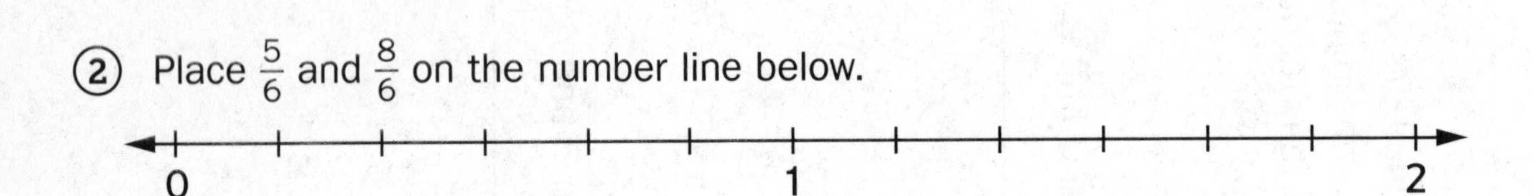

② Place $\frac{5}{6}$ and $\frac{8}{6}$ on the number line below.

③ Explain how you knew where to place $\frac{5}{6}$ and $\frac{8}{6}$ on the number line.

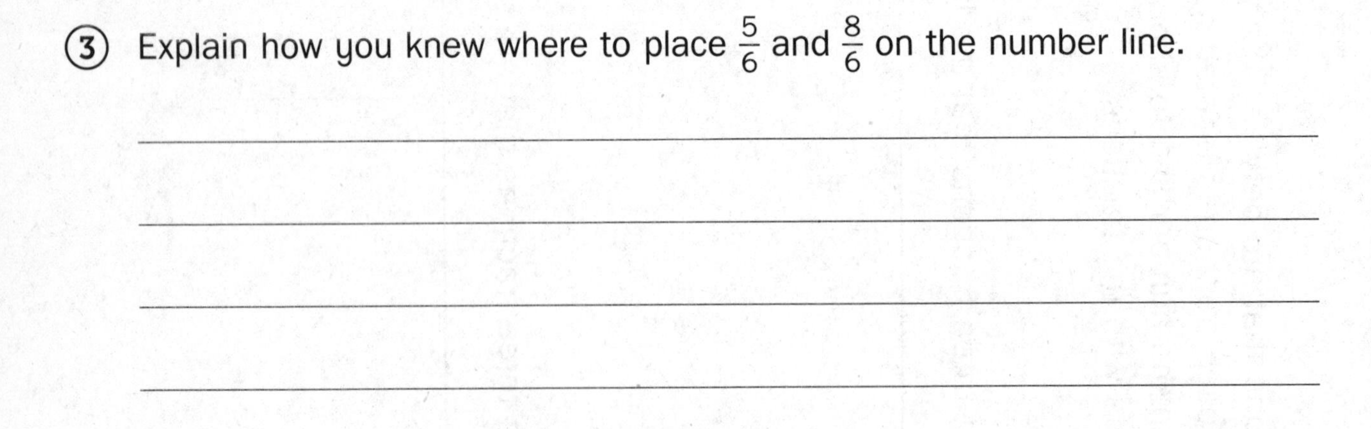

Fractions on Number Lines

Lesson 7-6

NAME DATE TIME

Number Line A

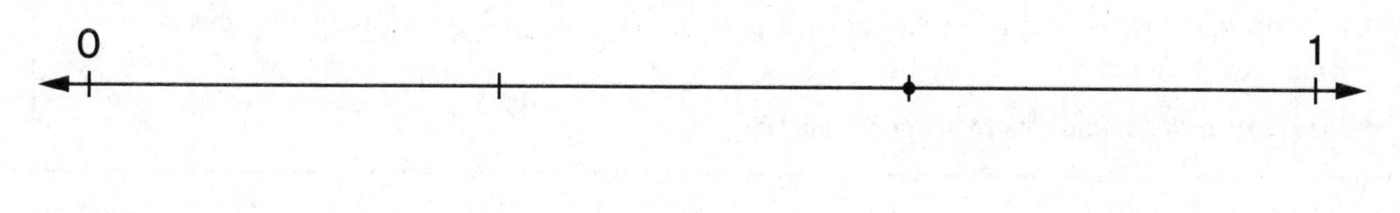

Number Line B

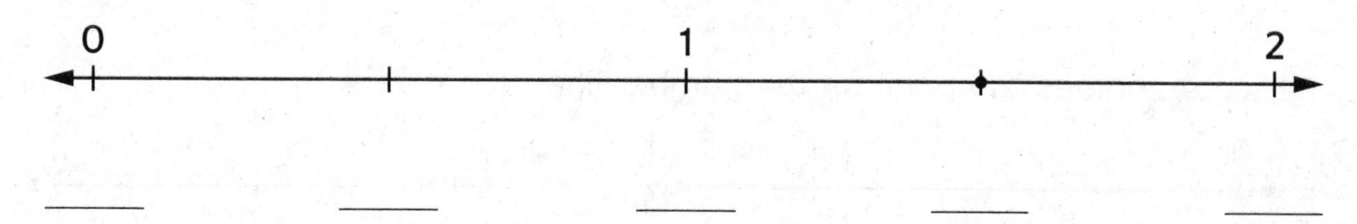

Number Line C

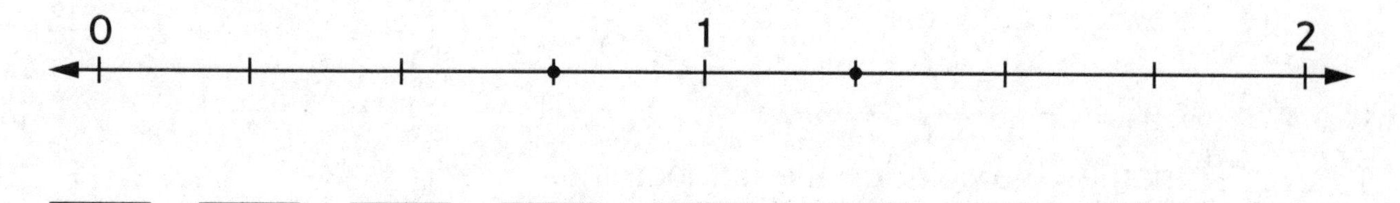

Number Line D

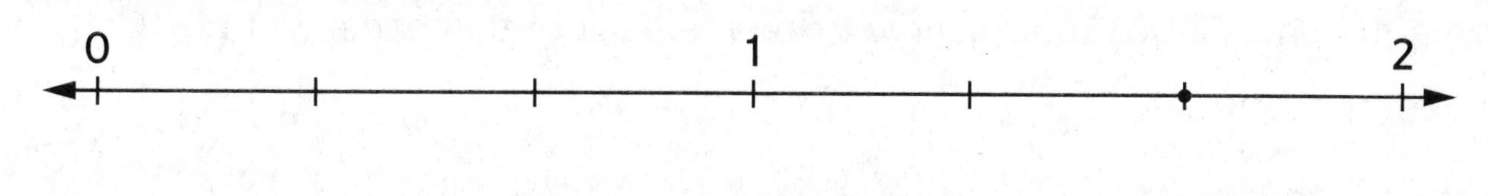

More Fractions on Number Lines

Home Link 7-6

NAME DATE TIME

Family Note Today your child identified fractions that are less than or greater than 1 on number lines. Help your child count the number of equal parts or distances between 0 and 1 and label each tick mark with a fraction.

Please return this Home Link to school tomorrow.

For each number line, fill in the missing numbers. Then name the fraction at each point.

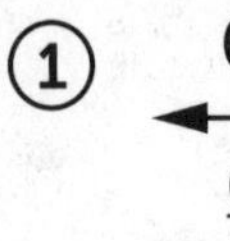

① 0 ... 1

$\frac{0}{3}$ ____ ____ $\frac{3}{3}$

______ names the point on the number line.

② 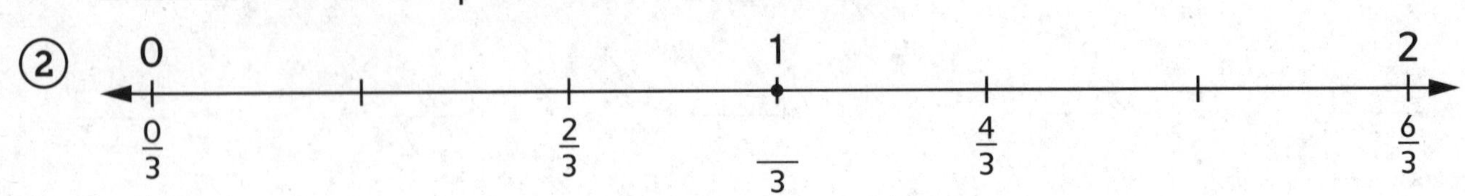

____ ____

______ names the point on the number line.

③ 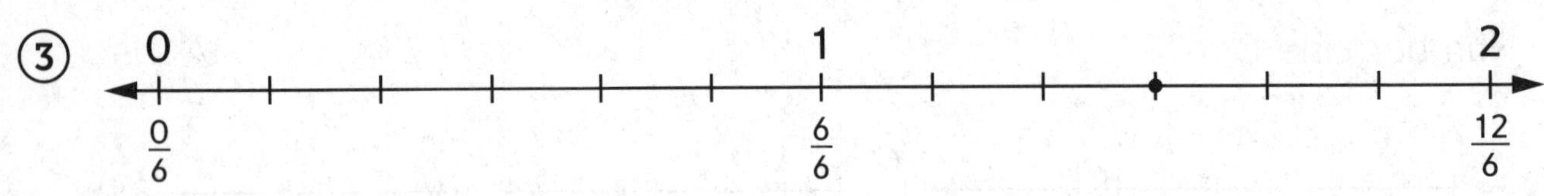

____ ____ ____ ____ ____ ____ ____ ____ ____ ____

______ names the point on the number line.

④ Look at the point on each number line. Which point names a fraction greater than 1? ______

Practice

Fill in the unit. Solve. Show your work on the back of this page.

⑤ 549 − 289 = ________ ⑥ 739 + 261 = ________

Comparing Fractional Distances

Lesson 7-7

NAME DATE TIME

Luke, Jenna, Dylan, and Alex are running 1 mile.

Here is how far they have gone:

Luke: $\frac{3}{4}$ of a mile

Jenna: $\frac{1}{2}$ of a mile

Dylan: $\frac{2}{4}$ of a mile

Alex: $\frac{1}{4}$ of a mile

Show how far each runner has gone on the number line below. Label the number line with fractions. Then write the initial of each runner below the correct fraction to show how far he or she has run.

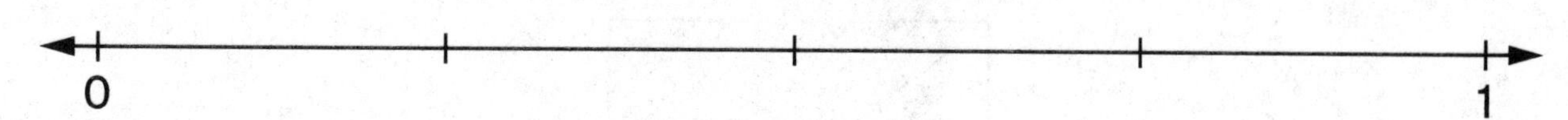

① Who has run the farthest so far? ______________

② Has Jenna run farther than Dylan? Explain your answer.

____

____

Try This

③ How much farther does Alex have to go to finish the race?

____

Estimating Regions of Fractions

Lesson 7-7

NAME DATE TIME

Math Message

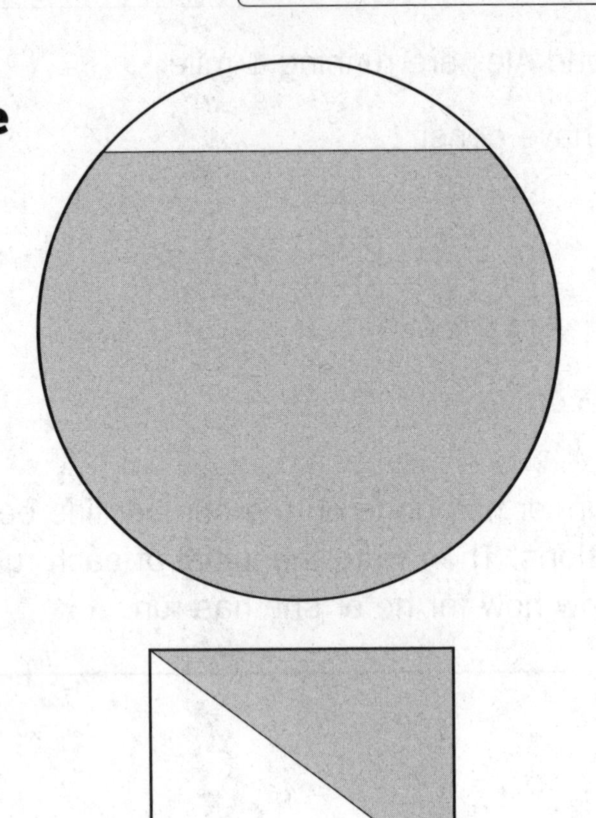

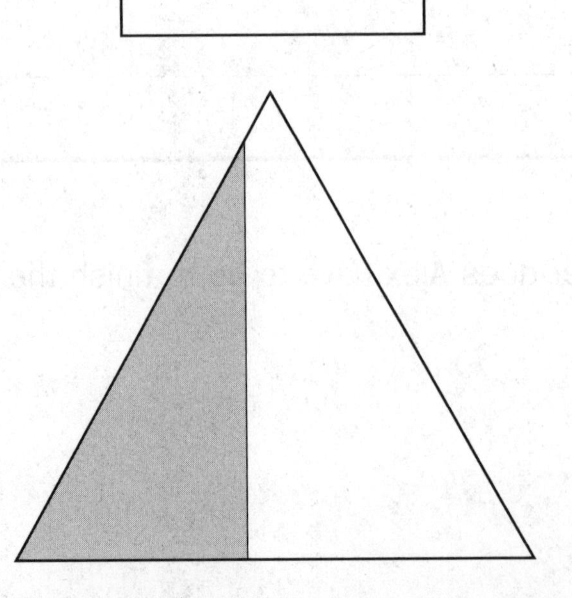

Comparing Fractions to $\frac{1}{2}$

Home Link 7-7

NAME DATE TIME

Family Note Today your child wrote number sentences comparing fractions shown with fraction circles and number lines. Ask your child to explain whether the fractions represented below are greater than (>), less than (<), or equal to (=) $\frac{1}{2}$.

Please return this Home Link to school tomorrow.

Shade each circle to match the fraction below it.

SRB 155-156

Example:

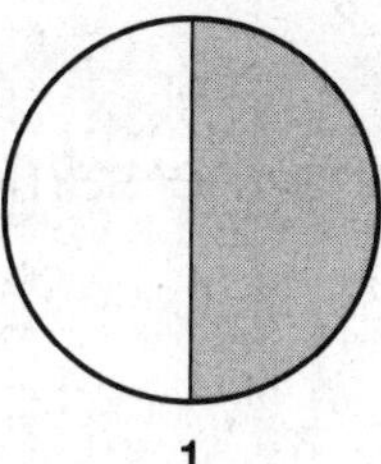

$\frac{1}{2}$

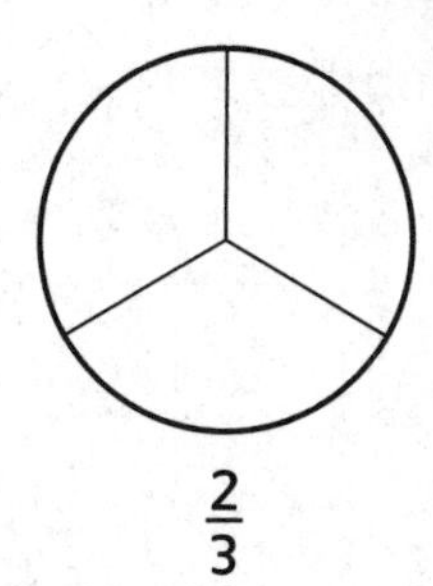

$\frac{2}{3}$

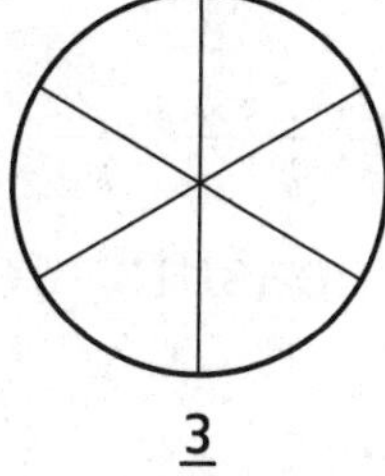

$\frac{3}{6}$

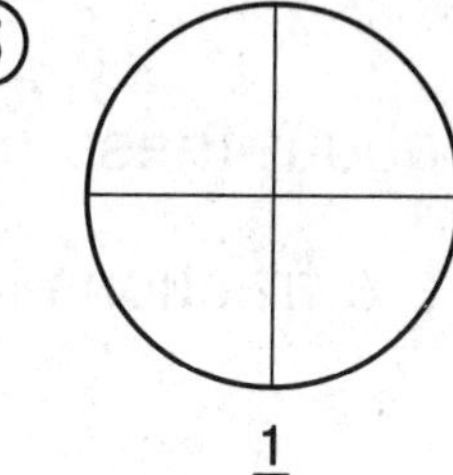

$\frac{1}{4}$

Look at the shaded parts of the circles.

4. Write the fraction above that is less than $\frac{1}{2}$.

   ____

5. Write the fraction above that is equal to $\frac{1}{2}$.

   ____

6. Write the fraction above that is greater than $\frac{1}{2}$.

   ____

7. Write <, >, or = to make the sentence true.
   You may draw a picture to help.

   $\frac{3}{4}$ ________ $\frac{1}{4}$

Finding Rules to Order Fractions

Lesson 7-8

NAME DATE TIME

For each problem, you may use any of your fraction tools to help (fraction circles, fraction strips, Fraction Number-Line Poster, fraction cards).

① Think about these fractions: $\frac{1}{6}, \frac{1}{8}, \frac{1}{10}, \frac{1}{3}, \frac{1}{4}, \frac{1}{2}, \frac{1}{5}$

Write the fractions in order from least to greatest:

What patterns do you notice? How are these fractions the same or different?

② Think about these fractions: $\frac{2}{6}, \frac{2}{3}, \frac{2}{8}, \frac{2}{4}, \frac{2}{10}, \frac{2}{5}$

Write the fractions in order from least to greatest:

What patterns do you notice? How are these fractions the same or different?

Finding Rules to Order Fractions

(continued)

Lesson 7-8

NAME DATE TIME

③ Write a rule for ordering fractions with the same numerator.

Sorting Fractions

Home Link 7-8

NAME DATE TIME

Family Note Today your child looked for patterns to help order fractions with the same numerator. Children recognized that as a fraction's denominator gets larger the fraction gets smaller. They were able to write this as a rule for ordering fractions with the same numerator. For Problem 1, your child will sort a set of fractions into two groups: fractions greater than 1 and fractions less than 1. In Problem 2, your child will look for patterns to help sort any set of fractions into these two groups.

Please return this Home Link to school tomorrow.

① Look at the fractions below and sort them into two groups: fractions less than 1 and fractions greater than 1.
Use the number lines on the following page to help you.

$\frac{1}{2}, \frac{2}{3}, \frac{6}{4}, \frac{3}{2}, \frac{7}{8}, \frac{5}{3}, \frac{6}{8}, \frac{7}{6}$

Less Than 1	Greater Than 1

② Look for a pattern in the fractions you sorted. Describe a pattern that can help you decide whether a fraction is less than 1 or greater than 1.

Sorting Fractions

(continued)

Home Link 7-8

NAME DATE TIME

Fraction Number Lines

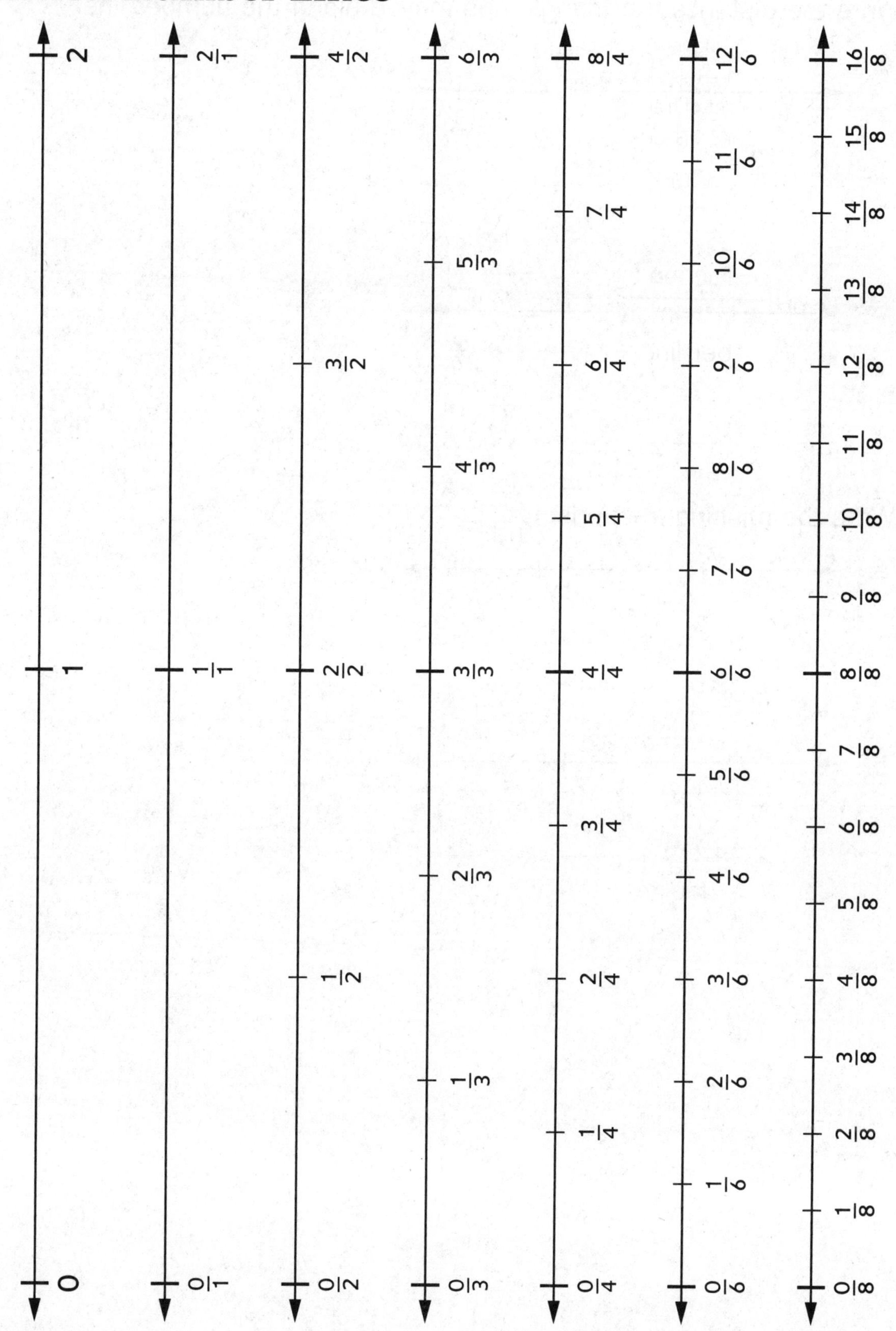

Locating and Representing Fractions

Lesson 7-9

NAME DATE TIME

① Write the distance the triangle has moved along the number line.

a.

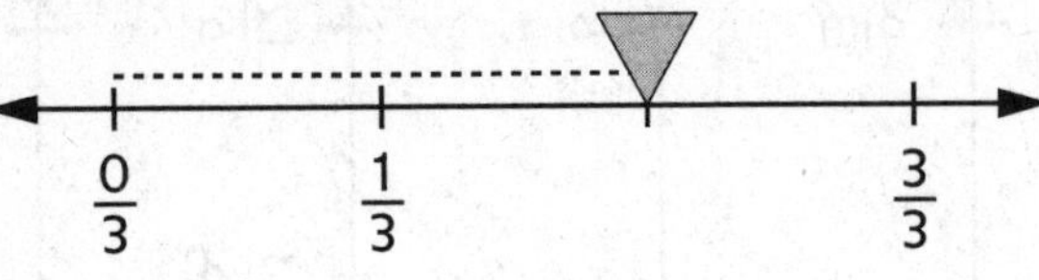

______

b.

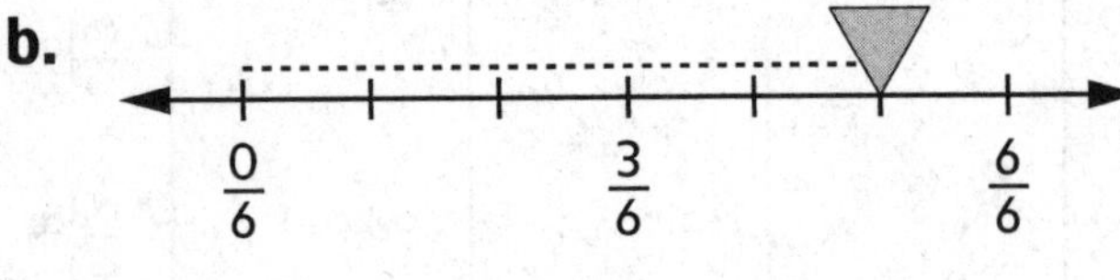

______

② Write the missing fractions.

a.

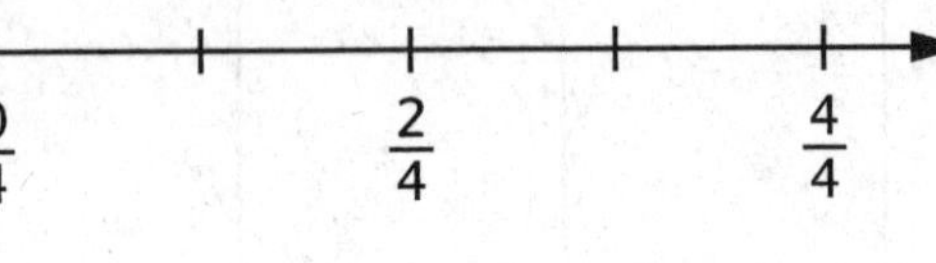

______ ______

b.

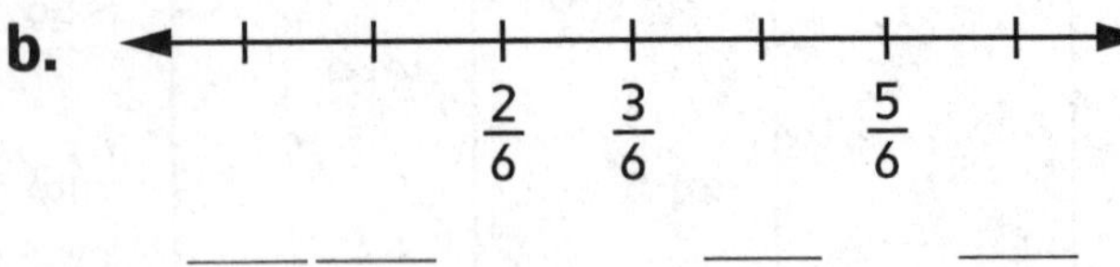

______ ______ ______ ______

Partitioning a Number Line

Lesson 7-9

NAME DATE TIME

Partition the number line into 12 equal parts with tick marks.

Hint: Think about how you partitioned a number line into 6 equal parts.

Then label each tick mark with a fraction $\frac{0}{12}$ through $\frac{12}{12}$.

0

1

Locating Equivalent Fractions

Lesson 7-9

NAME DATE TIME

Balen and Amy each have a different fraction card, but the fractions are equivalent. Use your fraction cards to find two fractions they could have. Write the fractions on the cards below.

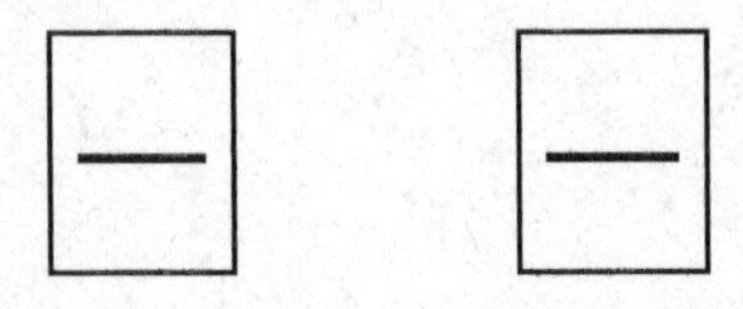

Balen's Card Amy's Card

Draw number lines below to show how you know the fractions are equivalent. Explain your work with a picture and words.

Picture:

Words: ____

____

____

____

Locating $\frac{1}{2}$ on Number Lines

Lesson 7-9

NAME DATE TIME

Math Message

The whole is the distance from 0 to 1.

- Label $\frac{1}{2}$ where you think it should go on the number line.
- Share your thinking with your partner.

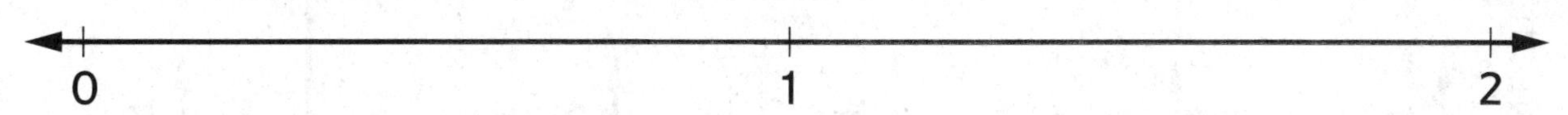

Listen to your teacher for directions for Problems 1 and 2.

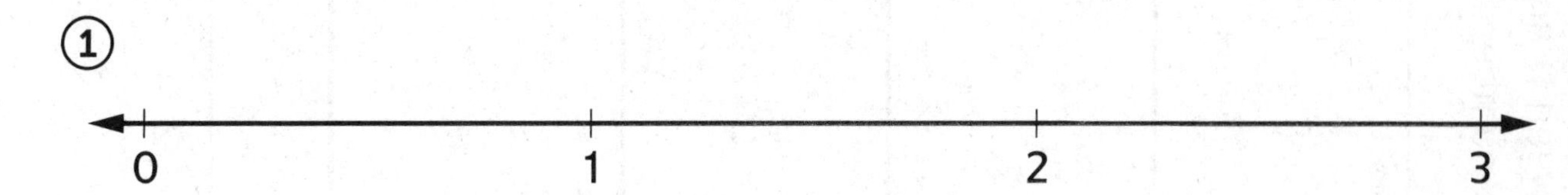

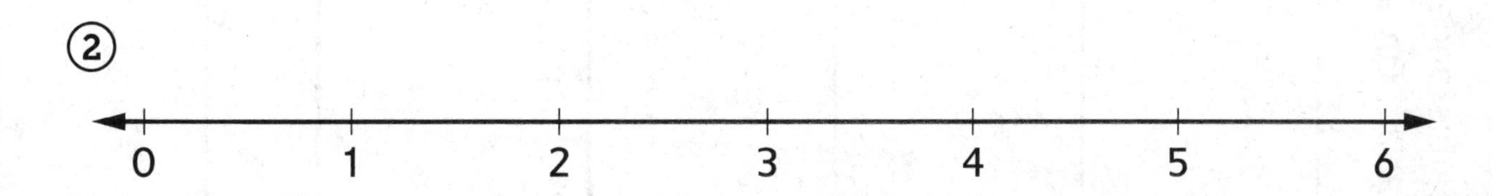

Fractions on Number Lines

Lesson 7-9

NAME DATE TIME

The whole on the number lines below is the distance between 0 and 1. Partition each number line to locate and then label the given fraction. *Hint:* On some number lines, there is more than 1 whole.

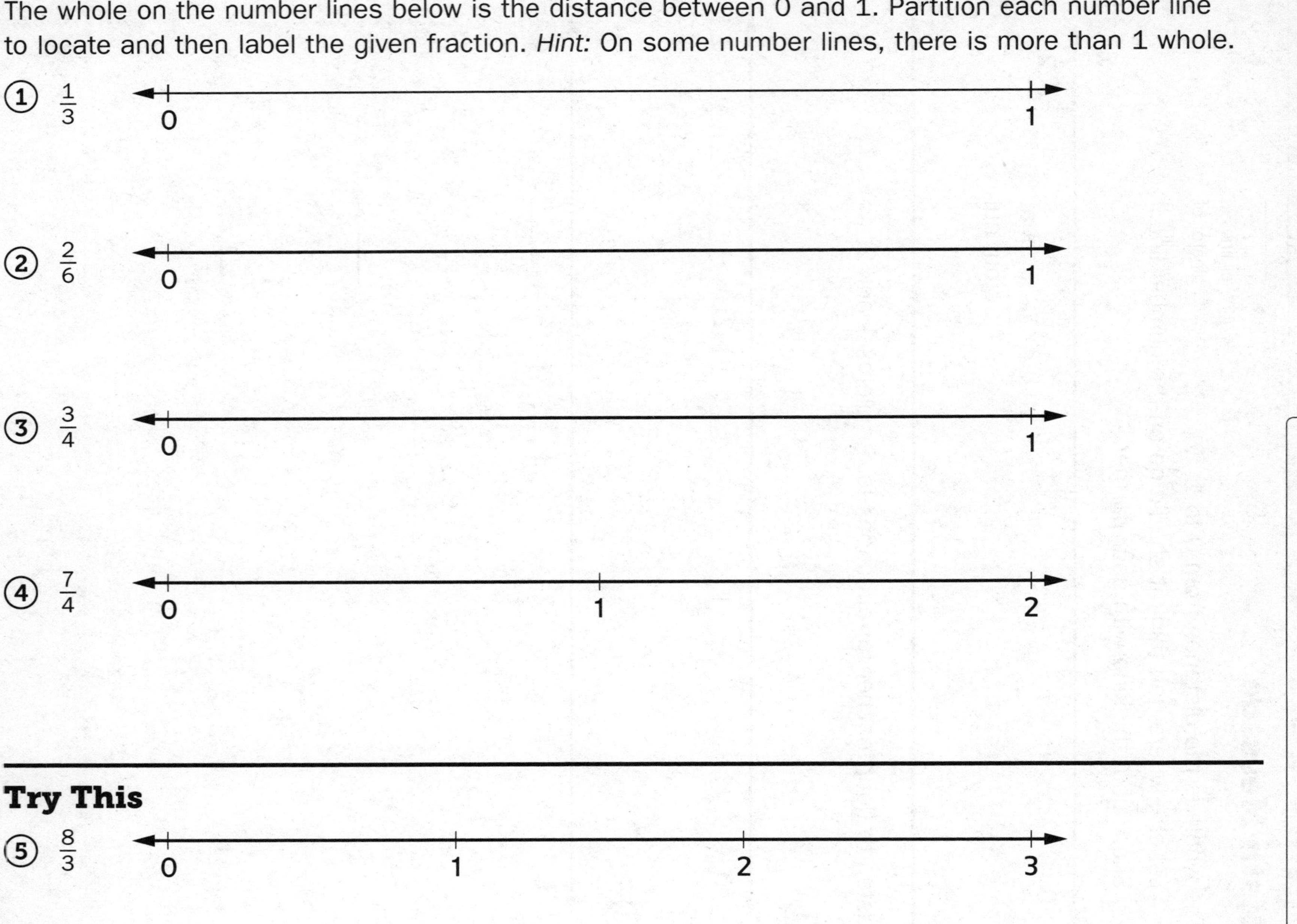

Locating Fractions on Number Lines

Home Link 7-9

NAME DATE TIME

Family Note Today your child learned to partition, or divide, number lines into equal parts and then locate and label fractions between whole numbers. The denominator of the given fraction describes the number and size of the equal parts. A whole on each number line below is equal to the distance between 0 and 1.

Please return this Home Link to school tomorrow.

Partition the wholes on each number line. Then locate and label the given fractions. Tell someone at home how you partitioned your number lines.

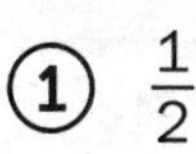

1. $\frac{1}{2}$

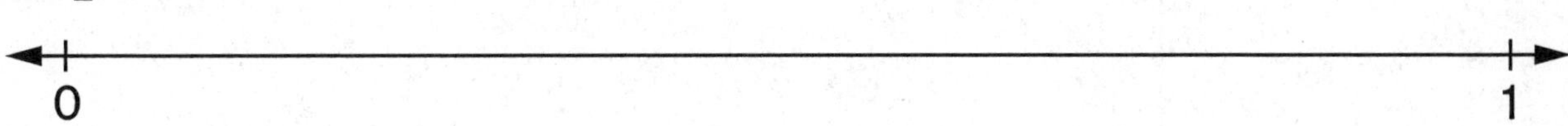

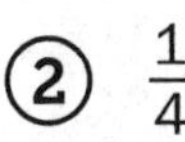

2. $\frac{1}{4}$

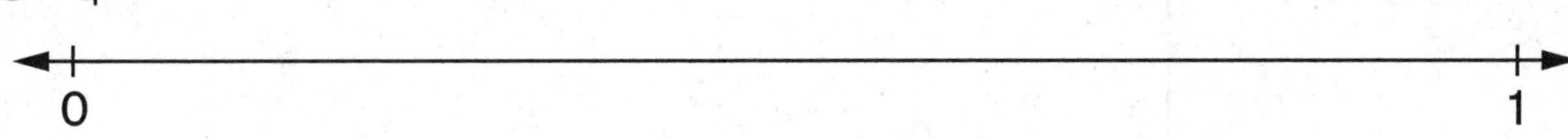

3. $\frac{2}{3}$

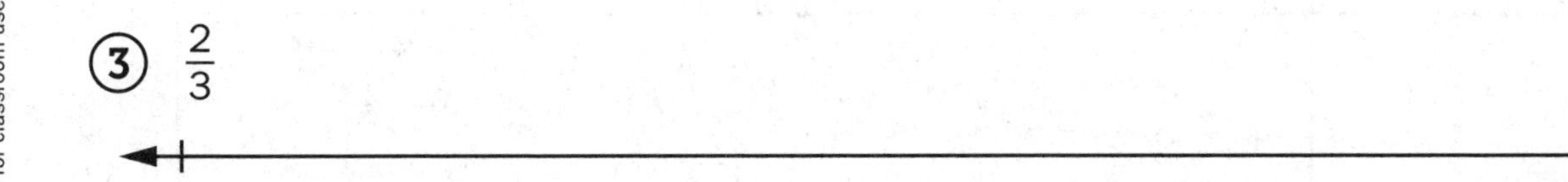

4. $\frac{5}{6}$

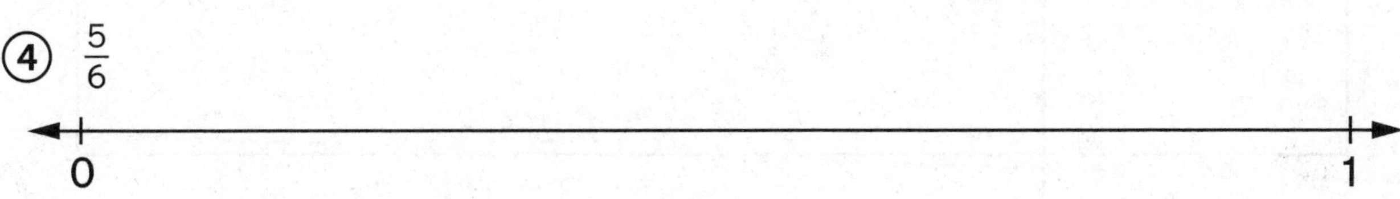

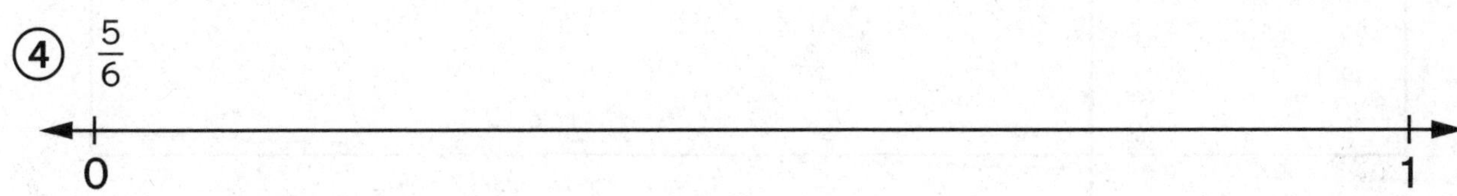

Try This

5. $\frac{3}{2}$

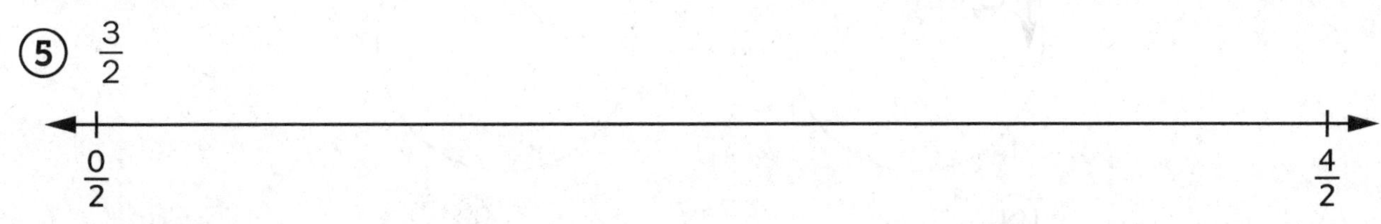

Modeling Equivalent Fractions

Lesson 7-10

NAME DATE TIME

Name a fraction that is equivalent to $\frac{2}{3}$. You may use your fraction tools to help.

$\frac{2}{3}$ = ______

Show that your number sentence is true with number lines, fraction strips, and fraction circle pieces. Sketch pictures to show what you did.

Number Lines

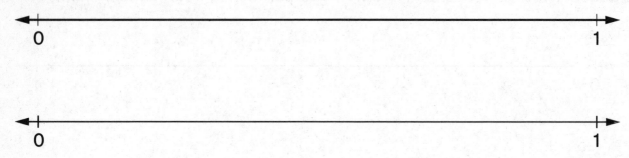

Fraction Strips

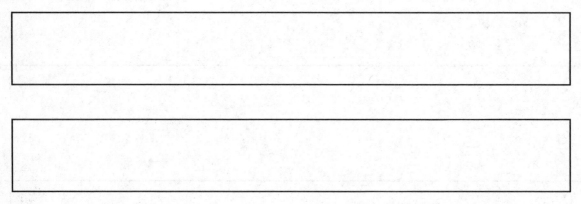

Fraction Circle Pieces

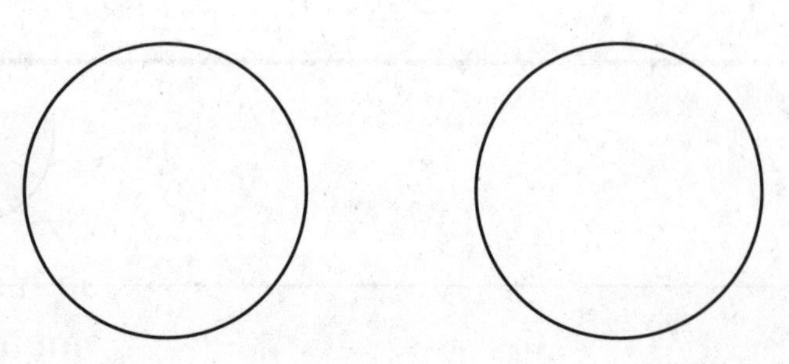

Matching Fraction Tools

Home Link 7-10

NAME DATE TIME

Family Note Your child has been using number lines, fraction circles, and fraction strips to learn about and represent fractions. Today your child used these tools to make and justify fraction comparisons. Encourage your child to explain how he or she matched each number sentence with a picture that represents the fraction comparison.

Please return this Home Link to school tomorrow.

Draw a line from each number sentence to the picture that represents it.

$\frac{1}{2} > \frac{1}{8}$

① $\frac{1}{2}$

$\frac{2}{6} < \frac{5}{6}$

② 0 1

0 1

$\frac{2}{3} = \frac{4}{6}$

③ 0 1

0 1

$\frac{3}{8} < \frac{3}{6}$

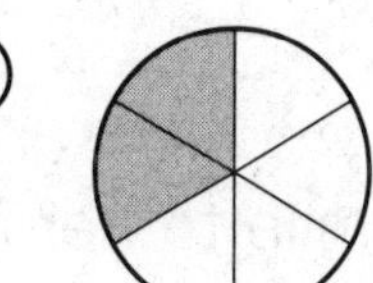

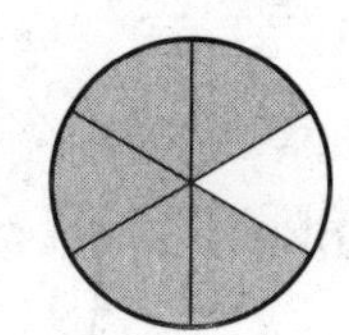

On the back of this page, write one of the above fraction number sentences. Sketch a different fraction tool that shows the same comparison.

Solving More Fraction Number Stories

Lesson 7-11

NAME DATE TIME

Use your fraction circles, fraction strips, number lines, or drawings to help you solve each problem.

1. Chantel used $\frac{2}{3}$ of a length of ribbon to wrap presents.

   Did she use more or less than $\frac{3}{4}$ of the ribbon? ________

2. Dorothy walks $1\frac{1}{2}$ miles to school.

   Jaime walks $\frac{3}{2}$ miles to school.

   Compare the two distances. What do you notice?

   ____

3. Anik read $\frac{1}{4}$ of a book on Monday.

   The next day he read $\frac{2}{4}$ of the book.

   What fraction of the book did he read on those 2 days?

   ______ of the book

4. Blake painted $\frac{3}{4}$ of his room.

   Did he paint more or less than $\frac{1}{2}$ of his room?

   ____

5. Greg ate $\frac{2}{8}$ of a small pizza.

   Carla ate $\frac{3}{6}$ of a small pizza.

   Who ate more pizza?

   ____

6. Lucia walked $\frac{2}{8}$ of a mile to her aunt's house.

   Then she walked $\frac{4}{8}$ of a mile to her friend's house.

   What fraction of a mile did she walk in all? ______ of a mile

Solving Art Class Fraction Stories

Lesson 7-11

NAME DATE TIME

Use your fraction circles, fraction strips, number lines, or drawings to help you solve each problem.

1. The children all started with the same amount of paint.

   Juliana used all of her paint, Chris used $\frac{1}{3}$ of his paint, and Olivia used $\frac{2}{3}$ of her paint. Who used the least amount of paint?

   ____

   What fraction of her paint did Juliana use? ____________

2. One-eighth of the class finished their art projects. The rest of the class is not finished with their projects. What fraction of the class is not finished?

   ________ of the class

3. The art teacher has 3 kilograms of clay. 4 children will equally share the clay. Write a fraction to show how much clay each child will get.

   ________ of a kilogram of clay

Try This

4. It took Nathan $1\frac{1}{4}$ hours to do his art project. How many minutes did he spend on his project? ________ minutes

Fraction Number Stories

Home Link 7-11

NAME DATE TIME

Family Note Today your child solved fraction number stories using a variety of fraction models, including pictures. Encourage your child to sketch a picture to represent each story.

Please return this Home Link to school tomorrow.

Solve these number stories. Show your answer as a fraction. You may draw pictures to show your work.

1. Ralph read $\frac{1}{8}$ of his book. What fraction of the book does he still have left to read?

   My sketch:

   _______ of his book

2. Four friends equally share two bottles of juice. How much juice will each friend get?

   My sketch:

   _______ of a bottle of juice

3. Nora rode her bike $\frac{2}{2}$ of a block. Brady rode his bike $\frac{4}{4}$ of the same block. Compare the distances each child rode. What do you notice? Explain your answer.

   My sketch:

Solving a Fraction Puzzle

Lesson 7-12

NAME DATE TIME

Use counters to solve the fraction puzzle. Draw and label a picture to show all of your work. Explain your thinking to a partner.

1. Emilio was playing a game with marbles.

   In the first round, he lost $\frac{1}{2}$ of his marbles.

   In the second round, he lost $\frac{1}{4}$ of his remaining marbles.

   In the third round, he lost $\frac{1}{3}$ of his remaining marbles.

   He gave one marble to his friend.

   He had one marble left.

   How many marbles did Emilio have when he started the game?

   ____

Try This

2. Create your own fraction puzzle, and write it on the back of this page.

Using Fractions to Name Parts of Sets

Lesson 7-12

NAME DATE TIME

Solve each problem. Use counters to help. Share solution strategies with others in your group.

1. Equally share a collection of 8 stickers between 2 people.

   How many stickers does each person get? ______________ (unit)

   What fraction of the stickers in the collection does each person get?

   ______________ of the stickers

2. Equally share a collection of 8 crackers among 4 people.

   How many crackers does each person get? ______________ (unit)

   What fraction of the crackers in the collection does each person get?

   ______________ of the crackers

3. Equally share a collection of 6 markers among 3 people.

   How many markers does each person get? ______________ (unit)

   What fraction of the markers in the collection does each person get?

   ______________ of the markers

4. Equally share a collection of 12 pennies among 2 people.

   How many pennies does each person get? ______________ (unit)

   What fraction of the pennies in the collection does each person get?

   ______________ of the pennies

Try This

5. Write an equivalent fraction for your answer to Problem 1.

   ______________ of the stickers

Fractions of Collections

Home Link 7-12

NAME DATE TIME

Family Note Today your child used fractions to name parts of collections of objects. As you help your child, encourage him or her to use sketches, pennies, or other tools to solve the number stories.

Please return this Home Link to school tomorrow.

Solve. Explain to someone at home how you figured out the numerator and the denominator for each fraction in Problems 1–3.

① 12 dogs are in the park. 2 of them are chasing a ball.

What fraction of the dogs are chasing a ball? _______

② 7 children are waiting for the school bus. 4 of them are girls.

What fraction of the children are girls? _______

③ There are 16 tulips in the garden. 4 of them are red.

What fraction of the tulips are not red? _______

④ Lisa and Carlie each have 6 cups. $\frac{2}{6}$ of Lisa's cups are yellow. $\frac{4}{6}$ of Carlie's cups are yellow. Who has more yellow cups?

Draw a picture to show your thinking.

_________ has more yellow cups.

Practice

Fill in the unit. Solve. Show your work on the back of this page.

Unit

⑤ $476 = 741 -$ _______

⑥ $558 =$ _______ $- 328$

NAME DATE TIME

Multiplication and Division

In this unit your child will deepen his or her understandings of measurement, multiplication and division, and geometric shapes.

In Unit 8, children will:

- Use a ruler to measure lengths to the nearest $\frac{1}{4}$ inch.

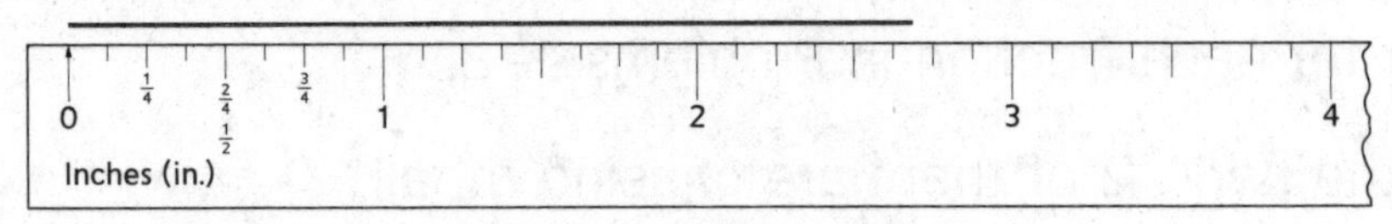

This line segment is about $2\frac{3}{4}$ inches long.

- Develop strategies for solving extended multiplication and division facts.
- Recognize and determine factor pairs of counting numbers within 100.
- Model equal-sharing situations involving money amounts.
- Apply understanding of factors while playing *Factor Bingo.*
- Extend work with fraction comparisons and equivalents.
- Examine features of rectangles with given area measurements.
- Explore the attributes of prisms.

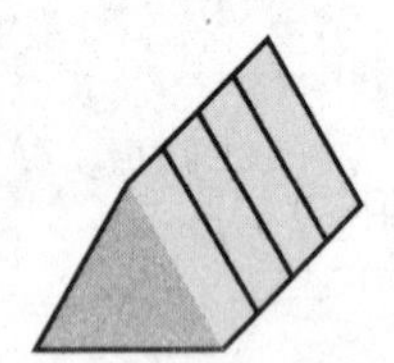

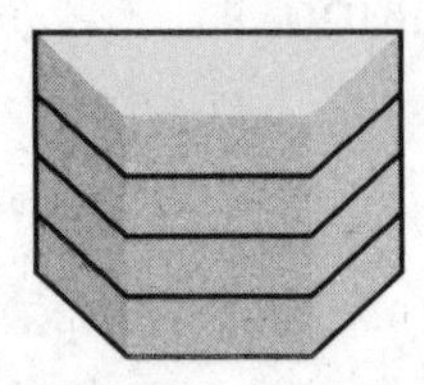
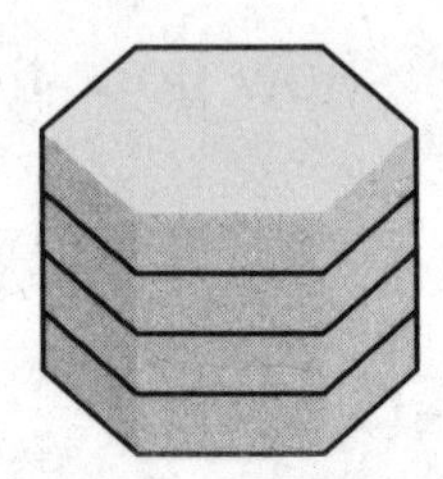

Children construct prisms with pattern blocks and explore common attributes.

Please keep this Family Letter for reference as your child works through Unit 8.

Vocabulary

Important terms in Unit 8:

base of a prism Either of the two parallel faces of a prism that are used to name it. (*See prism.*) *Example:* The base of a triangular prism is a triangle.

edge A line segment where two faces of a 3-dimensional shape meet.

extended fact A variation of a basic fact involving multiples of 10, 100, and so on. *Example:* The extended fact $3 \times 90 = 270$ is a variation of $3 \times 9 = 27$.

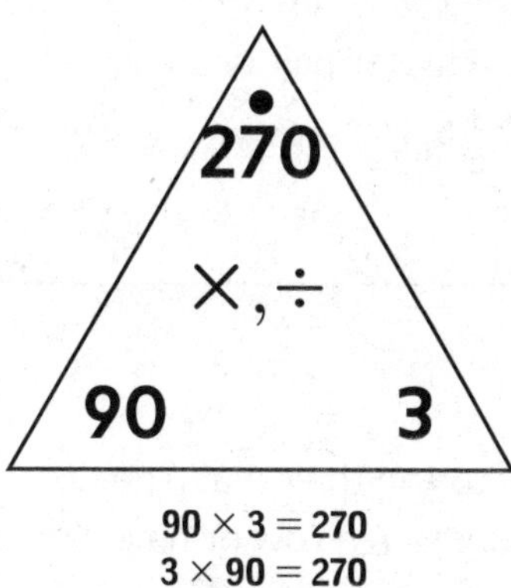

$90 \times 3 = 270$
$3 \times 90 = 270$
$270 \div 90 = 3$
$270 \div 3 = 90$

Extended Fact Triangle and extended fact family for 3, 90, and 270

face A flat surface that helps form the outside of a 3-dimensional shape.

factor pair Two counting numbers that multiply together to give a specified product. The specified product may have more than one factor pair. For example, the factor pairs for 18 are 1 and 18, 2 and 9, and 3 and 6.

multiple of 10 A product of 10 and a counting number. *Example:* 80 is a multiple of 10 because $10 \times 8 = 80$.

plot To mark a location on a number line, graph, map, or chart.

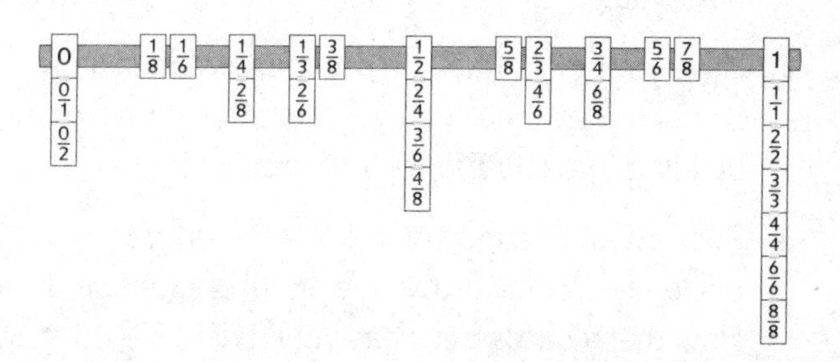

polyhedron A closed 3-dimensional figure whose surfaces are all flat and formed by polygons.

prism A polyhedron that has two parallel bases that are the same size and shape. The faces connecting the bases are all rectangles. Prisms take their names from the shape of their bases.

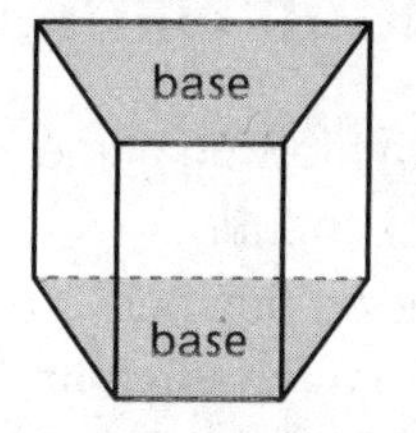

Trapezoidal prism

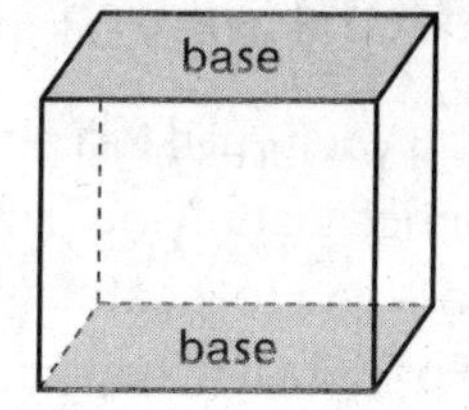

Rectangular prism

product The solution to a multiplication problem. *Example:* In $4 \times 7 = 28$, the product is 28.

3-dimensional (3-D) figure Solid shapes that have volume. Rectangular prisms and spheres are 3-dimensional figures.

2-dimensional (2-D) figure Flat shapes that have area but not volume. Rectangles and triangles are 2-dimensional figures.

vertex A point where edges of a polyhedron meet.

Do-Anytime Activities

The following activities provide practice for concepts taught in this and previous units.

1. Challenge your child to solve extended multiplication facts mentally by using a related basic fact. *Example:* $7 \times 50 = ?$ Use $7 \times 5 = 35$ and think 7×5 [10s] = 35 [10s] or 350.
2. Have your child find and use mathematical language (bases, faces, edges, vertices, and so on) to describe real-world examples of rectangular prisms. *Examples:* books, buildings, boxes, and other containers
3. Pose equal-sharing situations. Encourage your child to act out the situation using cards or pennies, or by drawing a picture. *Example:* 3 friends equally share 39 baseball cards. How many cards does each child get? $3 \times 10 = 30$. There are 9 left over and $3 \times 3 = 9$, so each child gets 3 more. $10 + 3 = 13$. Each child gets 13 baseball cards.
4. Ask your child to find factor pairs for a given number and say the resulting multiplication sentence. *Example:* 20. $4 \times 5 = 20$, so 4 and 5 are a factor pair of 20.
5. Use a ruler or tape measure to measure objects to the nearest $\frac{1}{4}$ inch.

Building Skills through Games

In Unit 8 your child will play the following games to practice identifying factors of counting numbers and to locate fractions on number lines. For detailed instructions, see the *Student Reference Book.*

Finding Factors Players use counters to mark factors on a strip. They multiply the factors together to find and circle products on the gameboard.

Factor Bingo Players choose products to write in their game mat. They turn over a number card to generate a factor and find products with that factor on their mat. The first player to cover 5 products in a row or have 12 products covered wins.

Fraction Number-Line Squeeze The leader thinks of a mystery fraction. Two players place brackets over each end of the number line. As players guess the mystery fraction, the leader states whether his or her fraction is greater or less than the guess and moves a bracket accordingly. Play continues until one player guesses the mystery fraction or the fraction is "squeezed" between the brackets.

As You Help Your Child with Homework

As your child brings home assignments, you may want to go over the instructions together, clarifying them as necessary. The answers listed below will guide you through this unit's Home Links.

Home Link 8-1

Answers vary.

Home Link 8-2

1. $8 \times 20 = 160$
   $20 \times 8 = 160$
   $160 \div 8 = 20$
   $160 \div 20 = 8$
2. $9 \times 30 = 270$
   $30 \times 9 = 270$
   $270 \div 9 = 30$
   $270 \div 30 = 9$
3. $6 \times 40 = 240$
   $40 \times 6 = 240$
   $240 \div 6 = 40$
   $240 \div 40 = 6$
4. Answers vary.

Home Link 8-3

1. Sample answers: 1 row with 18 chairs, 3 rows with 6 chairs, 6 rows with 3 chairs, 2 rows with 9 chairs; 1, 18; 2, 9; The number of rows and the number of chairs in each row are factors of 18.
2. Sample answers: 5, 8; 2, 20
3. Sample answers: 1, 72; 8, 9
4. Sample answers: 1, 150; 3, 50

Home Link 8-4

1. Sample answers:

X X X X X X
X X X X X X
X X X X X X
X X X X X X

X X X X X X X X
X X X X X X X X
X X X X X X X X

2. Answers vary.

Home Link 8-5

Sample answers: 10, 9, 60, 8, 21, 12, 16, 18, 15, 20, 24

Home Link 8-6

1. Sample answer: *D*; the number of dollars for each person; $\$76 \div 4 = D$; $4 \times D = \$76$; \$19
2. 16
3. 14
4. Sample answer: If I have the same amount of money shared with more people, each person would have to get less. So $\$90 \div 5$ is more than $\$90 \div 6$.

Home Link 8-7

1.

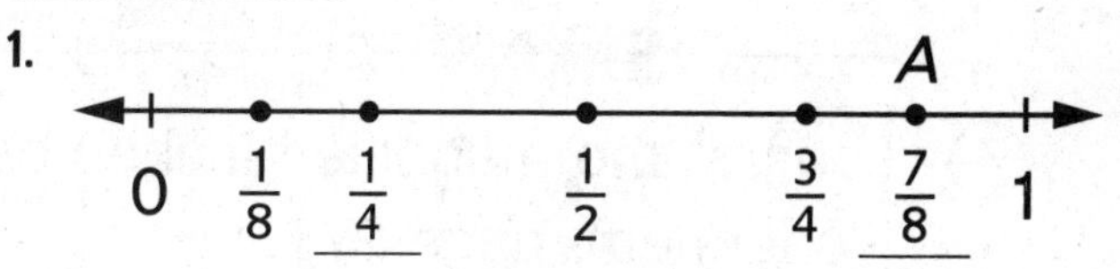

2. Sample answers: I saw the distance from fraction *A* to 1 was about the same distance as 0 to $\frac{1}{8}$. So, I knew fraction *A* was $\frac{7}{8}$. I know that $\frac{3}{4}$ is equivalent to $\frac{6}{8}$, so the next point is $\frac{7}{8}$.
3. 28
4. 48
5. 7
6. 8

Home Link 8-8

1. Triangles
2. Rectangles
3. 3
4. Triangular prism

Comparing Rulers and Number Lines

Lesson 8-1

NAME DATE TIME

① Look at your ruler and the Class Number Line.

How is a ruler like a number line?

____

____

② Look at the small lines between 0 and 1 on the inch ruler. What do these small lines mean?

____

____

③ Give examples of numbers that come between 0 and 1.

____

④ Look at the magnified inches on Comparing Rulers and Number Lines (continued), page 271.

Fill in the blanks below the ruler with the correct fractions.

How did you know which fractions to write?

____

____

____

____

Comparing Rulers and Number Lines

(continued)

Lesson 8-1

NAME DATE TIME

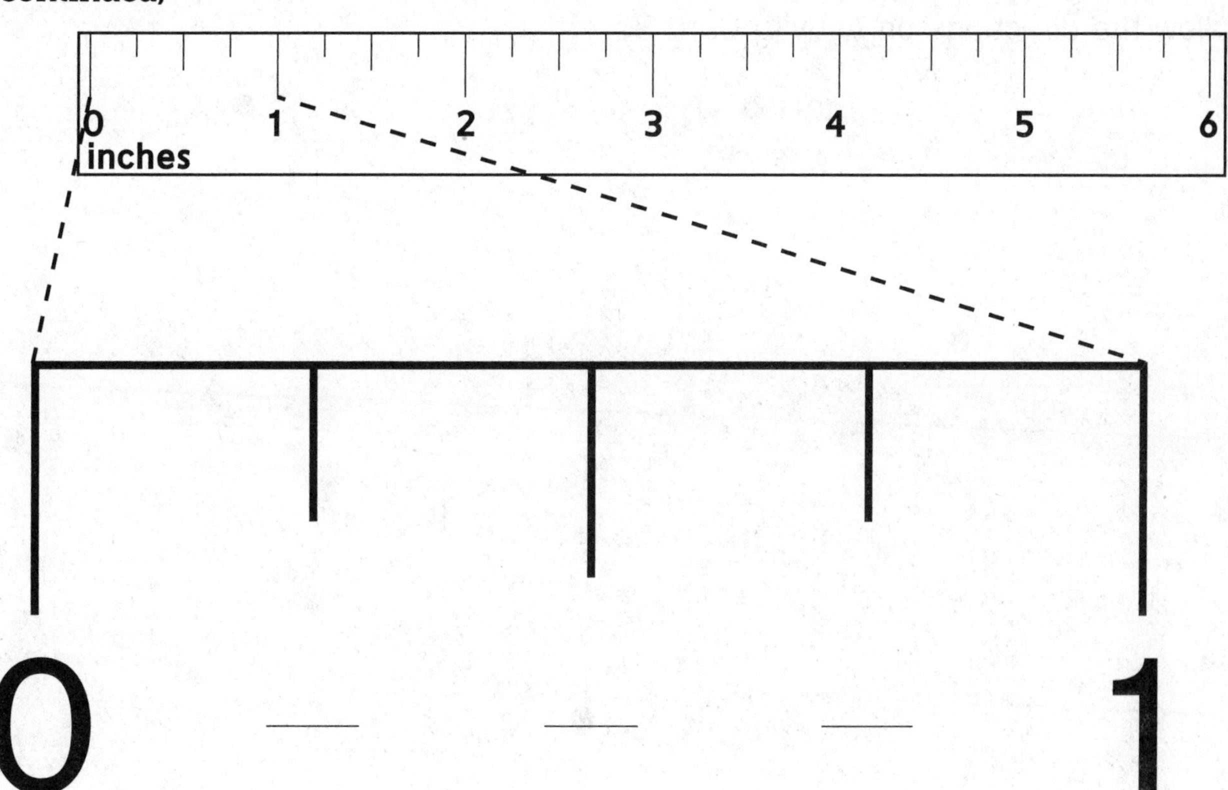

0 ___ ___ ___ 1

Drawing a Path to Treasure

Lesson 8-1

NAME DATE TIME

Follow the directions on Activity Card 85.

Start

0 1 mile

Scale:
1 inch represents 1 mile

Add your measures to find the length of the entire path.

My path is about ______ inches long.

My partner's path is about ______ inches long.

Whose path is longer? ____________

How much longer? About ______ inches

Completing a Story with Measures

Lesson 8-1

NAME ______ DATE ______ TIME ______

Make a list of 8 measurements of length. Use each of the following units at least once: inch, centimeter, foot, and meter. You may look at *Student Reference Book,* page 288 for units of length.

A. ____________ (number) ____________ (unit)

B. ____________ (number) ____________ (unit)

C. ____________ (number) ____________ (unit)

D. ____________ (number) ____________ (unit)

E. ____________ (number) ____________ (unit)

F. ____________ (number) ____________ (unit)

G. ____________ (number) ____________ (unit)

H. ____________ (number) ____________ (unit)

Match the letters in the story with the letters of your measurements. Fill in each blank in the story with the measurement that has the same letter.

This morning, after walking (A) ____________ to school, I saw my friend Emma looking up into a tree. She was staring at a spot about (B) ____________ above the ground. How in the world did her backpack get up there? I ran (C) ____________ to my friend Henry's house. He had a stepladder (D) ____________ tall. We raced back to school and set it up under the tree. I climbed up as far as I could go. If only I were (E) ____________ taller, I could have reached the backpack! Isaac came running with a stick that was (F) ____________ long. I held it up, caught the backpack, and jiggled it free. It crashed to the ground with a thud!

Luckily, my locker is (G) ____________ tall and (H) ____________ wide. The ladder just fit. At the end of the day, we returned the ladder to Henry's house. Emma and I decided she was getting too big and strong to be tossing backpacks just for fun.

- Read the story. Is it silly or sensible? How do you know?
- Discuss how you would change the measurements and distances to make the story more sensible.

Ants at a Picnic

Lesson 8-1

NAME DATE TIME

Two ants visit a picnic and notice a strawberry. They each take a different crooked path to get to the strawberry. Use Ruler D to measure each part of their paths to the nearest $\frac{1}{4}$ inch. Then label the parts with your measures.

Which ant took the shorter path? ___________

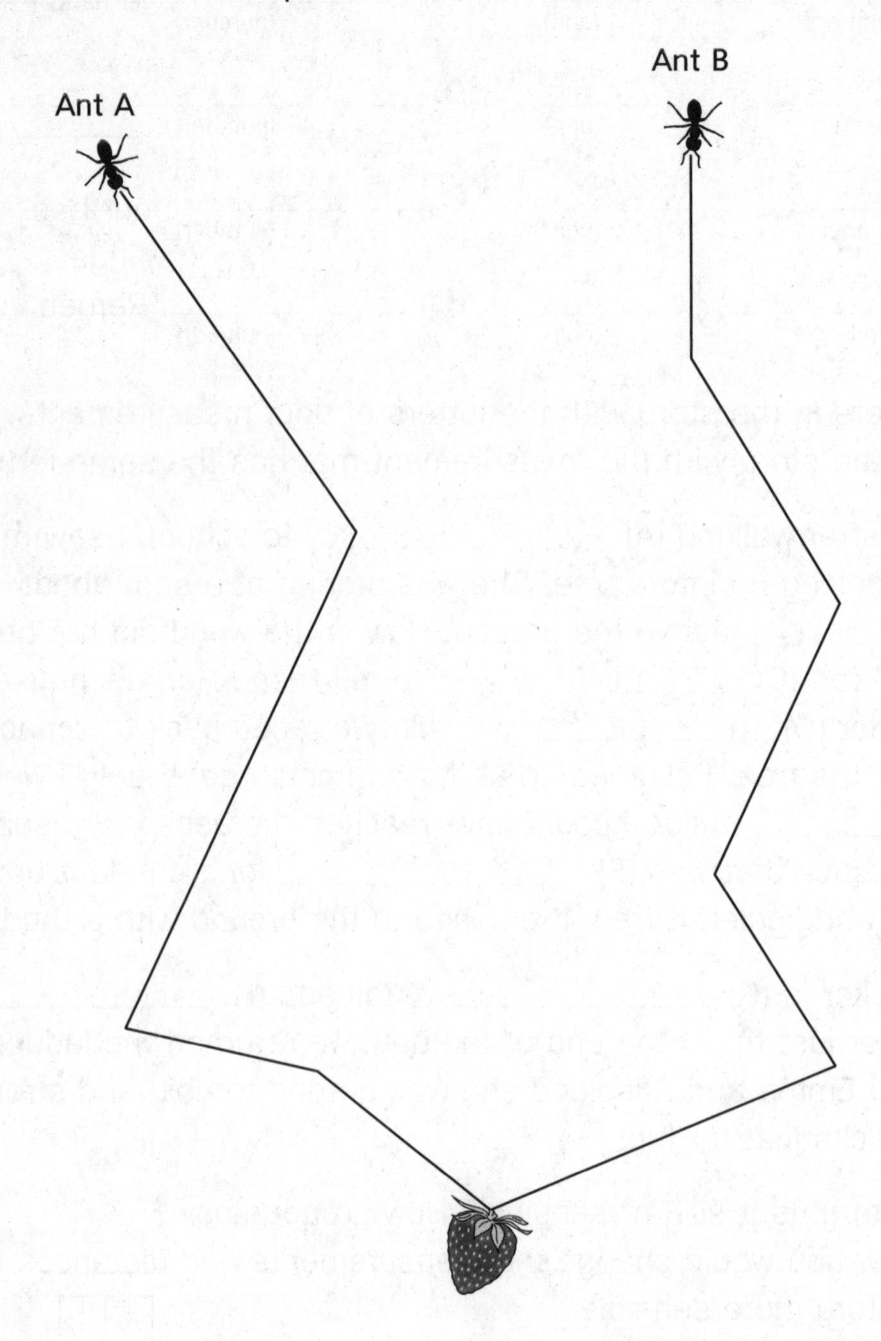

Measuring Fingers

Home Link 8-1

NAME DATE TIME

Family Note Today your child measured objects to the nearest $\frac{1}{4}$ inch by using $\frac{1}{8}$-inch markings on a ruler to determine which $\frac{1}{4}$-inch mark was closer to the end of the object. When measuring an item, if one end is lined up with the 0 mark and the other end is to the right of the $\frac{1}{8}$-inch mark, the measure is rounded to the next larger $\frac{1}{4}$ inch. If the end of the item is to the left of the $\frac{1}{8}$-inch mark, the measure is rounded to the next smaller $\frac{1}{4}$ inch. Help your child trace his or her hand and use the $\frac{1}{8}$-inch marks on the ruler to measure finger lengths to the nearest $\frac{1}{4}$ inch.

Please return this Home Link to school tomorrow.

SRB 171-172

① Cut out the ruler below. Carefully trace around one of your hands in the space below. Measure the length of each traced finger to the nearest $\frac{1}{4}$ inch. Write the measurement on each finger. Remember to record the unit.

② Have someone at home trace his or her hand on the back of this page. Measure the lengths of the traced fingers to the nearest $\frac{1}{4}$ inch. Write the measurements above each finger.

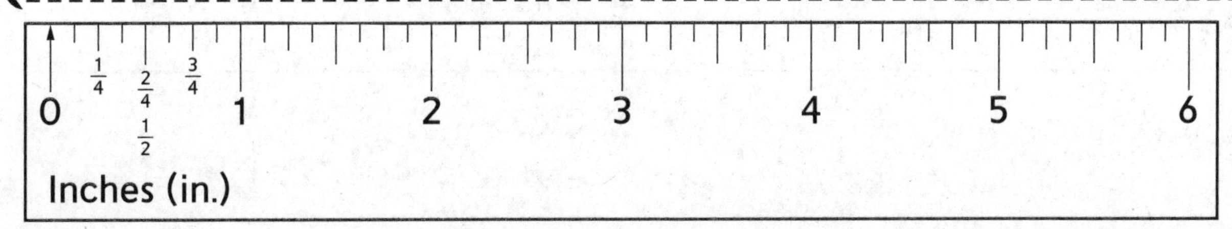

Using Multiples of 10

Lesson 8-2

NAME DATE TIME

Work with a partner.

One partner uses a calculator to skip count by 1s, 10s, and 100s. Record the numbers in the correct place in the chart as they are displayed in the calculator. The other partner models each count with base-10 blocks. Use base-10 shorthand cubes, longs, and flats (. | □) to record the counts in the chart.

The first two numbers are done for you.

Count by 1s.									
one 1		two 1s		three 1s		four 1s		five 1s	
1	.	2	..						

Count by 10s.									
one 10		two 10s		three 10s		four 10s		five 10s	
10	|								

Count by 100s.									
one 100		two 100s		three 100s		four 100s		five 100s	

Solving a Number Story with Extended Facts

Lesson 8-2

NAME DATE TIME

Solve the number story.

A 30-**minute** television show has:

- two 60-**second** commercials at the beginning.
- two 60-**second** commercials at the end.
- four 30-**second** commercials in the middle.

a. How many **minutes** of commercials are there? ______
(unit)

Record number models to show how you figured it out:

b. How many **minutes** long is the show, not including commercials?

______
(unit)

How did you figure it out?

______

______

Extended Facts: Multiplication and Division

Home Link 8-2

NAME DATE TIME

Family Note Today your child learned to use basic multiplication facts, such as $4 \times 6 = 24$, to solve extended multiplication facts, such as 4×60, by thinking of groups of ten. For example, 4×60 can be thought of as 4×6 [10s]. If you know that $4 \times 6 = 24$, then you also know that 4×6 [10s] = 24 [10s] or 240. The same approach works for extended division facts like $120 \div 3 = 40$. If you know that $12 \div 3 = 4$, then you also know that 12 [10s] $\div 3 = 4$ [10s] or 40. The extended Fact Triangles below work the same way as the basic Fact Triangles.

Please return this Home Link to school tomorrow.

Fill in the extended Fact Triangles. Write the extended fact families.

SRB 57

1.

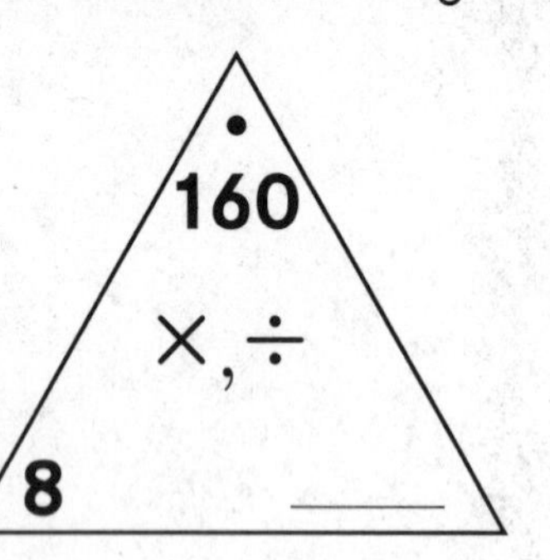

____ × ____ = ____

____ × ____ = ____

____ ÷ ____ = ____

____ ÷ ____ = ____

2.

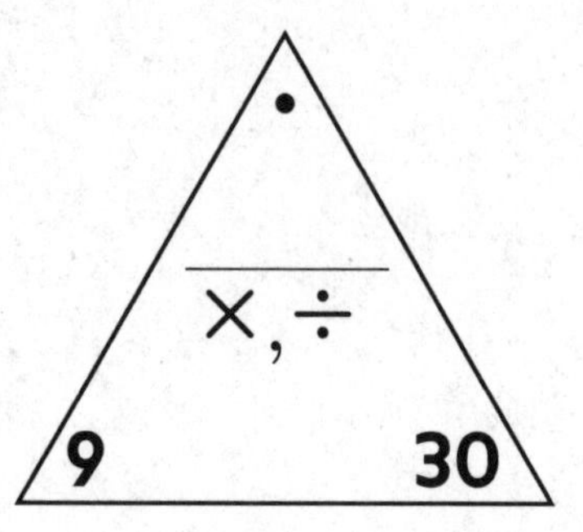

____ × ____ = ____

____ × ____ = ____

____ ÷ ____ = ____

____ ÷ ____ = ____

3.

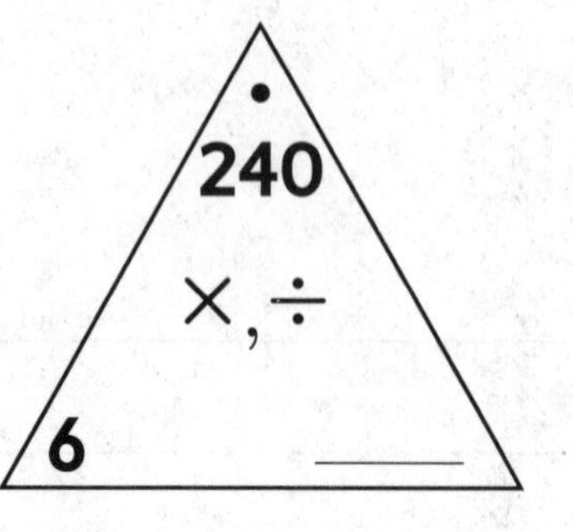

____ × ____ = ____

____ × ____ = ____

____ ÷ ____ = ____

____ ÷ ____ = ____

4. Write your own.

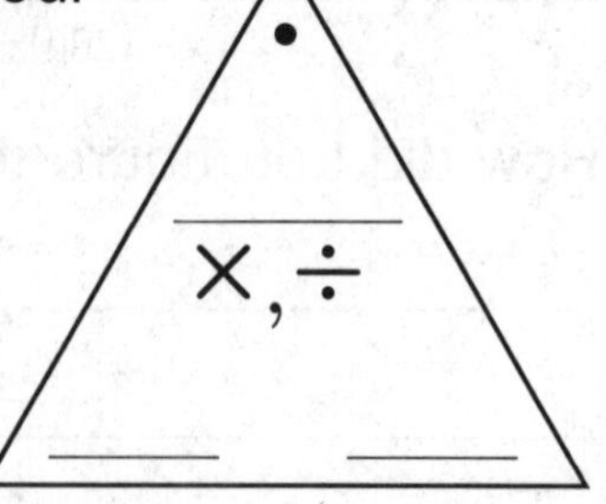

____ × ____ = ____

____ × ____ = ____

____ ÷ ____ = ____

____ ÷ ____ = ____

Finding Factor Pairs

Lesson 8-3

NAME DATE TIME

The covered number on the number grid is your product for each round.

Product: ______	Product: ______	Product: ______
______ × ______ ______ × ______ ______ × ______ ______ × ______ ______ × ______	______ × ______ ______ × ______ ______ × ______ ______ × ______ ______ × ______	______ × ______ ______ × ______ ______ × ______ ______ × ______ ______ × ______
Product: ______	Product: ______	Product: ______
______ × ______ ______ × ______ ______ × ______ ______ × ______ ______ × ______	______ × ______ ______ × ______ ______ × ______ ______ × ______ ______ × ______	______ × ______ ______ × ______ ______ × ______ ______ × ______ ______ × ______

Multiplication Practice

Lesson 8-3

NAME DATE TIME

Suppose you want to solve 6 × 7 using the break-apart method. Show four different ways you can do this by following the steps below.

1. Cut out one of the 6-by-7 rectangles below.
2. Make one cut to break apart the rectangle into two smaller rectangles.
3. Tape or glue the smaller rectangles on another paper. Record the product of each smaller rectangle and the sum of both.
4. Repeat Steps 1–3 for the other 3 rectangles.

Example: You can cut apart a 6-by-7 rectangle into a 6-by-6 rectangle and a 6-by-1 rectangle:

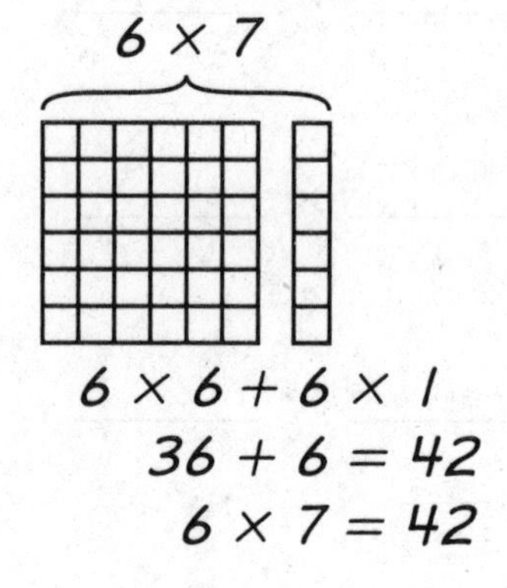

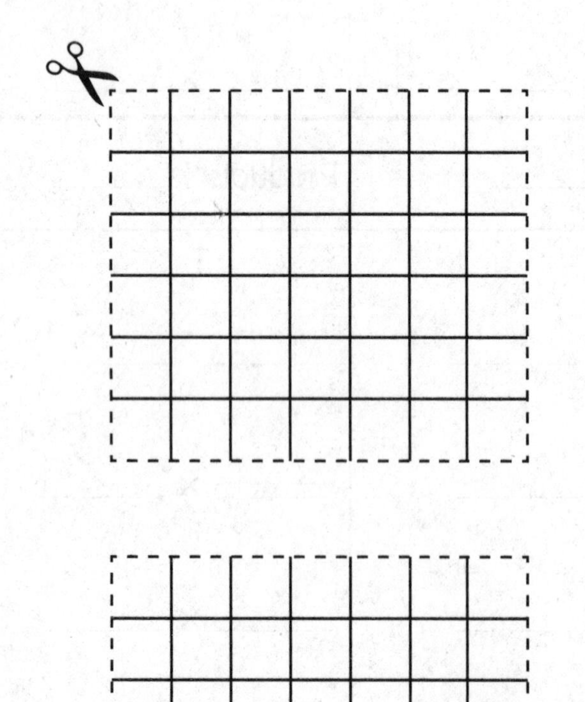

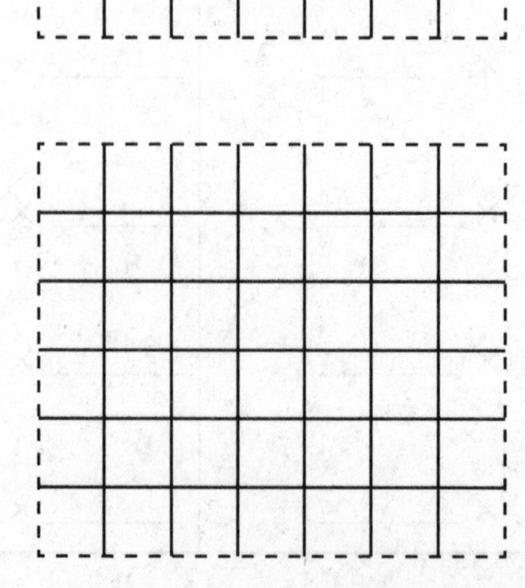

Factor Pairs

Home Link 8-3

NAME DATE TIME

Family Note Today your child found factor pairs for numbers by using basic facts, pictures, and arrays. For example, 2 and 8 are a factor pair for 16 because $2 \times 8 = 16$.

Please return this Home Link to school tomorrow.

Use facts, counters, or drawings to help you solve the problems.

① The third-grade class is putting on a play. They have 18 chairs for the audience. Jayla and Kevin are in charge of arranging the chairs in equal rows with no chairs left over.

Describe ways that Jayla and Kevin can arrange the chairs.

____

____

List two factor pairs for 18:

_____ × _____ = 18

_____ × _____ = 18

How does knowing ways to arrange 18 chairs in equal rows help you find factors of 18?

____

____

② 40 = _____ × _____ 40 = _____ × _____

③ 72 = _____ × _____ 72 = _____ × _____

Try This

④ 150 = _____ × _____ 150 = _____ × _____

Setting Up Chairs: My Arguments

Lesson 8-4

NAME DATE TIME

Here is the problem: **How many chairs did Ms. Soto set up?**

Here are the clues:

Clue A: When there are 2 chairs in each row, there is 1 leftover chair.

Clue B: When there are 3 chairs in each row, there is 1 leftover chair.

Clue C: When there are 4 chairs in each row, there is still 1 leftover chair.

Clue D: When there are 5 chairs in each row, there are no leftover chairs.

How many chairs do you think Ms. Soto set up? **Make two conjectures.**

Remember, the room holds no more than 35 chairs.

First conjecture: Ms. Soto set up ______ chairs.

Second conjecture: Ms. Soto set up ______ chairs.

1. Use the clues to make a mathematical argument for or against your **first conjecture** of ______ chairs. Show your reasoning.

Setting Up Chairs: My Arguments

(continued)

Lesson 8-4

NAME DATE TIME

② Use the clues to make a mathematical argument for or against your **second conjecture** of ______ chairs. Show your reasoning.

Making Conjectures and Arguments

Home Link 8-4

NAME DATE TIME

Family Note Today your child learned how conjectures and arguments are related. In mathematics, a **conjecture** is a statement that is thought to be true, and an **argument** is the mathematical reasoning used to show whether a conjecture is true or false. In the problem below, children are asked to find two different ways band members can be arranged for marching. Then they are asked to choose which arrangement they think is better. When children are asked to explain the reasoning for their choice, they are being asked to make an argument. Encourage your child to show the mathematical reasoning he or she used in the explanation for which arrangement is better.

Please return this Home Link to school tomorrow.

1. There are 24 members in the school band. The band director wants them to march in rows with the same number of band members in each row. Find two different ways that the band members can be arranged. Draw a sketch that shows each arrangement.

2. Which way do you think is better? Explain your reasoning.

Identifying Multiples

Lesson 8-5

NAME DATE TIME

① Elena turns over a 6 when playing *Factor Bingo.* She thinks she can cover 36, 16, 18, or 24 on her game mat. Do you agree? Explain your answer.

② Use the *Factor Bingo* game mat to the right. List at least three products that Elena could cover if she draws a 9.

38	5	19	24	23
31	40	16	53	20
21	27	15	89	14
66	41	70	36	25
9	32	18	29	50

Explain how you know that 9 is part of a factor pair for one of the products above.

③ List the products on Elena's game mat that have 5 as a factor.

Try This

④ Which of the products that you wrote for Problems 2 and 3 would you include on a new *Factor Bingo* game mat? Why?

Factor Bingo

Home Link 8-5

NAME DATE TIME

Family Note Today your child learned to play *Factor Bingo* to practice identifying factors of products. When the circled products on the game mat form a row, column, or diagonal, your child can call *Bingo!*

Please return this Home Link to school tomorrow.

SRB 240-241

Look for a product for each factor in the table below on the *Factor Bingo* game mat. Circle the product on the game mat and record it next to the factor in the table. You can only use each product on the game mat one time. Explain to someone at home how you chose that product. For example, 2 is a factor of 6 because 2 × 3 equals 6. Call out *Bingo!* if you get five products in a row, column, or diagonal.

Factor	Product
2	6
5	_____
3	_____
10	_____
4	_____
7	_____
3	_____
2	_____
9	_____
5	_____
4	_____
8	_____

***Factor Bingo* Game Mat**

10	8	11	24	23
38	40	(6)	35	27
21	20	15	90	75
28	17	31	36	45
16	12	18	9	60

Sharing Money with a Partner

Lesson 8-6

NAME DATE TIME

Record each round.

Factor	Factor	Product	Amount of money each partner gets	Leftover money	Try This Amount of money each partner gets with leftovers

Sharing Money with a Partner

Lesson 8-6

NAME DATE TIME

Record each round.

Factor	Factor	Product	Amount of money each partner gets	Leftover money	Try This Amount of money each partner gets with leftovers

Copyright © McGraw-Hill Education. Permission is granted to reproduce for classroom use.

Buying Tickets

Lesson 8-6

NAME DATE TIME

1

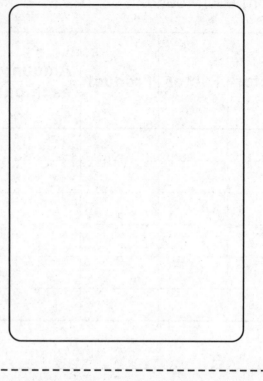

Buying Tickets

Lesson 8-6

NAME DATE TIME

1

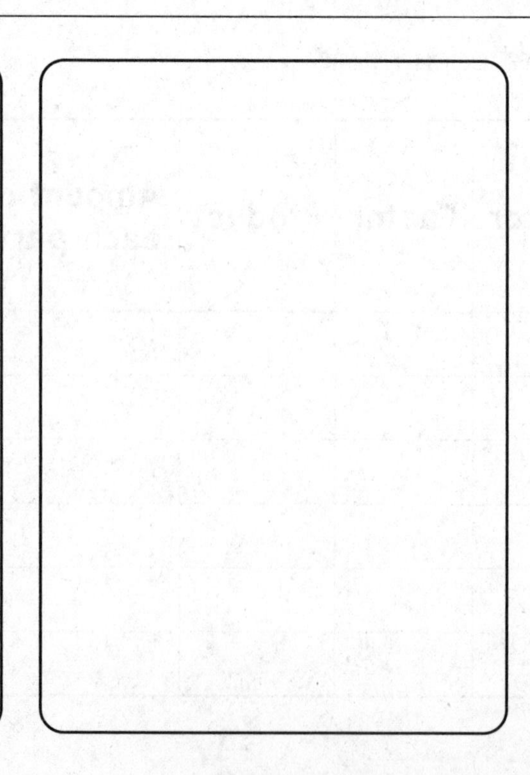

Sharing Money with Friends

Home Link 8-6

NAME DATE TIME

Family Note Today your child modeled equal sharing by distributing money amounts into equal groups. Equal sharing is one way to think about division. Work with more formal division algorithms will begin in *Fourth Grade Everyday Mathematics.* In the meantime, encourage your child to solve the following problems in his or her own way and to explain the strategy to you. Have your child model these problems with play money or with slips of paper labeled $10 and $1.

Please return this Home Link to school tomorrow.

1. Four friends share $76. They have seven $10 bills and six $1 bills. They can go to the bank to get smaller bills.

   The letter _______ represents ______________________________.

   ____
   (number model with letter)

   Use numbers or pictures to show how you solved the problem:

   Answer: Each friend gets a total of $_______.

Model each sharing problem below. Record your answer.

2. $48 ÷ 3 = $_______

3. $56 ÷ 4 = $_______

Try This

4. Without calculating, explain how you know that $90 ÷ 5 would be larger than $90 ÷ 6.

____

____

____

Completing the Whole

Lesson 8-7

NAME DATE TIME

The red circle is the whole.

Use your fraction circle pieces to complete the *missing* part of each whole with at least two different-size pieces. Write a number sentence to show how your pieces are equal to the missing part.

①

$\frac{1}{3}$ OR

What fraction of the whole is missing? ______

Number sentence:

____

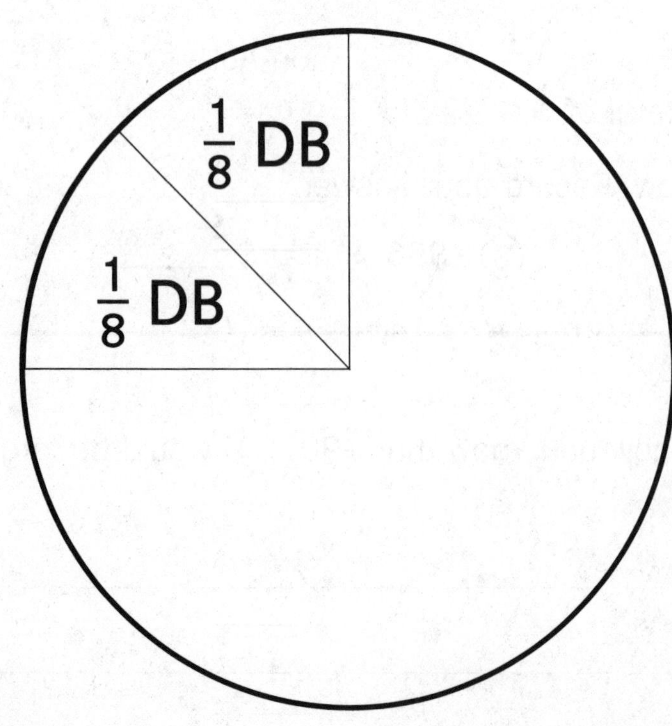

What fraction of the whole is missing? ______

Number sentence:

____

More Geoboard Areas

Lesson 8-7

NAME DATE TIME

Follow the "next time you work on this card" directions on Activity Card 91 to build possible rectangles with the given areas. For each rectangle, record the lengths of two sides that touch.

Area	Side 1	Side 2
8 square units	______ units	______ units
8 square units	______ units	______ units
20 square units	______ units	______ units
20 square units	______ units	______ units
24 square units	____________	____________
24 square units	____________	____________

More Geoboard Areas

Lesson 8-7

NAME DATE TIME

Follow the "next time you work on this card" directions on Activity Card 91 to build possible rectangles with the given areas. For each rectangle, record the lengths of two sides that touch.

Area	Side 1	Side 2
8 square units	______ units	______ units
8 square units	______ units	______ units
20 square units	______ units	______ units
20 square units	______ units	______ units
24 square units	____________	____________
24 square units	____________	____________

Copyright © McGraw-Hill Education. Permission is granted to reproduce for classroom use.

Exploring Equivalent Fractions

Lesson 8-7

NAME DATE TIME

Follow the "next time you work on this card" directions on Activity Card 92.

The red circle is the whole.

① $\frac{1}{2}$ PK

What fraction of the whole is missing? ______

$\frac{\square}{\square} = \frac{\square}{\square}$

② $\frac{1}{4}$ Y $\frac{1}{4}$ Y $\frac{1}{4}$ Y

What fraction of the whole is missing? ______

$\frac{\square}{\square} = \frac{\square}{\square}$

Exploring Equivalent Fractions (continued)

Lesson 8-7

NAME DATE TIME

③

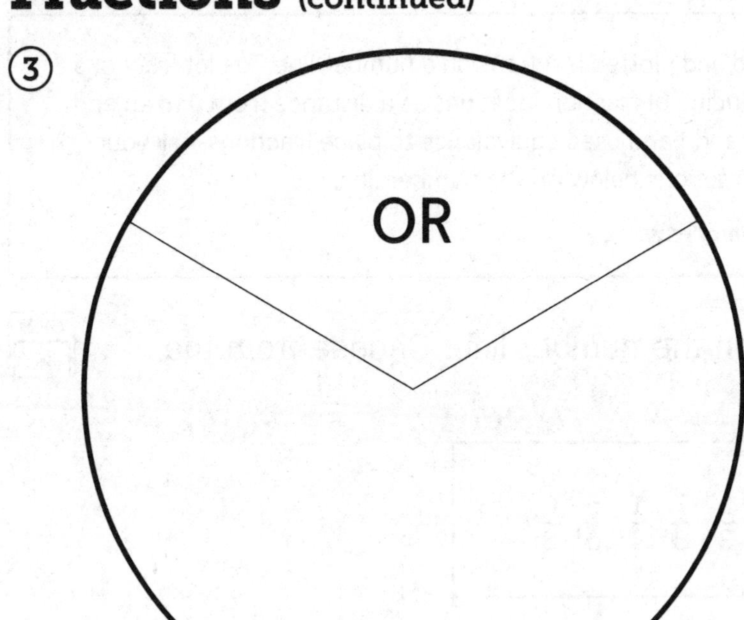

What fraction of the whole is missing? ______

$\frac{\square}{\square} = \frac{\square}{\square}$

④

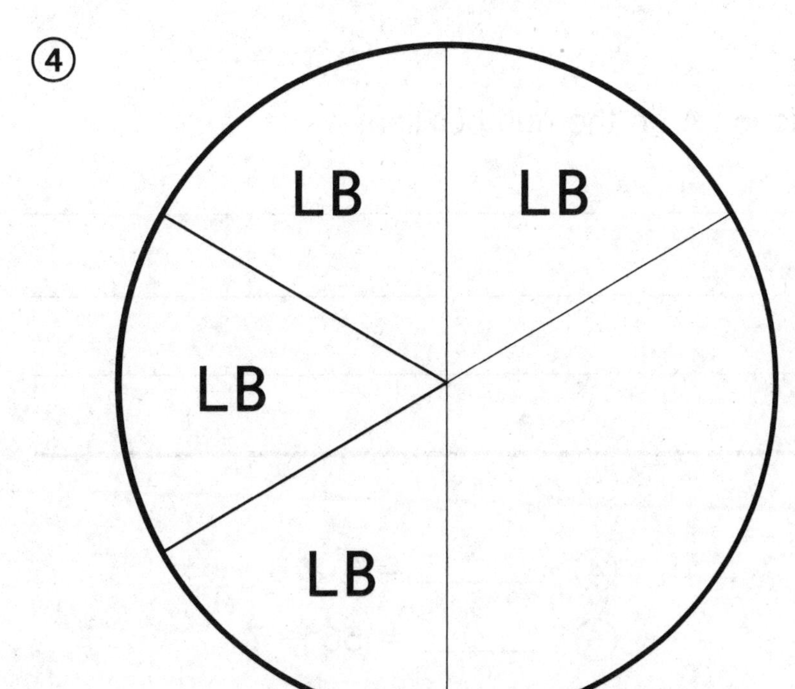

What fraction of the whole is missing? ______

$\frac{\square}{\square} = \frac{\square}{\square}$

⑤ How do you know when fractions are equivalent?

____

____

Locating Fractions

Home Link 8-7

NAME DATE TIME

Family Note Today your child located and plotted fractions on a number line. To plot fractions accurately, children applied their understanding of fraction locations as a distance from 0 to an end point. They also made comparisons to 0, $\frac{1}{2}$, and 1 and used equivalence to place fractions. Ask your child to explain how he or she placed the fractions below on the number line.

Please return this Home Link to school tomorrow.

(1) Fill in the missing fractions on the number line. Choose from the fractions in the box below.

$\frac{1}{3}, \frac{7}{8}, \frac{1}{4}, \frac{5}{6}, \frac{6}{8}$

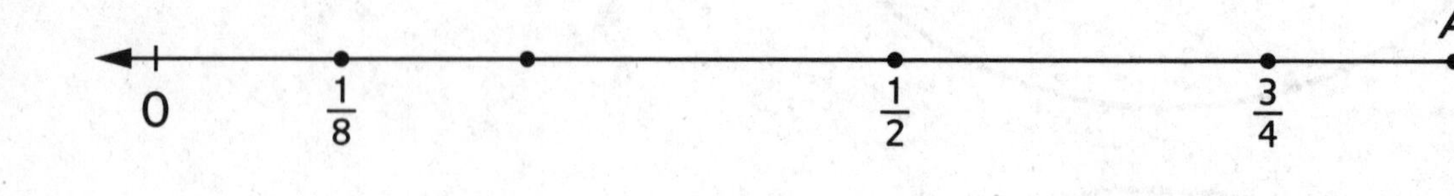

______ ______

(2) Explain how you placed fraction *A* on the number line.

____

____

____

Practice

(3) $4 \times 7 =$ ______

(4) ______ $= 6 \times 8$

(5) $49 \div 7 =$ ______

(6) ______ $= 56 \div 7$

Constructing a Pentagonal Prism

Lesson 8-8

NAME DATE TIME

Cut on the dashed lines. Fold on the dotted lines. Tape or paste each TAB inside or outside the shape.

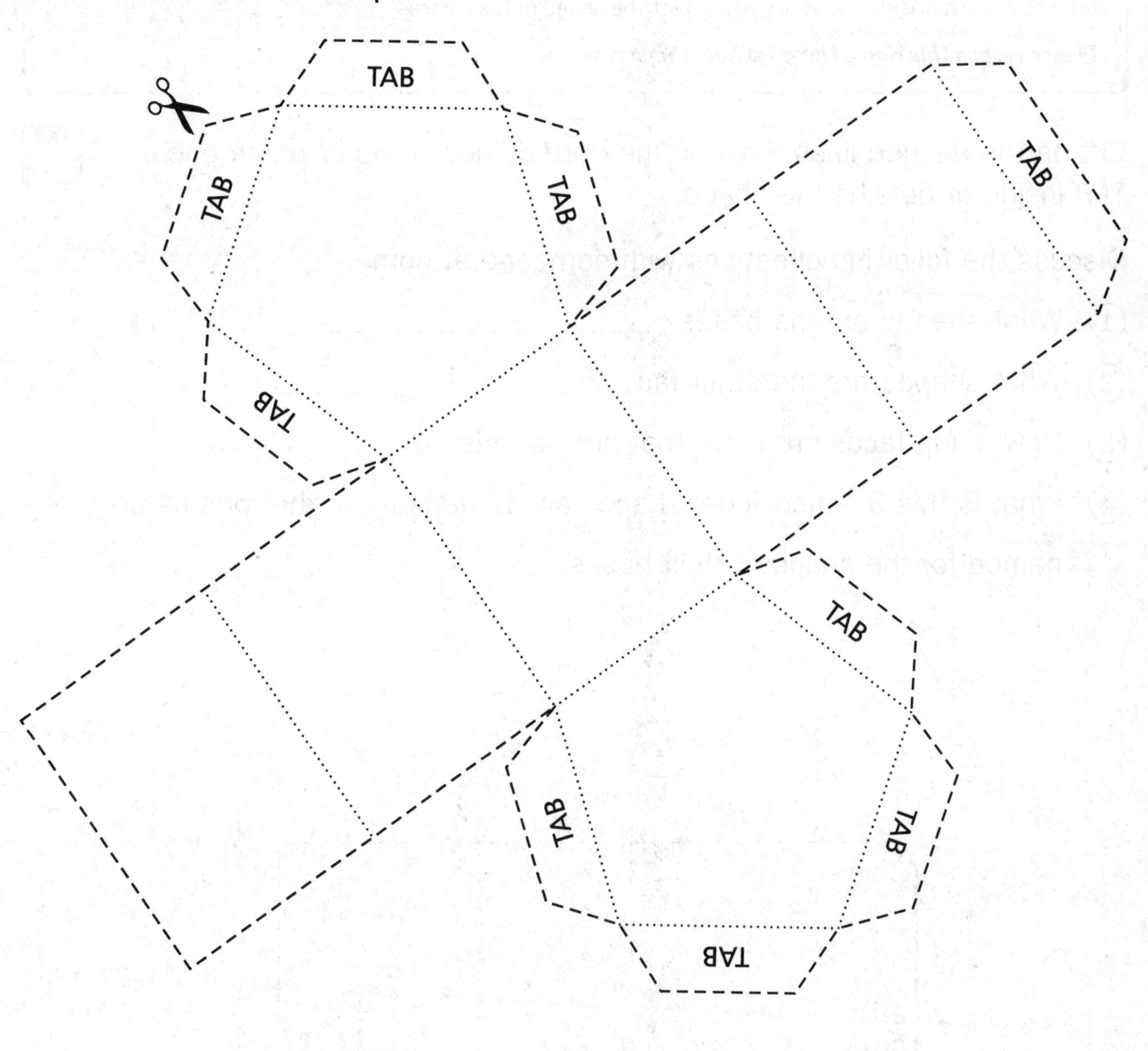

Making a Prism

Home Link 8-8

NAME DATE TIME

Family Note Today your child explored attributes of prisms. The pattern on this page can be used to make a prism. Prisms are named for the shape of their bases.

Please return this Home Link to school tomorrow.

Cut on the dashed lines. Fold on the dotted lines. Tape or paste each TAB inside or outside the shape.

Discuss the following questions with someone at home.

① What shapes are the bases? ____________________

② What shapes are the other faces? ____________________

③ How many faces are there that are not bases? ______

④ What is this 3-dimensional shape called? Remember that prisms are named for the shape of their bases. ____________________

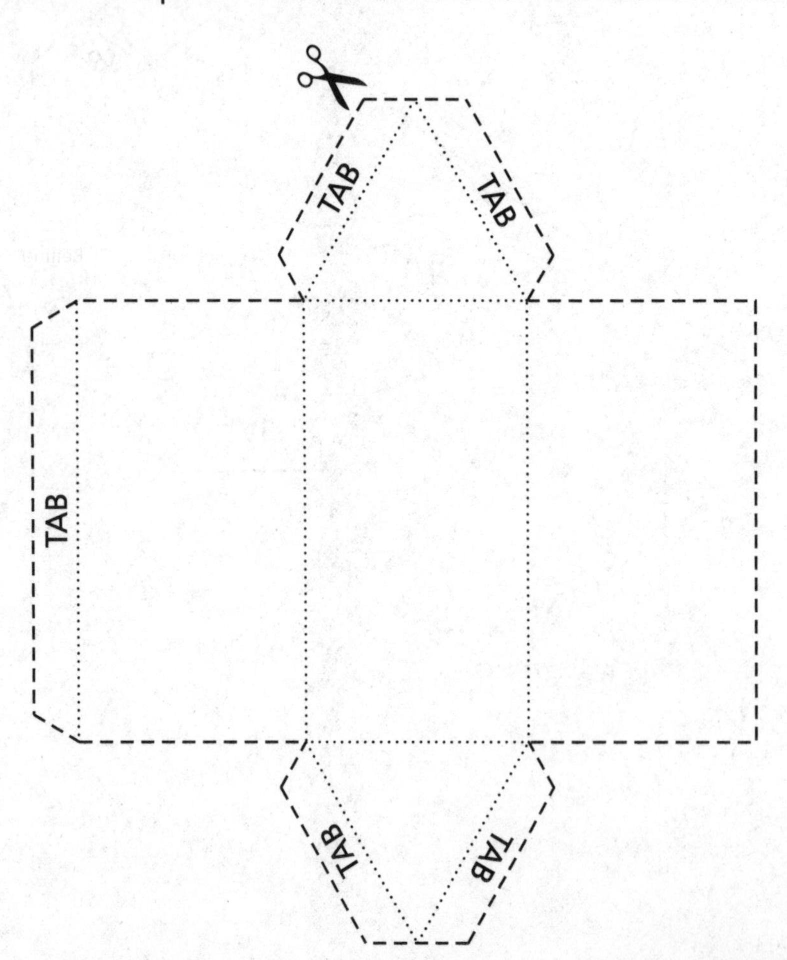

Unit 9: Family Letter

Home Link 8-9

NAME DATE TIME

Multidigit Operations

In Unit 9, children make sense of and solve problems involving multiplication and division, units of mass, geometry, and elapsed time. They play a new multiplication game that encourages automaticity with all multiplication facts, which is an end-of-year goal. Using mental math and area models, children apply basic fact strategies to solve multiplication and division number stories with larger factors. They revisit the Length-of-Day Project and calculate elapsed time.

In Unit 9, your child will:

- Use multiplication-fact knowledge while playing *Product Pile-Up.*
- Make sense of number stories and solve them by multiplying and dividing with multiples of 10.
- Use mental arithmetic to multiply problems involving larger factors.
- Solve multidigit multiplication problems using area models.
- Analyze bar graphs that show the class length-of-day data.
- Calculate the length of day for locations around the world.

Sunrise and Sunset Data for June 21, 2016

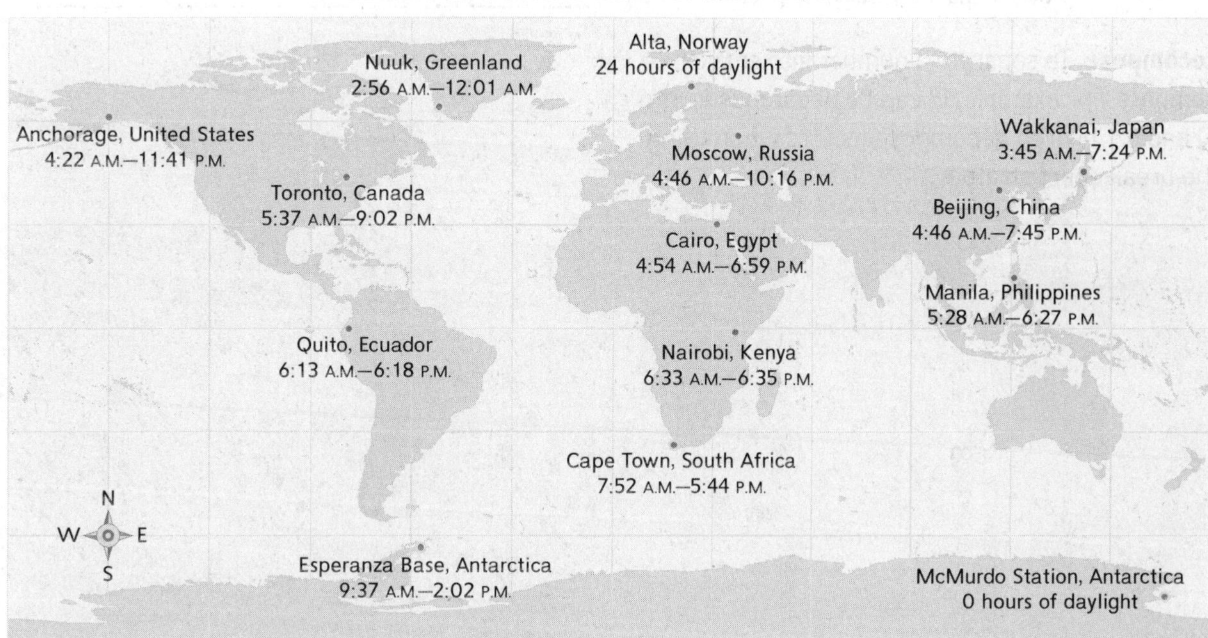

Vocabulary

Important terms in Unit 9:

basic multiplication and division facts Multiplication facts with whole-number factors between 0 and 10 and the corresponding division facts, except there can be no division by 0. For example, $4 \times 6 = 24$ and $24 \div 6 = 4$ are basic facts.

break-apart strategy A multiplication strategy in which one or both factors are broken into two or more smaller parts, resulting in easier-to-solve multiplication problems. Partitioning a rectangular area model is a way to represent this strategy.

	10	6
9	$9 \times 10 = 90$	$9 \times 6 = 54$

$90 + 54 = 144$

$9 \times 16 = 144$
A rectangular area model can be used to model the break-apart strategy.

decompose To separate a number into smaller numbers. For example, 16 can be decomposed into 10 and 6. Children decompose factors when using the break-apart strategy.

elapsed time The amount of time that has passed from one point to the next. For example, the elapsed time between 12:45 P.M. and 1:30 P.M. is 45 minutes.

extended fact Variations of basic arithmetic facts involving multiples of 10, 100, and so on. For example, the extended fact $40 \times 5 = 200$ is related to the basic fact $4 \times 5 = 20$.

length of day Total number of hours and minutes between sunrise and sunset.

multiplication/division diagram A diagram for modeling situations with equal-size groups. The diagram has a number of groups, a number in each group, and a total number.

number of birds	grams per bird	grams in all
6	20	?

Do-Anytime Activities

The following activities provide practice for concepts taught in this unit and previous units.

1. Continue to work toward automaticity with all multiplication facts using Fact Triangles or by playing games such as *Product Pile-Up, Multiplication Top-It,* and *Salute!*.
2. Practice using basic facts to solve extended multiplication and division facts, such as using $3 \times 7 = 21$ to solve $3 \times 70 = 210$, or $18 \div 6 = 3$ to solve $180 \div 6 = 30$.
3. Calculate how long daily activities take. For example: *Shawna arrives at school at 8:45 A.M. and leaves at 3:00 P.M. How many hours and minutes is she at school? Your dentist appointment is at 3:15 P.M. It takes 20 minutes to drive to the dentist. If we leave at 2:45 P.M., will we arrive on time?*

Building Skills through Games

Product Pile-Up Players are dealt eight number cards. They take turns selecting two cards and multiplying the numbers to generate a product that is greater than the product of the last two cards played. The winner is the first player to run out of cards or the player with the fewest cards when there are no more cards to draw. For detailed instructions, see the *Student Reference Book*.

As You Help Your Child with Homework

As your child brings home assignments, you may want to go over the instructions together, clarifying them as necessary. The answers listed below will guide you through this unit's Home Links.

Home Link 9-1

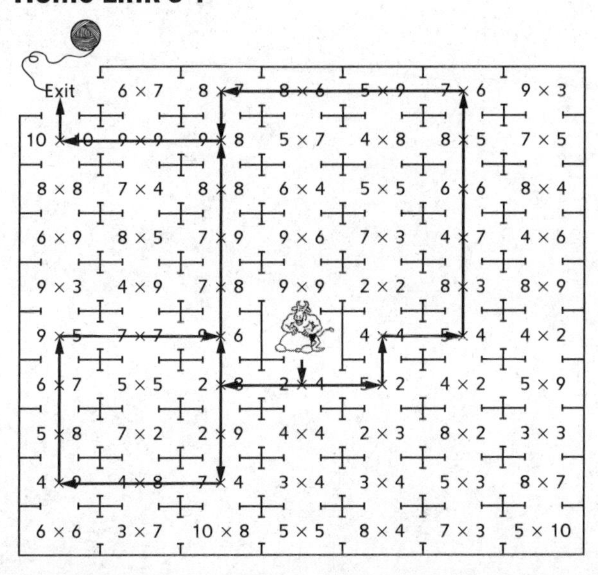

Home Link 9-2

1.

number of flamingos	mass of 1 flamingo in kg	total mass in kg
40	2	?

Sample answer: $2 \times 40 = ?$; 80 kg

2.

number of bluebirds	mass of 1 bluebird in g	total mass in g
9	?	270

Sample answer: $9 \times ? = 270$; 30 g

3. Sample answer: I thought of 9 × what number is 270. I know 9 × 3 = 27, so 9 × 30 = 270. One bluebird has a mass of about 30 g.

Home Link 9-3

1. Sample answer: I broke apart 12 into 6 and 6. I know 6 × 9 = 54, so I can double that to get 12 × 9. So 12 × 9 = 54 + 54 = 108; 108 kilograms
2. Sample answers: I broke apart 7 into 4 and 3. I know 25 × 4 = 100 and 25 × 3 = 75. So 7 × 25 = 100 + 75 = 175; I used the break-apart strategy and thought 25 × 7 = 20 × 7 + 5 × 7 = 140 + 35 = 175. So 25 × 7 = 175; 175 grams
3. Answers vary.

Home Link 9-4

1. 50, 60, 20, 75 **2.** 240 **3.** 210

4. 480 **5.** 720

Home Link 9-5

1. 210

	40	2	
5	5 × 40 = 200	5 × 2 = 10	200 + 10 = 210

(42 = 40 + 2)

2. 324

	50	4	
6	6 × 50 = 300	6 × 4 = 24	300 + 24 = 324

(54 = 50 + 4)

3. Answers vary.

Home Link 9-6

1. 15; Sample answer: I knew that the number of cartons had to be more than 10 because 10 × 12 = 120 and the class needed 180 eggs. So I tried 12 × 12 on the calculator, but that was only 144. So I tried 12 × 13, 12 × 14, and 12 × 15. 12 × 15 = 180, so 15 is the number of cartons they need.
2. Sample answer: 1, 8, 0, –, 12, =, =, =, =, =, =, =, =, =, =, =, =, =, =, =; I had to push the = key 15 times to reach 0, so the number of cartons is 15.

Home Link 9-7

1. San Francisco: 9 hours 33 minutes; Minneapolis: 8 hours 46 minutes; Miami: 10 hours 32 minutes
2. Miami
3. Minneapolis

Writing A Guide to Playing Math Games

Lesson 9-1

NAME DATE TIME

Use these game titles and topics to help you create *A Guide to Playing Math Games.*

Name That Number	***Factor Bingo***
How 0s can help you pick up more cards How 1s can help you pick up more cards	Which products are the best to use on a game mat and why How to set up a winning game mat
Finding Factors	***Shuffle to 100***
How to decide which factor to cover to win a square How to decide which factor to cover to block your opponent	How to use combinations of 10 to decide which digits to use in the 10s place
The Area and Perimeter Game	***Product Pile-Up***
How to choose whether to find the area or the perimeter How to choose whether to have your partner find the area or the perimeter	How to block players from using their last two cards Which cards to hold in your hand until later in the game Which cards to play early in the game

Comparing Products

Home Link 9-1

NAME DATE TIME

Family Note Today your child learned a game that involves finding a multiplication product greater than the one just played. The activity below provides practice with this skill. Have your child start at the picture of the Minotaur and use a pencil so that he or she can erase wrong turns.

Please return this Home Link to school tomorrow.

SRB 65-66

According to Greek mythology, there was a monster called the Minotaur that was half bull and half human. The king had a special mazelike dwelling built, from which the Minotaur could not escape. The dwelling, called a **labyrinth** (la buh rinth), had many rooms and passageways that formed a puzzle. No one who went in could find their way out without help. One day, a Greek hero named Theseus decided to slay the monster. To find his way out of the labyrinth, Theseus's friend Ariadne gave him a very, very long ball of string to unwind as he walked through the passageways. After Theseus slew the Minotaur, he followed the string to escape.

Pretend you are Theseus. To find your way out of the maze, each room you enter must have a product greater than the product in the room you are leaving. Start at the Minotaur's chambers in the middle and draw a path to the exit.

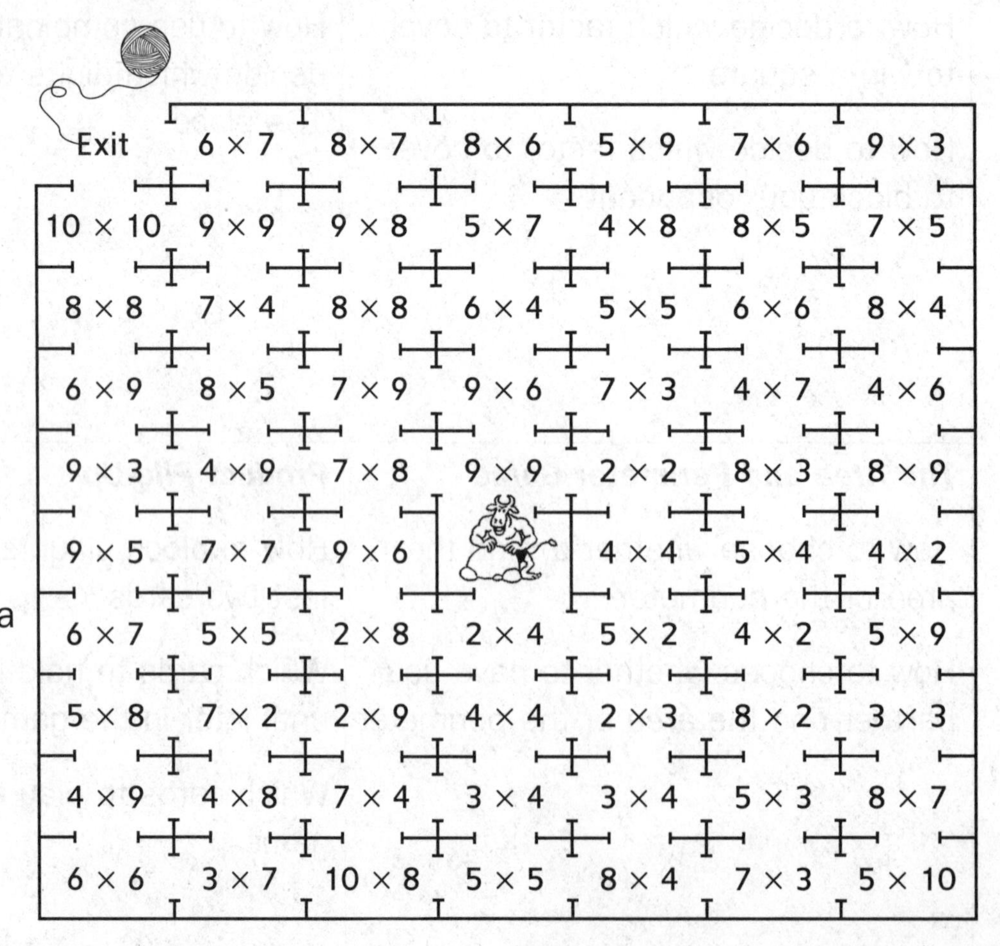

Modeling Extended Multiplication Facts

Lesson 9-2

NAME DATE TIME

Represent each pair of problems using base-10 blocks.
Then write the product.

① $2 \times 4 =$ ______

$2 \times 40 =$ ______

② $3 \times 3 =$ ______

$3 \times 30 =$ ______

③ $7 \times 2 =$ ______

$7 \times 20 =$ ______

④ $6 \times 4 =$ ______

$6 \times 40 =$ ______

Modeling Extended Multiplication Facts

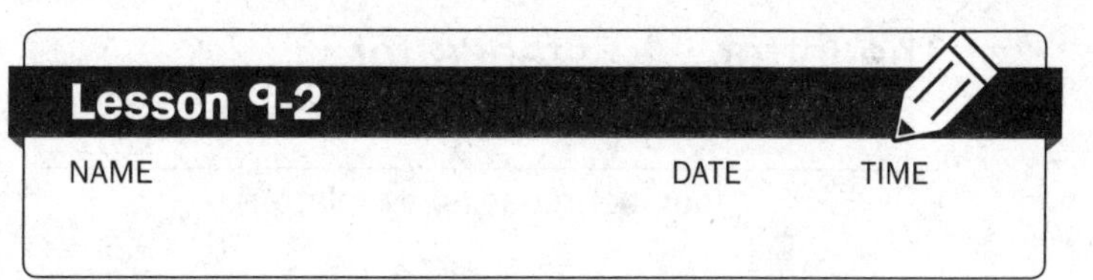

Represent each pair of problems using base-10 blocks.
Then write the product.

① $2 \times 4 =$ ______

$2 \times 40 =$ ______

② $3 \times 3 =$ ______

$3 \times 30 =$ ______

③ $7 \times 2 =$ ______

$7 \times 20 =$ ______

④ $6 \times 4 =$ ______

$6 \times 40 =$ ______

Copyright © McGraw-Hill Education. Permission is granted to reproduce for classroom use.

Feeding Hummingbirds

Lesson 9-2

NAME DATE TIME

Solve. Write number models with a letter for the unknown. You may draw pictures or diagrams.

1. The aviary has 80 hummingbirds with a mass of about 5 grams each. Each bird eats about half of its mass in sugar every day. How many grams of sugar do all the hummingbirds need each day?

   The letter ___ stands for ____.

   ____
   (number model(s) with letter)

   Answer: about ______ grams of sugar

2. A 1-liter bottle of hummingbird nectar has 200 grams of sugar. How many liters of nectar do the hummingbirds need for one week?

   The letter ___ stands for ____.

   ____
   (number model(s) with letter)

   Answer: about ______ liters of nectar

3. The aviary moves 40 of the hummingbirds to a different area for four days. How many grams of sugar will those hummingbirds need for four days?

   The letter ___ stands for ____.

   ____
   (number model(s) with letter)

   Answer: about ______ grams of sugar

Comparing Masses of Birds

Lesson 9-2

NAME DATE TIME

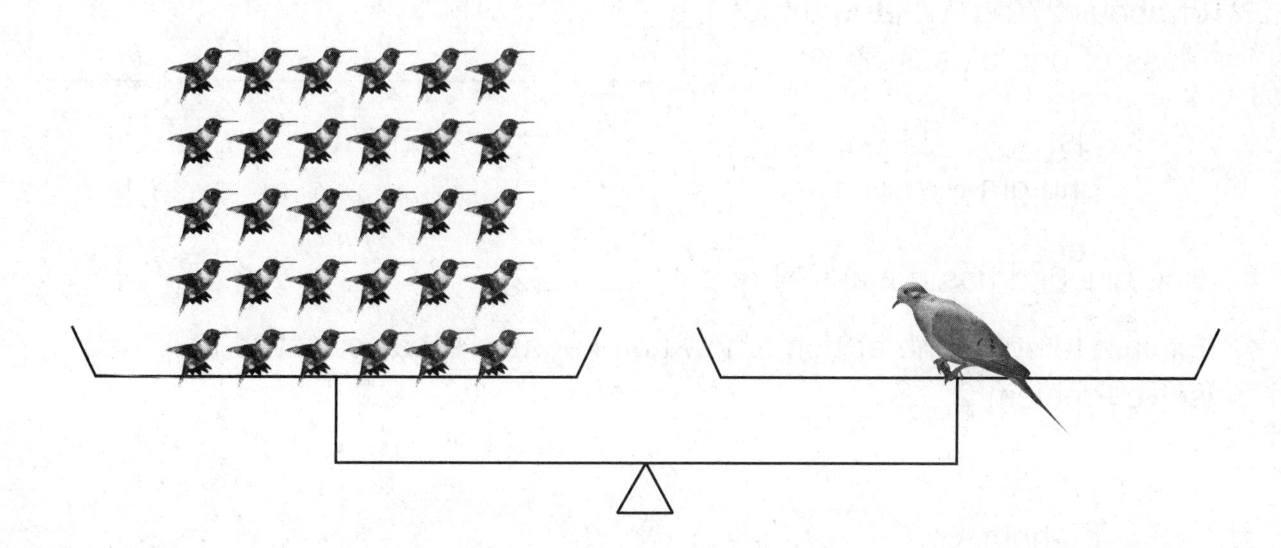

Multiplication and Division Number Stories

Home Link 9-2

NAME DATE TIME

Family Note Today your child solved number stories involving multiples of 10. The class examined a map displaying the masses of adult North American birds to make sense of the stories and used multiplication/division diagrams to organize information. For the problems below, encourage your child to use a known basic fact to help solve the number models with extended facts involving multiples of 10.

SRB 61-64

Please return this Home Link to school tomorrow.

Write a number model. Then solve each number story. You may draw a picture or use the multiplication/division diagram.

1. One American flamingo has a mass of about 2 kg. What is the mass of 40 American flamingos that each have a mass of about 2 kg?

number of flamingos	mass of 1 flamingo in kg	total mass in kg

_______________ (number model with ?)

40 flamingos have a mass of about _______ kg.

2. There are 9 bluebirds that each have about the same mass. Together they have a mass of about 270 g. What is the mass of one bluebird?

number of bluebirds	mass of 1 bluebird in g	total mass in g

_______________ (number model with ?)

One bluebird has a mass of about _______ g.

3. Explain to someone at home how you can use a basic fact to help you solve Problem 2.

Using Mental Multiplication

Lesson 9-3

NAME DATE TIME

Solve each number story in your head. Use number models and words to show your thinking.

1. The mass of one black-capped chickadee is about 12 grams. What is the mass of eight 12-gram black-capped chickadees?

   My thinking:

   Answer: about ______________ (unit)

2. The mass of one mountain bluebird is about 30 grams. What is the mass of six 30-gram mountain bluebirds?

   My thinking:

   Answer: about ______________ (unit)

3. Which has more mass, fifteen 5-kilogram American white pelicans or thirteen 6-kilogram bald eagles? Show your thinking below.

   Answer: ______________

Using Mental Math to Multiply

Home Link 9-3

NAME DATE TIME

Family Note Today your child practiced applying efficient fact strategies to solve multiplication problems with larger factors. Your child broke apart factors into easier numbers to mentally solve problems involving masses of North American birds.

Please return this Home Link to school tomorrow.

Solve each problem in your head. Use number models and words to show your thinking.

1. The mass of one California condor is 9 kilograms. What is the mass of twelve 9-kilogram California condors?

   My thinking:

   Answer: ______________ (unit)

2. The mass of one mountain bluebird is 25 grams. What is the mass of seven 25-gram bluebirds?

   My thinking:

   Answer: ______________ (unit)

3. Explain to someone at home how you can use the break-apart and doubling strategies to solve problems with larger factors.

Solving Number Stories about Time and Mass

Lesson 9-4

NAME DATE TIME

Use the schedule to answer Problems 1 and 2. You may draw an open number line to help.

① How much longer is the prairie study than the marsh walk?

Watershed Nature Center Activities	
Activity	**Time**
Marsh walk	10:00 A.M.–11:00 A.M.
Observation tower	11:00 A.M.–11:30 A.M.
Prairie study	11:30 A.M.–12:45 P.M.

Answer: _______ minutes

② Sujie arrived at 10:00 A.M. for the marsh walk and left after the prairie study. She went to the observation tower between the marsh walk and the prairie study. What is the total length of time Sujie spent at the nature center?

Answer: _______ hours _______ minutes

Solve.

③ Jasmine saw a northern cardinal from the observation tower. She knows that a cardinal can have a mass of about 45 grams. What is the total mass of three 45-gram cardinals?

about _______________
(unit)

Finding Field Day End Time

Lesson 9-4

NAME DATE TIME

Math Message

Jamar wants to know what time Field Day activities will end. He looks at the schedule to find out how long each activity lasts. Then he draws an open number line. Help Jamar finish his open number line.

Field Day Activities beginning at 9:15 A.M.	
Relay races	20 minutes
Long jump	10 minutes
Hula hoop	15 minutes
Double dutch	20 minutes
Two water breaks	5 minutes each

Field Day end time: ______________

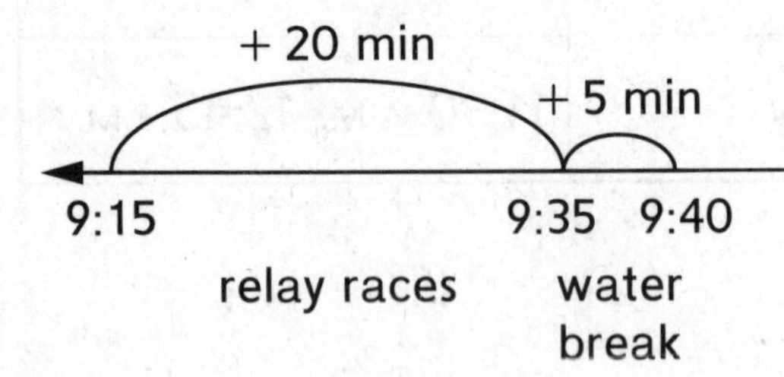

Finding Field Day End Time

Lesson 9-4

NAME DATE TIME

Math Message

Jamar wants to know what time Field Day activities will end. He looks at the schedule to find out how long each activity lasts. Then he draws an open number line. Help Jamar finish his open number line.

Field Day Activities beginning at 9:15 A.M.	
Relay races	20 minutes
Long jump	10 minutes
Hula hoop	15 minutes
Double dutch	20 minutes
Two water breaks	5 minutes each

Field Day end time: ______________

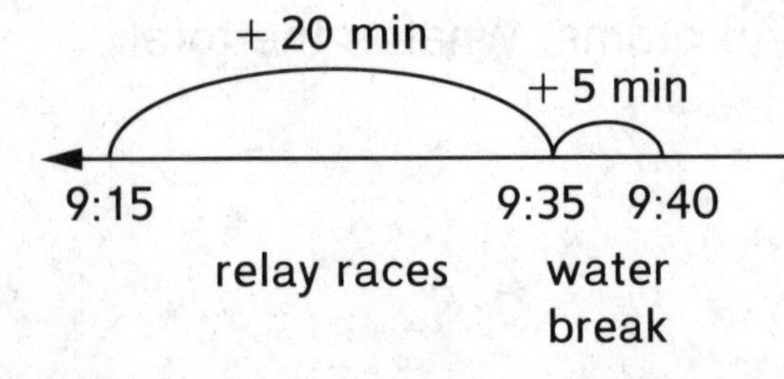

City Zoo Schedule

Lesson 9-4

NAME DATE TIME

City Zoo Schedule		
Activity	**Start Time**	**Length***
Predator or Prey? Show	9:25 A.M. 9:55 A.M. 10:25 A.M. 10:55 A.M. 11:25 A.M.	30 min
Rainforest Animals Show	9:00 A.M. 9:45 A.M. 10:30 A.M. 11:15 A.M. 12:00 P.M.	45 min
African Journey (long tour)	Any time	90 min
African Journey (short tour)	Any time	45 min
Big Cats Habitat	Any time	25 min
Great Ape House	Any time	30 min
Sea Lion Pool	Any time	20 min
Endangered Reptiles Exhibit	Any time	60 min

*The length of each activity includes travel time and bathroom breaks.

Two Squares

Lesson 9-4

NAME DATE TIME

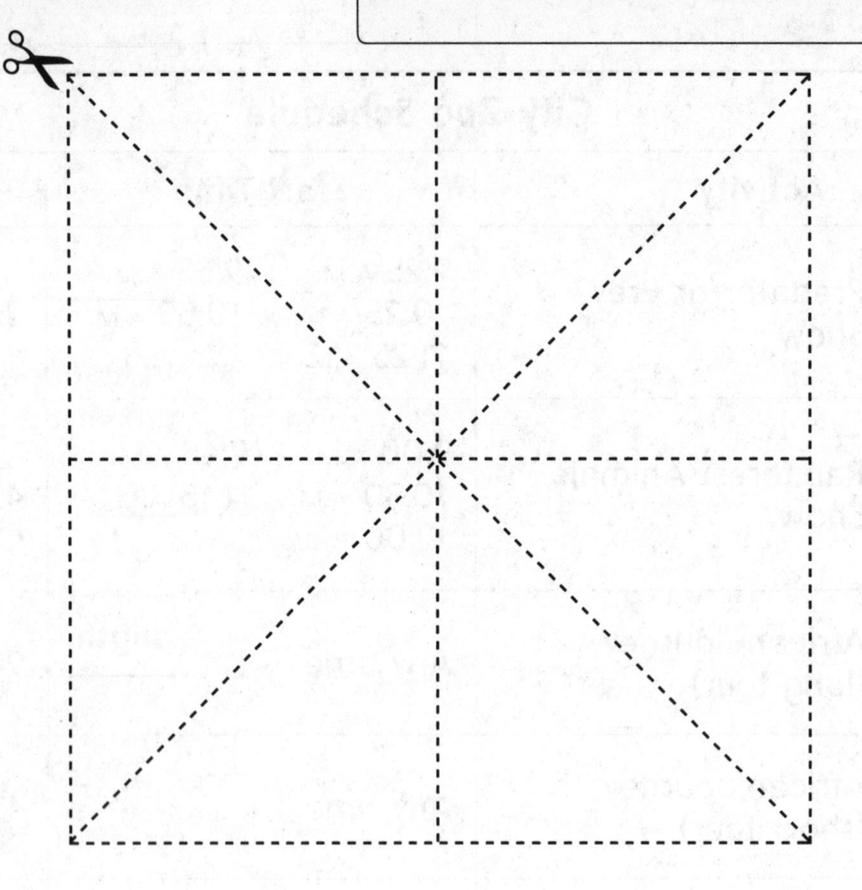

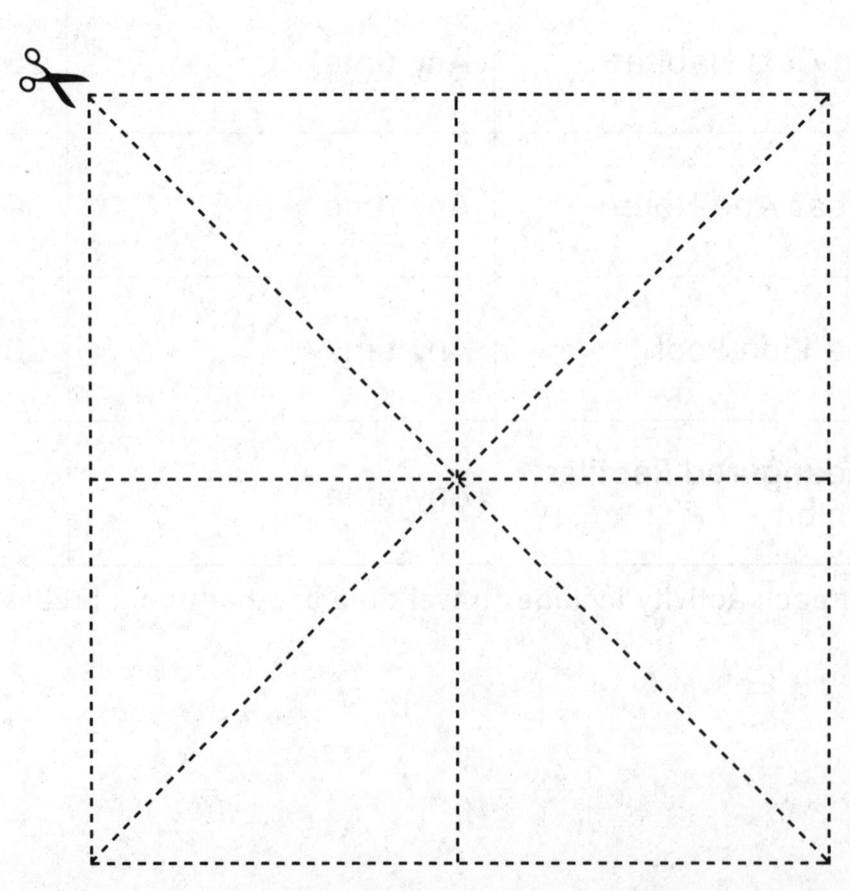

Measuring the Lengths of Activities

Home Link 9-4

NAME DATE TIME

Family Note Today your child practiced measuring time intervals by planning a schedule for a field trip. After completing Problem 1, have your child explain how he or she figured out the length of each activity.

Please return this Home Link to school tomorrow.

① Isabella wants to know how long each camp activity lasts. Use the table below to find the length of each activity. You may use open number lines, clocks, or another strategy.

Camp Activities		
Activity	**Schedule**	**Length, in minutes**
Art	8:30 A.M.–9:20 A.M.	
Swimming	9:20 A.M.–10:20 A.M.	
Snack	10:35 A.M.–10:55 A.M.	
Nature walk	10:55 A.M.–12:10 P.M.	

Practice

Solve.

② $4 \times 60 =$ ______

③ $70 \times 3 =$ ______

④ ______ $= 60 \times 8$

⑤ ______ $= 80 \times 9$

Breaking Apart Two Factors

Lesson 9-5

NAME DATE TIME

Study the example problem and solution. *Think:* How are the factors broken apart? What does the picture of the rectangle show?

Then use the same strategy to solve Problems 1 and 2.

Example:

13 × 42 = 546

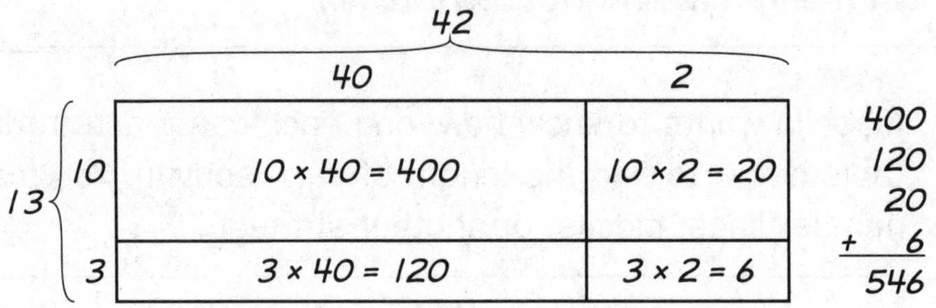

	40	2
10	10 × 40 = 400	10 × 2 = 20
3	3 × 40 = 120	3 × 2 = 6

(42 = 40 + 2; 13 = 10 + 3)

400
120
20
\+ 6
546

① 52 × 12 = ______

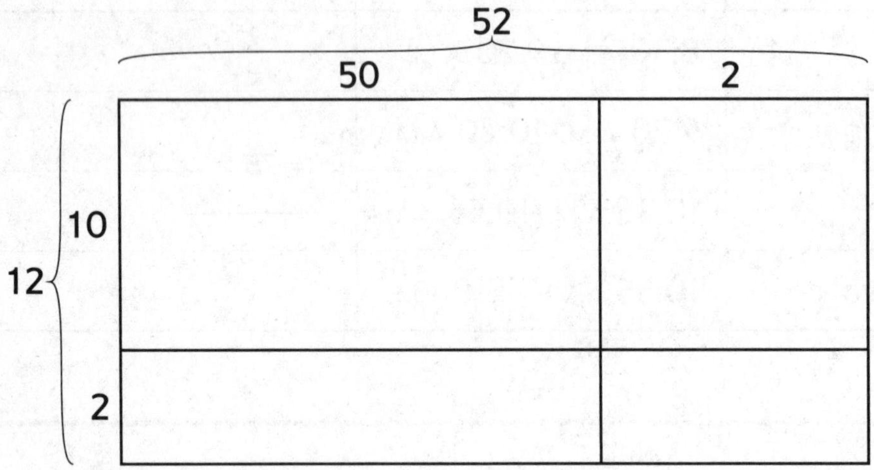

	50	2
10		
2		

(52 = 50 + 2; 12 = 10 + 2)

② 33 × 15 = ______

Multidigit Multiplication

Home Link 9-5

NAME DATE TIME

Family Note Today your child multiplied 2-digit numbers by 1-digit numbers using area models. Children drew a rectangle to represent the multiplication problem and then broke apart the larger factor into smaller, easier-to-multiply numbers.

Please return this Home Link to school tomorrow.

Use the break-apart strategy to solve the multiplication problems. Draw and partition a rectangle. Then record number sentences to show how you broke apart the factor.

Example:

3 × 28 = 84

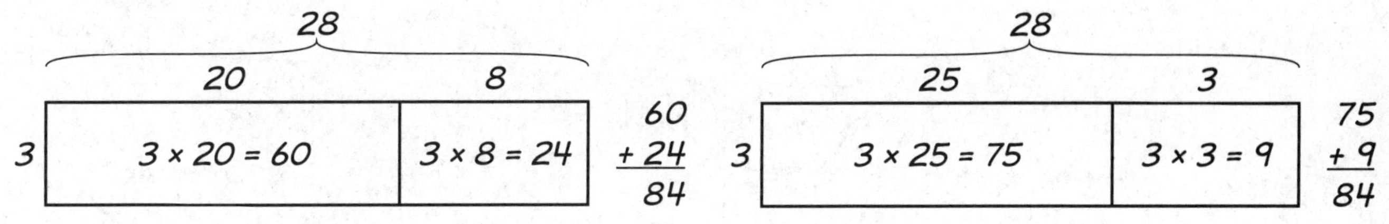

Two ways to break apart 28 to help solve 3 × 28.

① 5 × 42 = ______

② 6 × 54 = ______

③ Explain to someone at home how you broke apart the larger factors.

Packing Apples

Lesson 9-6

NAME DATE TIME

Pretend the division key (÷) on your calculator is broken. Solve this problem using other keys on your calculator.

Emma picks 176 apples to sell at the farmers' market. She packs them into boxes that each hold 16 apples. How many boxes does she need to take all of her apples to the market?

Show or tell how you used your broken calculator to solve this problem.

Answer: ______ boxes of apples

Using Tools Effectively

Home Link 9-6

NAME DATE TIME

Family Note Today your child pretended to use a calculator with a broken division key to solve a number story. In the problem below, your child is asked to solve a similar problem with a broken calculator. Ask your child to explain why both strategies work and how they are different.

Please return this Home Link to school tomorrow.

Ask someone at home for a calculator you can use to solve this problem. 

A third-grade class is planning to buy eggs for the school's pancake breakfast. They need 180 eggs for the breakfast. The teacher reminded the class that eggs come in cartons of 12 and asked them to figure out how many cartons they need. Lucy wants to use her calculator to solve the problem, but the (+) and (÷) keys are both broken. Help Lucy find a way to use her broken calculator to solve the problem.

1. Show or tell how to use Lucy's broken calculator to find the number of cartons of eggs the class needs to buy.

   The class needs to buy ______ cartons of eggs.

2. Show or tell another way for Lucy to use her broken calculator to solve the problem.

Measuring Time with an Open Number Line

Lesson 9-7

NAME DATE TIME

① Maggie's swim meet began at 9:30 A.M. and ended at 1:18 P.M. She sketched an open number line to help her figure out how long the swim meet lasted.

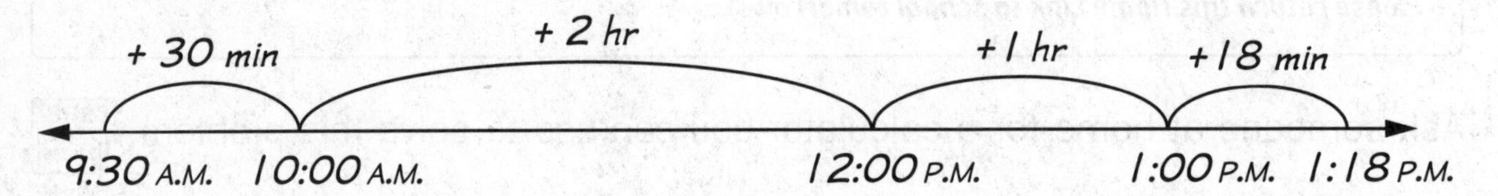

Use Maggie's open number line to calculate the length of the swim meet.

_______ hours _______ minutes

② After the swim meet, Maggie went to lunch with her grandmother. They started lunch at 1:27 P.M. and finished at 2:45 P.M. How long were they at lunch?

Show your thinking on an open number line.

_______ hour(s) _______ minute(s)

Finding Length-of-Day Trends

Lesson 9-7

NAME DATE TIME

Find each place listed below on the map on *Student Reference Book,* page 281. Use the June 21, 2016, sunrise and sunset times on the map to calculate the length of day. Next to each place on the chart, record the length of day.

	Place	Length of Day in Hours and Minutes
North	Alta, Norway	______ hours ______ minutes
↓	Nuuk, Greenland	______ hours ______ minutes
	Anchorage, Alaska, USA	______ hours ______ minutes
	Moscow, Russia	______ hours ______ minutes
	Wakkanai, Japan	______ hours ______ minutes
	Toronto, Canada	______ hours ______ minutes
	Beijing, China	______ hours ______ minutes
	Cairo, Egypt	______ hours ______ minutes
	Manila, Philippines	______ hours ______ minutes
	Quito, Ecuador	______ hours ______ minutes
	Nairobi, Kenya	______ hours ______ minutes
	Cape Town, South Africa	______ hours ______ minutes
	Esperanza Base, Antarctica	______ hours ______ minutes
South	McMurdo Station, Antarctica	______ hours ______ minutes

Finding Length-of-Day Trends

(continued)

Lesson 9-7

NAME DATE TIME

Use your length-of-day calculations in the chart and the map on *Student Reference Book*, page 281 to answer the questions.

1. Circle the names of the cities in the chart with lengths of day that are about 12 hours. What do you notice about their locations on the map?

____

____

2. What happens to the length of day on June 21, 2016, as you move from south (Antarctica) to north (Alaska or Norway)?

____

3. What happens to the sunrise time on June 21, 2016, as you move from north (Greenland) to south?

____

Try This

4. Look at your class Length-of-Day Graph. Find where you live on a map. Then make predictions about the length of day on December 21, 2016, for where you live.

____

____

Length-of-Day Graph for Chicago, Illinois 2015–2016

Lesson 9-7

NAME DATE TIME

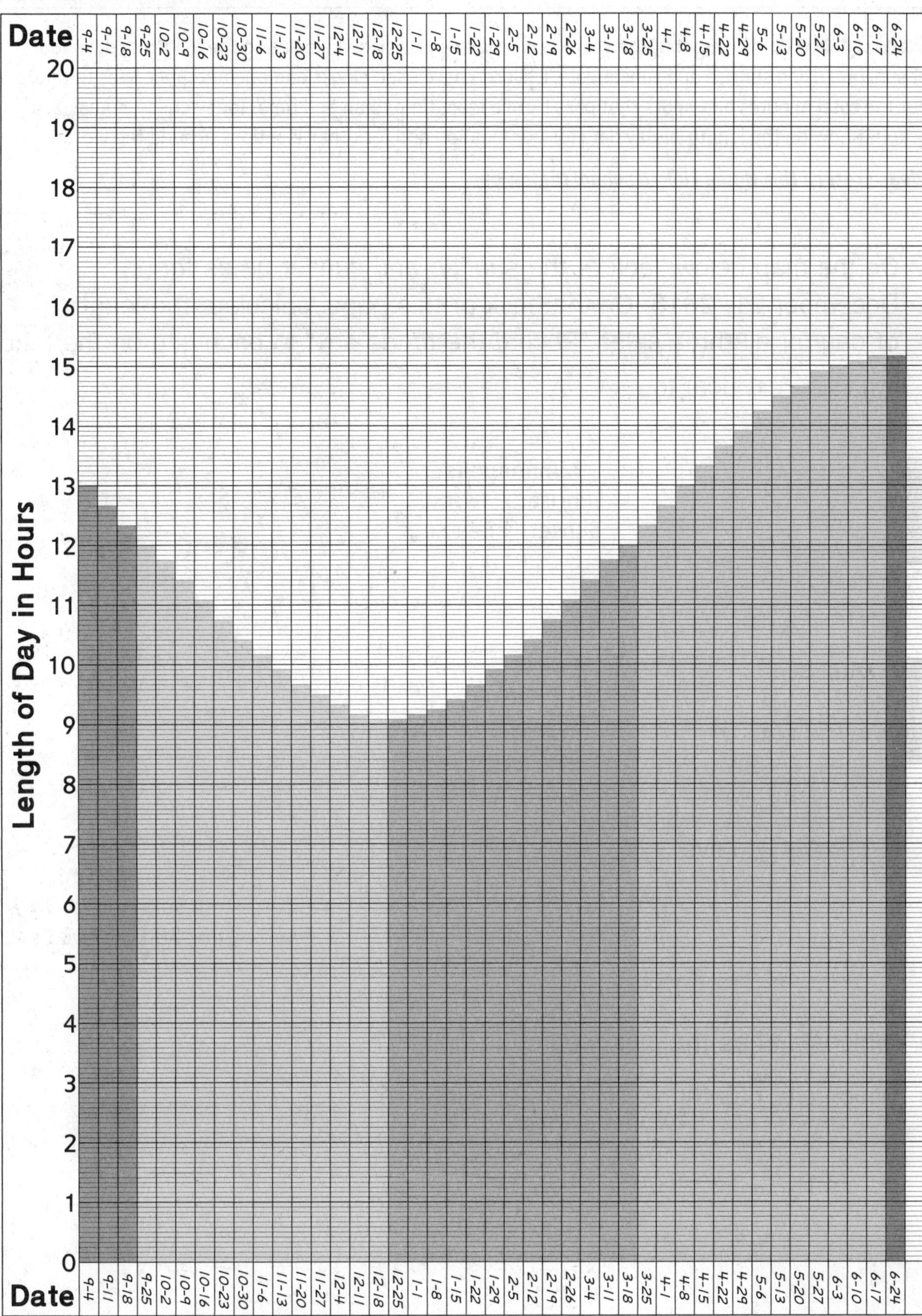

Calculating Elapsed Time

Home Link 9-7

NAME DATE TIME

Family Note Throughout the year, your child has practiced calculating the length of day (hours of sunlight) using sunrise and sunset data. Children have used clocks and open number lines to figure out the total minutes and hours that pass from a start time to an end time. Today children analyzed graphs showing the length-of-day data for our location and for other locations around the world.

Please return this Home Link to school tomorrow.

SRB 18-19, 187-188

1. On the map below, look at the sunrise and sunset times for December 21, 2016. On the back of this page, calculate the length of day for all three cities. Record the times next to each city on the map.

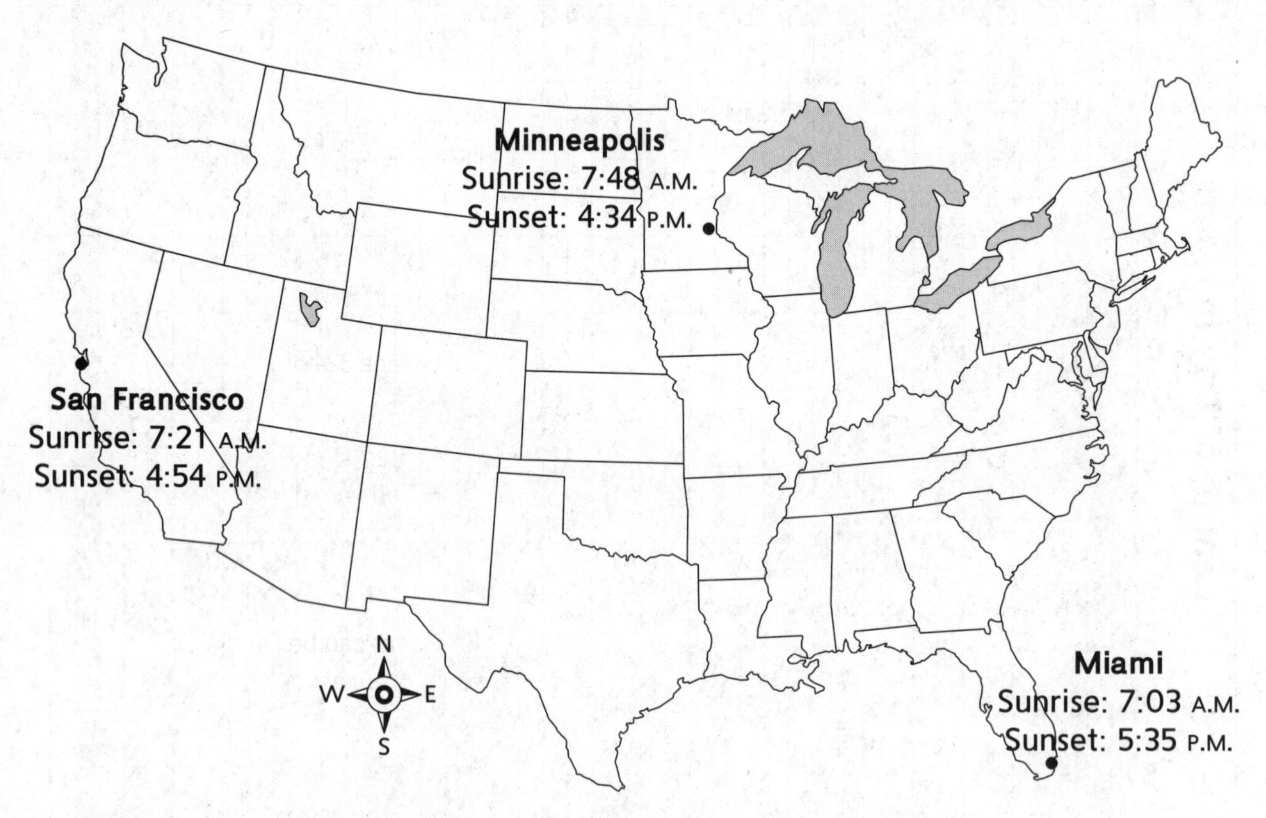

2. Which city has the *most* hours of sunlight? ____________________

3. Which city has the *least* hours of sunlight? ____________________

End-of-Year Family Letter

Home Link 9-8

NAME DATE TIME

Congratulations! By completing *Third Grade Everyday Mathematics* your child has accomplished a great deal. Thank you for all of your support.

This Family Letter is provided for you as a resource throughout your child's school vacation. It includes a list of Do-Anytime Activities, game directions, titles of mathematics-related books, and a sneak preview of what your child will be learning in *Fourth Grade Everyday Mathematics.*

Enjoy your summer!

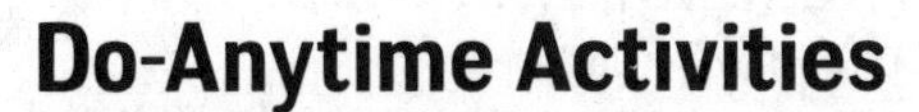

Do-Anytime Activities

The following activities are for you and your child to do together during the summer to help review concepts your child learned in third grade. These activities build on the skills from this year and help prepare your child for *Fourth Grade Everyday Mathematics.*

1. Pose number stories with single-digit numbers or multiples of 10 that can be solved using multiplication and division. Examples: *I have 12 crackers to share equally among you and your two sisters. How many crackers will you each get? 12 ÷ 3 = 4 crackers. 30 children can ride in one bus. How many can ride in 3 buses? 30 × 3 = 90 children.*
2. Have your child review and practice multiplication and division facts. Your child can use Fact Triangles or play a multiplication game (as described on the page 324 of this letter).
3. Pose simple fraction stories for your child to solve and encourage him or her to draw pictures to show his or her thinking. Examples: *The pizza is the whole. If you eat $\frac{1}{4}$ of the pizza and your sister eats $\frac{1}{2}$ of the pizza, who eats more? How do you know? My sister eats more because 1 out of 2 parts is larger than 1 out of 4 parts of the same pizza.*
4. Have your child practice reading analog clocks to the nearest minute.
5. Encourage your child to identify and describe geometric shapes that can be seen in the world. Example: *The window is a rectangle and a parallelogram because it has four sides and two sets of parallel sides.*
6. Examine food and drink labels to prompt discussions about mass and liquid volume units in the real world, such as milliliters, liters, grams, and kilograms. At the grocery store, look for items that have a mass of about 1 gram (blueberry) or 1 kilogram (pineapple).
7. Measure real-world objects or distances to the nearest inch, $\frac{1}{2}$ inch, and $\frac{1}{4}$ inch. Talk about when you need to make more precise measurements. Example: Hold a watermelon-seed spitting contest and measure the distances the seeds travel.
8. Find the areas and perimeters of rooms at home. Have your child estimate which room could have the largest area and largest perimeter (they might not be the same).
9. Practice multidigit addition and subtraction.
10. Practice extended facts using basic facts for multiplication and division. Example: If you know 3 × 3 = 9, then it helps you know that 3 × 30 = 90.

Building Skills through Games

Multiplication Top-It

Materials Number cards 0–10* (4 of each)
Players 2 to 4
Skill Multiplication facts 0 to 10
Object of the Game To collect the most cards

Directions

Shuffle the cards and place them facedown in a pile. Each player turns over 2 cards and calls out the product of the numbers. The player with the largest product wins the round and takes all the cards. In case of a tie for the largest product, each tied player turns over 2 more cards and calls out the product of the numbers. The player with the largest product then takes all the cards from both plays. The game ends when there are not enough cards left for each player to have another turn. The player with the most cards wins.

Variation: *Extended Multiplication Top-It*

Turn over 2 cards and make the second card a multiple of 10. For example, if you turn over 2 and 4, the 4 becomes 40. Multiply the single digit by the multiple of 10.

Name That Number

Materials Number cards 0–20* (4 of each card 0–10, and 1 of each card 11–20)
Players 2 to 4 (the game is more interesting when played by 3 or 4 players)
Skill Finding equivalent names for numbers
Object of the Game To collect the most cards

Directions

1. Shuffle the deck and place 5 cards number-side up on the table. Leave the rest of the deck number-side down. Then turn over the top card of the deck and lay it down next to the deck. The number on this card is the target number.
2. Players take turns. When it is your turn, try to name the target number by adding, subtracting, multiplying, or dividing the numbers on 2 or more of the 5 cards that are number-side up. A card may be used only once for each turn. If you can name the target number, take the target number card and the cards you used to name it. Then replace all the cards you took by drawing from the top of the deck. If you cannot name the target number, your turn is over. Turn over the top card of the deck and lay it down on the target-number pile. The number on this card becomes the new target number to be named.
3. Play continues until all of the cards in the deck have been turned over. The player who has taken the most cards wins.

Salute!

Materials	Number cards 1–10* (4 of each)
Players	3
Skill	Practicing multiplication and division facts
Object of the Game	To solve for a missing factor

Directions

One person begins as the Dealer. The Dealer gives one card to each of the other two Players. Without looking at their cards, the Players hold them on their foreheads with the numbers facing out. The Dealer looks at both cards and says the product of the two numbers. Each Player looks at the other Player's card. They use the number they see and the product said by the Dealer to figure out the number on their card (the missing factor). They say that number out loud. Once both Players have said their numbers, they can look at their own cards to check their answers. Rotate roles clockwise and repeat the game. Play continues until everyone has been the Dealer five times.

*The number cards used in some games can be made from index cards or from a regular deck of playing cards. (Use jacks for zeros, aces for ones, and other face cards for teen numbers.)

Product Pile-Up

Materials	Number cards 1–10 (4 of each)
Players	3
Skill	Practicing multiplication facts 1 to 10
Object of the Game	To play all of your cards

Directions

1. Take turns being the dealer. Shuffle and deal 8 cards to each player. Place the rest of the deck number-side down.
2. The player to the left of the dealer begins. This player selects 2 cards from his or her hand, places them number-side up on the table, multiplies the numbers, and says the product aloud.
3. Play continues with each player playing 2 cards with a product *greater than* the product of the last 2 cards played. If a player states an incorrect product, other players may suggest a helper fact or strategy to help find the correct product.
4. If a player is not able to play 2 cards with a greater product, the player draws 2 cards from the deck.
   - If the player is now able to make a greater product, those cards are played and the game continues. If the player still cannot make a greater product, the player keeps the cards and says "Pass." The game continues to the next person.
   - If all players must pass, the player who laid down the last 2 cards starts a new round beginning with Step 2 above.
5. The winner is the first player to run out of cards, or the player with the fewest cards when there are no more cards to draw.

Vacation Reading with a Mathematical Twist

Books can contribute to children's learning by presenting mathematics in a combination of real-world and imaginary contexts. The titles below were recommended by teachers who use *Everyday Mathematics.* Visit your local library and check out these mathematics-related books with your child.

Geometry

A Cloak for the Dreamer by Aileen Friedman

Fractals, Googols, and Other Mathematical Tales by Theoni Pappas

Measurement

How Tall, How Short, How Far Away by David Adler

Math Curse by Jon Scieszka

Measuring on Penny by Loren Leedy

Numeration

Fraction Fun by David Adler

How Much Is a Million? by David Schwartz

Operations

The Grapes of Math by Gregory Tang

The King's Chessboard by David Birch

The I Hate Mathematics! Book by Marilyn Burns

A Remainder of One by Elinor J. Pinczes

Anno's Mysterious Multiplying Jar by Masaichiro Anno and Mitsumasa Anno

Patterns, Functions, and Algebra

Eight Hands Round: A Patchwork Alphabet by Ann Whitford Paul

A Million Fish . . . More or Less by Patricia C. McKissack

Reference Frames

Pigs on a Blanket by Amy Axelrod

Three Days on a River in a Red Canoe by Vera B. Williams

Looking Ahead: *Fourth Grade Everyday Mathematics*

Next year your child will:

- Solve multistep problems involving the four operations.
- Explore multiples, factors, and prime and composite numbers.
- Explore multidigit multiplication and division methods.
- Add and subtract fractions with like denominators and multiply fractions by whole numbers.
- Read, write, compare, and order fractions and decimals.
- Convert between metric and U.S. customary units of measure.
- Apply formulas to find the area and the perimeter of rectangles.
- Identify geometric properties in a variety of shapes.
- Collect and interpret data.
- Identify lines of symmetry and symmetric shapes.
- Explore partial quotients for division.
- Solve number stories involving different units of measurement.

Again, thank you for your support this year. Have fun continuing your child's mathematical adventures throughout the vacation!

Teaching Aid Masters

Number Grids

NAME DATE TIME

−9	−8	−7	−6	−5	−4	−3	−2	−1	0
1	2	3	4	5	6	7	8	9	10
11	12	13	14	15	16	17	18	19	20
21	22	23	24	25	26	27	28	29	30
31	32	33	34	35	36	37	38	39	40
41	42	43	44	45	46	47	48	49	50
51	52	53	54	55	56	57	58	59	60
61	62	63	64	65	66	67	68	69	70
71	72	73	74	75	76	77	78	79	80
81	82	83	84	85	86	87	88	89	90
91	92	93	94	95	96	97	98	99	100
101	102	103	104	105	106	107	108	109	110
111	112	113	114	115	116	117	118	119	120

−9	−8	−7	−6	−5	−4	−3	−2	−1	0
1	2	3	4	5	6	7	8	9	10
11	12	13	14	15	16	17	18	19	20
21	22	23	24	25	26	27	28	29	30
31	32	33	34	35	36	37	38	39	40
41	42	43	44	45	46	47	48	49	50
51	52	53	54	55	56	57	58	59	60
61	62	63	64	65	66	67	68	69	70
71	72	73	74	75	76	77	78	79	80
81	82	83	84	85	86	87	88	89	90
91	92	93	94	95	96	97	98	99	100
101	102	103	104	105	106	107	108	109	110
111	112	113	114	115	116	117	118	119	120

Copyright © McGraw-Hill Education. Permission is granted to reproduce for classroom use.

Number Grid

NAME DATE TIME

−9	−8	−7	−6	−5	−4	−3	−2	−1	0
1	2	3	4	5	6	7	8	9	10
11	12	13	14	15	16	17	18	19	20
21	22	23	24	25	26	27	28	29	30
31	32	33	34	35	36	37	38	39	40
41	42	43	44	45	46	47	48	49	50
51	52	53	54	55	56	57	58	59	60
61	62	63	64	65	66	67	68	69	70
71	72	73	74	75	76	77	78	79	80
81	82	83	84	85	86	87	88	89	90
91	92	93	94	95	96	97	98	99	100
101	102	103	104	105	106	107	108	109	110
111	112	113	114	115	116	117	118	119	120

Paper Clock

NAME DATE TIME

① Cut out the clock face, the minute hand, and the hour hand.

② Punch a hole through the center of the clock face and through the X on the minute hand and the hour hand.

③ Fasten the hands to the clock face with a brad.

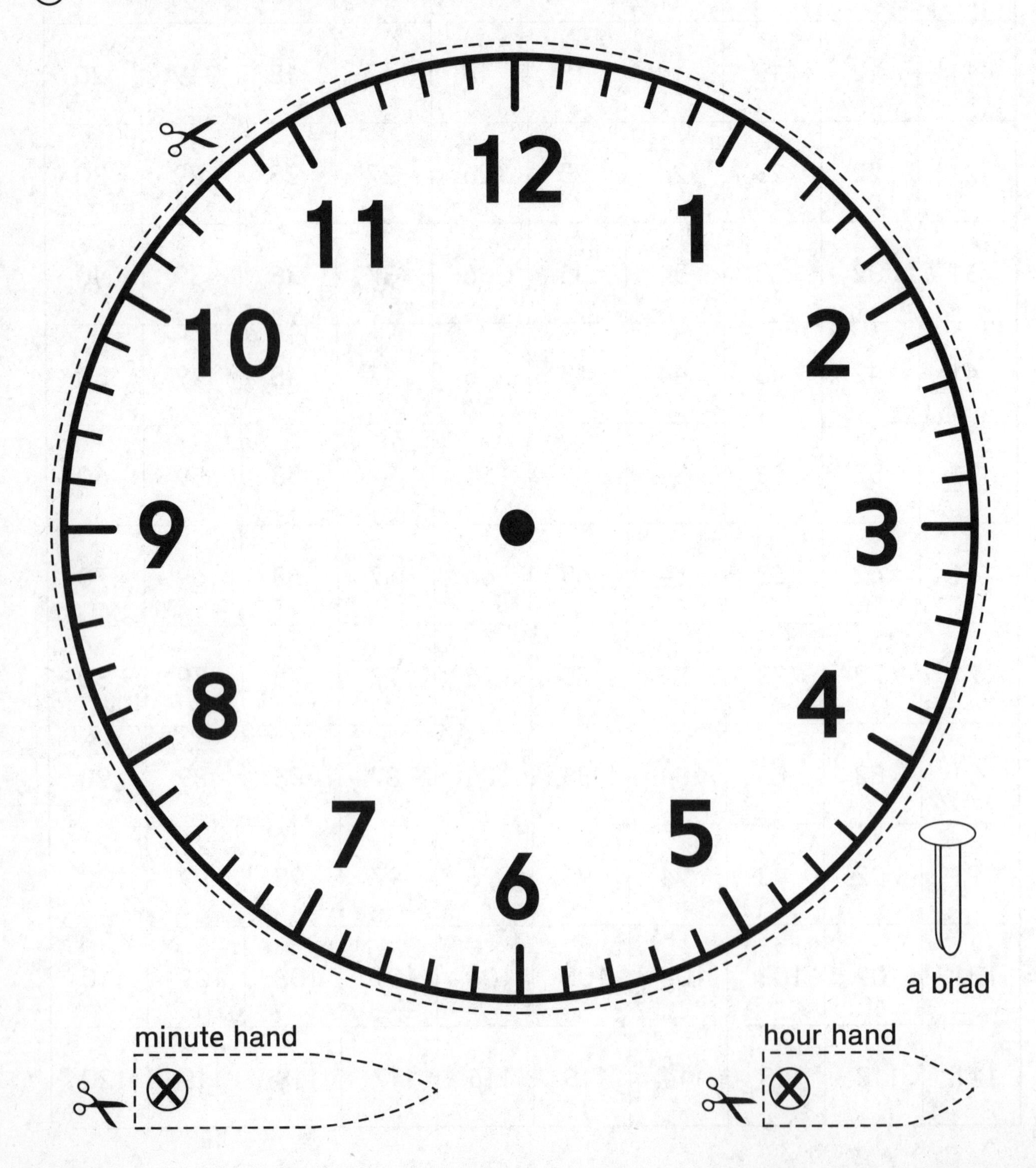

Math Boxes

NAME DATE TIME

①	②
③	④
⑤	⑥

Reengagement Planning Form

Common Core State Standard (CCSS):

Goal for Mathematical Practice (GMP):

Strengths and understandings:

Weaknesses and misconceptions:

Planning the Reengagement Discussion

Issue to address	Work samples that illustrate this issue	Questions to ask about the sample student work

A Bar Graph

NAME DATE TIME

Make a bar graph for your set of data.

Title: ______________________

A Number Story

NAME DATE TIME

Multiplication/ Division Fact Triangle

NAME DATE TIME

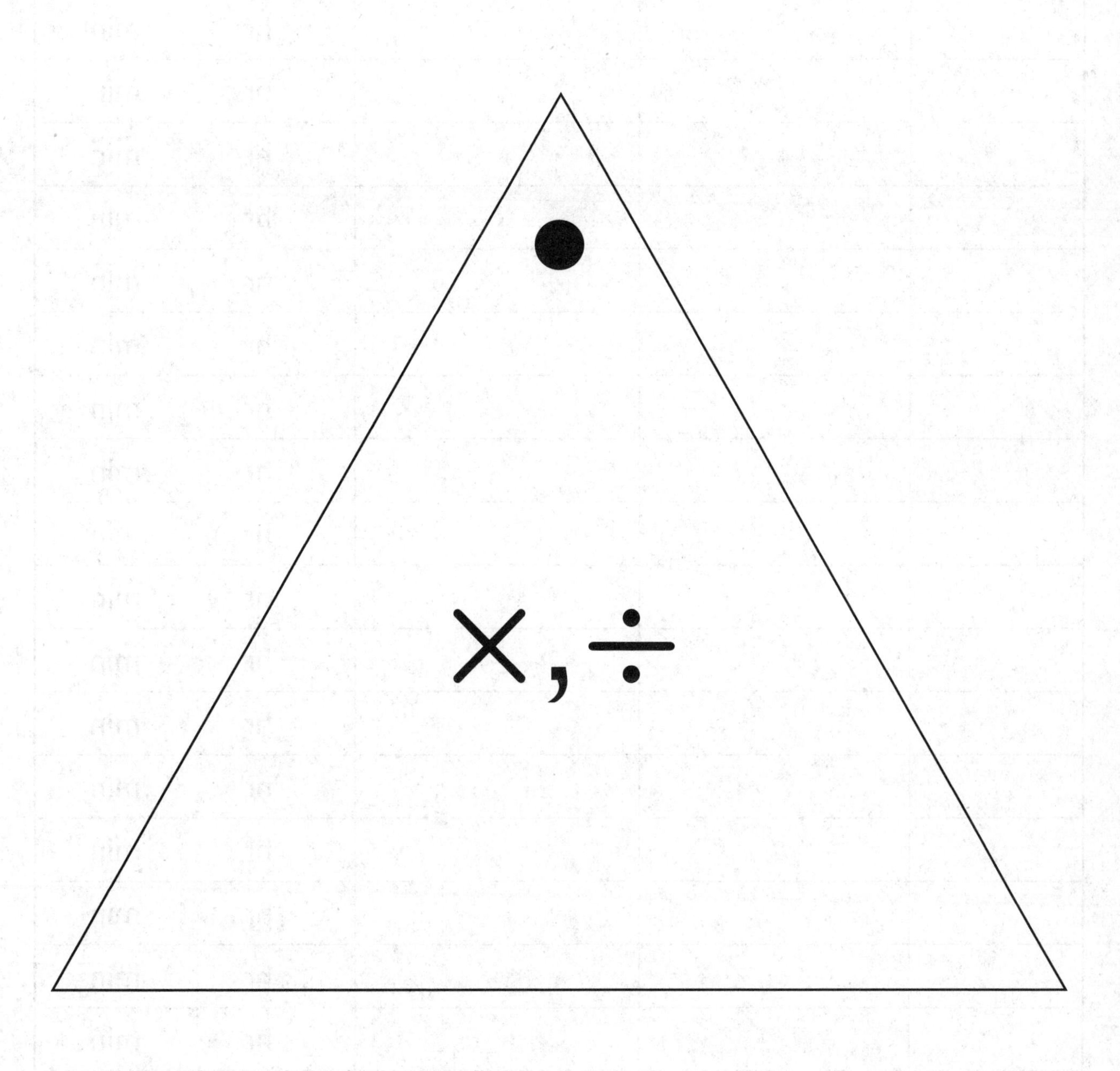

Sunrise and Sunset Data

NAME DATE TIME

Date	Time of Sunrise	Time of Sunset	Length of Day
			hr min
			hr min
			hr min
			hr min
			hr min
			hr min
			hr min
			hr min
			hr min
			hr min
			hr min
			hr min
			hr min
			hr min
			hr min
			hr min
			hr min
			hr min
			hr min
			hr min
			hr min

Elapsed Time Cards (Deck A)

NAME DATE TIME

5 minutes later	10 minutes later	15 minutes later	20 minutes later
30 minutes later	45 minutes later	1 hour later	2 hours later
5 minutes earlier	10 minutes earlier	15 minutes earlier	20 minutes earlier
30 minutes earlier	45 minutes earlier	1 hour earlier	2 hours earlier

Elapsed Time Cards (Deck B)

NAME DATE TIME

3 minutes later	9 minutes later	22 minutes later	39 minutes later
55 minutes later	1 hour and 25 minutes later	3 hours and 35 minutes later	5 hours and 40 minutes later
6 minutes earlier	12 minutes earlier	28 minutes earlier	49 minutes earlier
50 minutes earlier	1 hour and 55 minutes earlier	4 hours and 10 minutes earlier	7 hours and 45 minutes earlier

Paper Pancakes

NAME DATE TIME

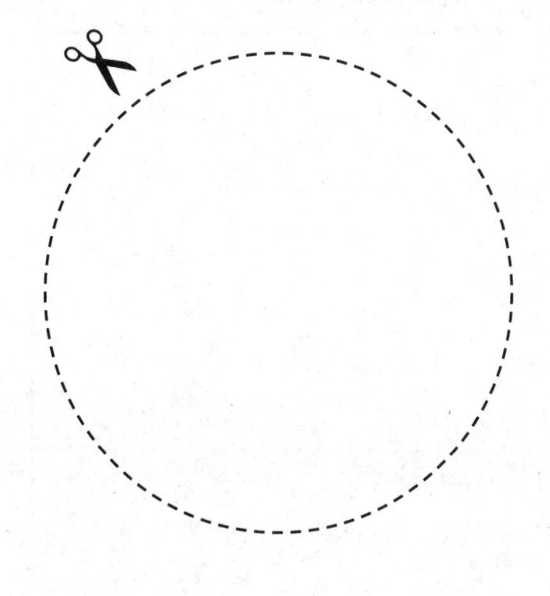
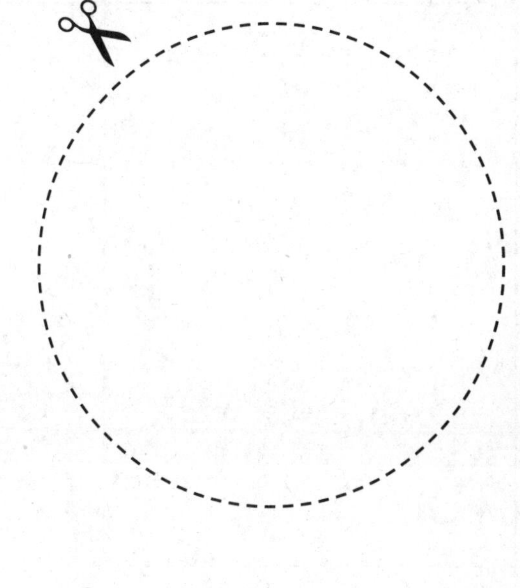
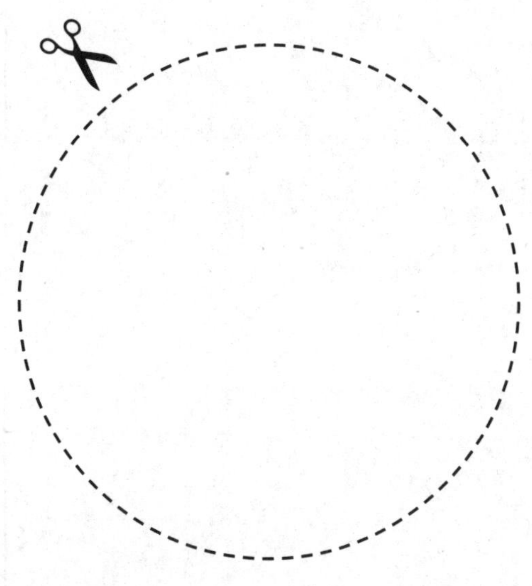
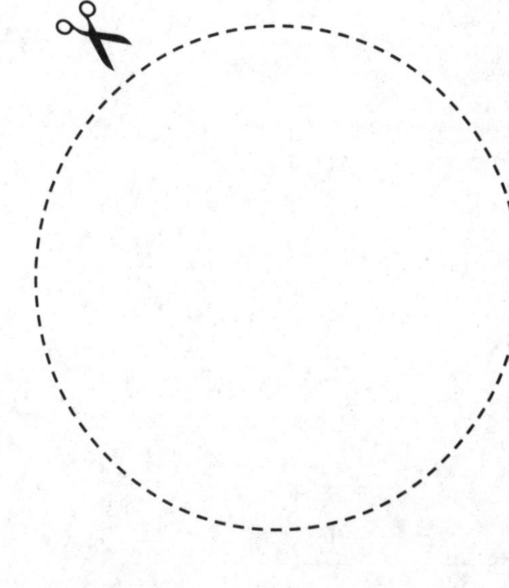
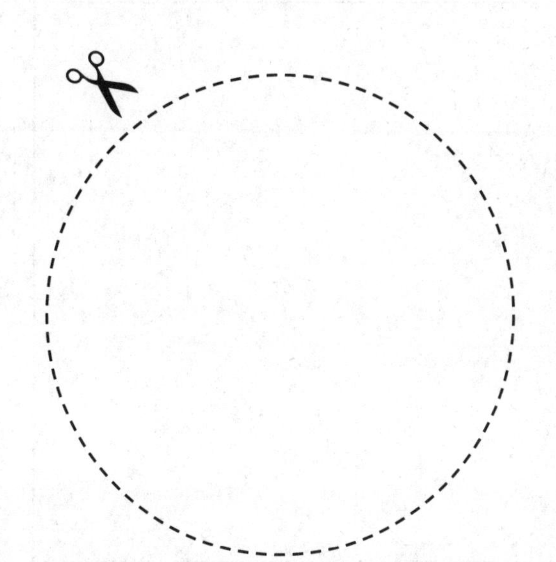
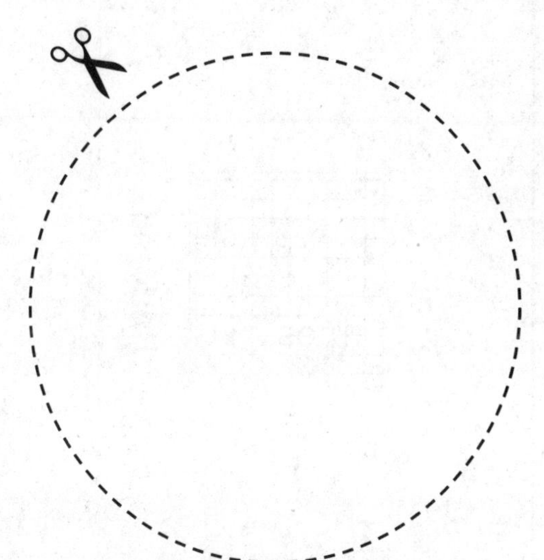

3-Digit Place-Value Mat

NAME DATE TIME

Ones	
Tens	
Hundreds	

Guide to Solving Number Stories

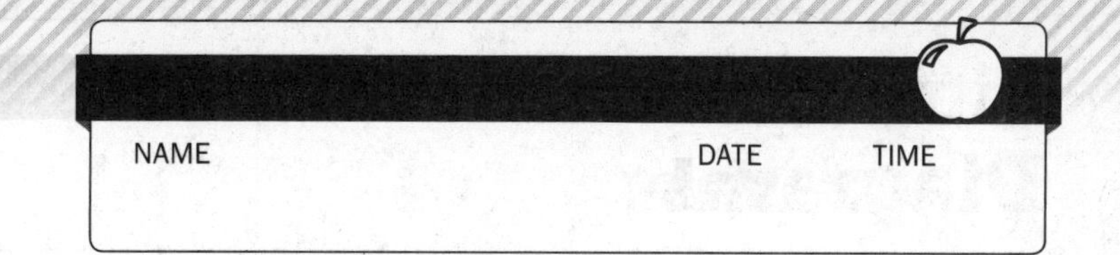

Make sense of the problem.

- Read the problem. What do you understand?
- What do you know?
- What do you need to find out?

Make a plan.

What will you do?

- Add?
- Subtract?
- Multiply?
- Divide?
- Draw a picture?
- Draw a diagram?
- Make a table?
- Make a graph?
- Write a number model?
- Use counters or base-10 blocks?
- Make tallies?
- Use a number grid or number line?

Solve the problem.

- Show your work.
- Keep trying if you get stuck.
  → Reread the problem.
  → Think about the strategies you have already tried and try new strategies.
- Write your answer with the units.

Check.

Does your answer make sense? How do you know?

Situation Diagrams

NAME DATE TIME

Parts-and-Total

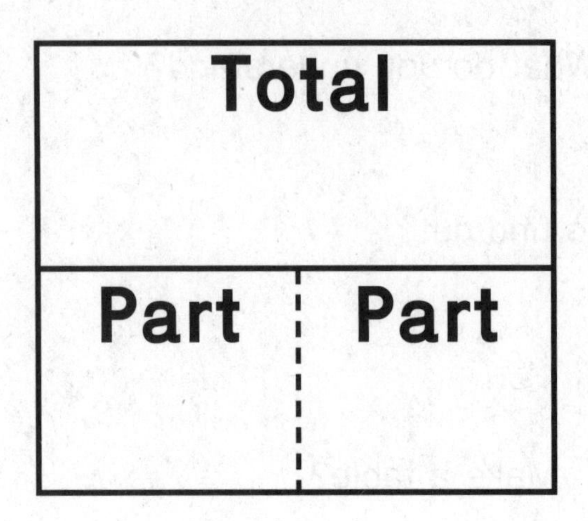

Change

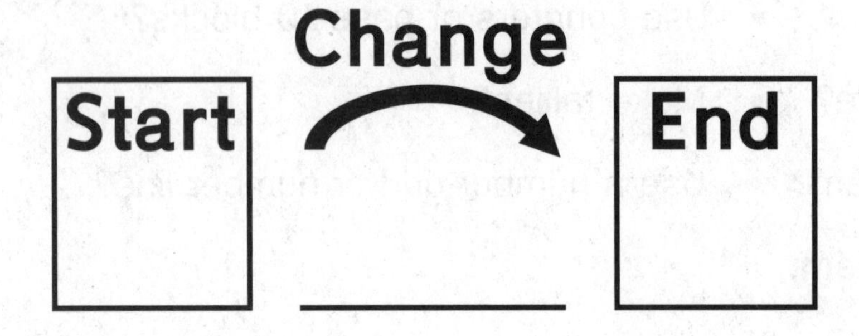

Comparison

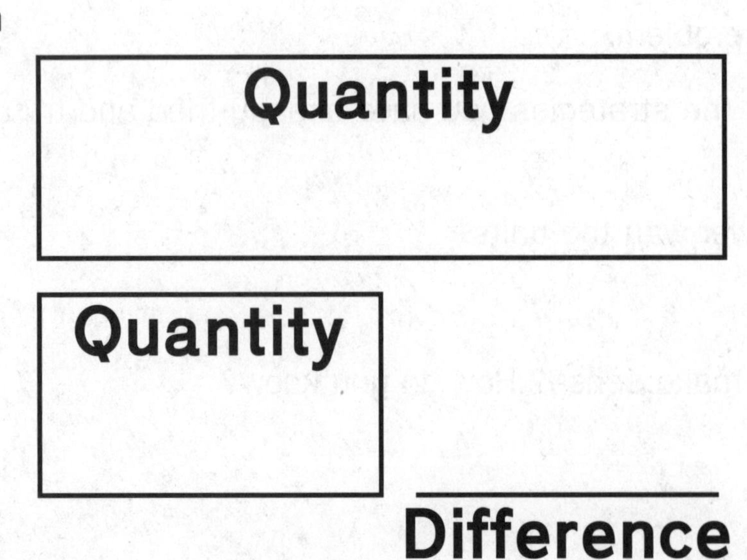

$\frac{1}{4}$-inch Grid Paper

NAME DATE TIME

4 × 3 Grid Dot Paper

NAME DATE TIME

Centimeter Grid Paper

NAME DATE TIME

4-Square Graphic Organizer

NAME DATE TIME

Word

More Frames and Arrows

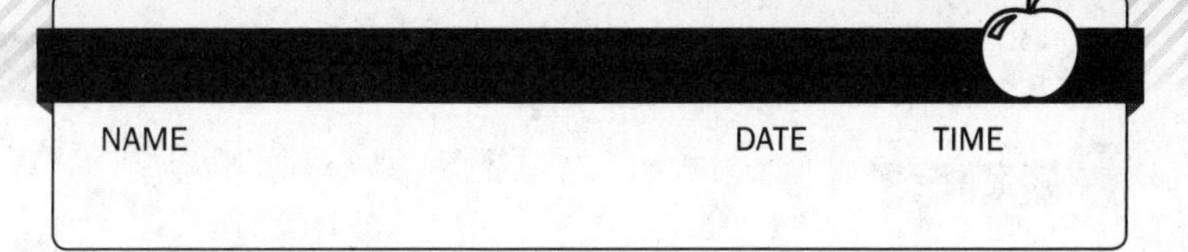

① Rule

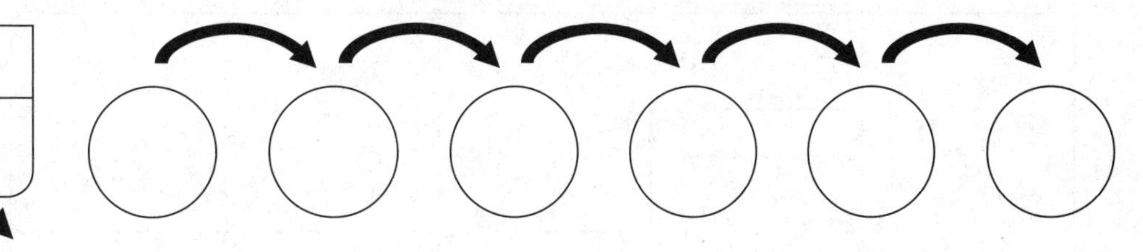

② Rule

③ Rule

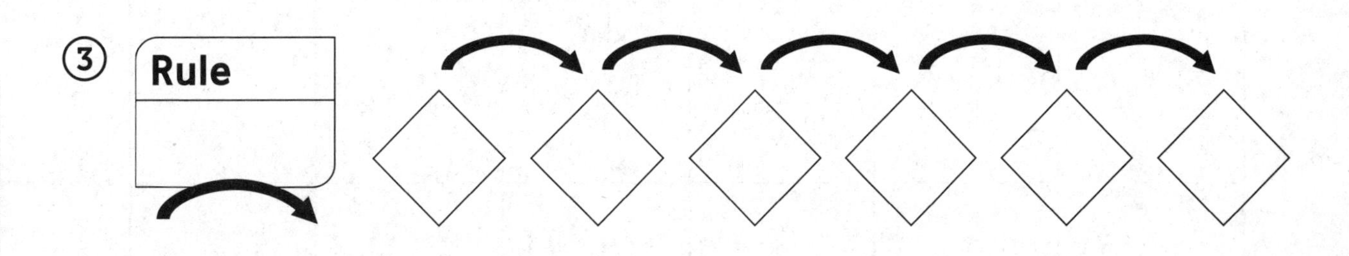

④ Rule

One-Inch Grid Paper

NAME DATE TIME

"What's My Rule?" Table

NAME DATE TIME

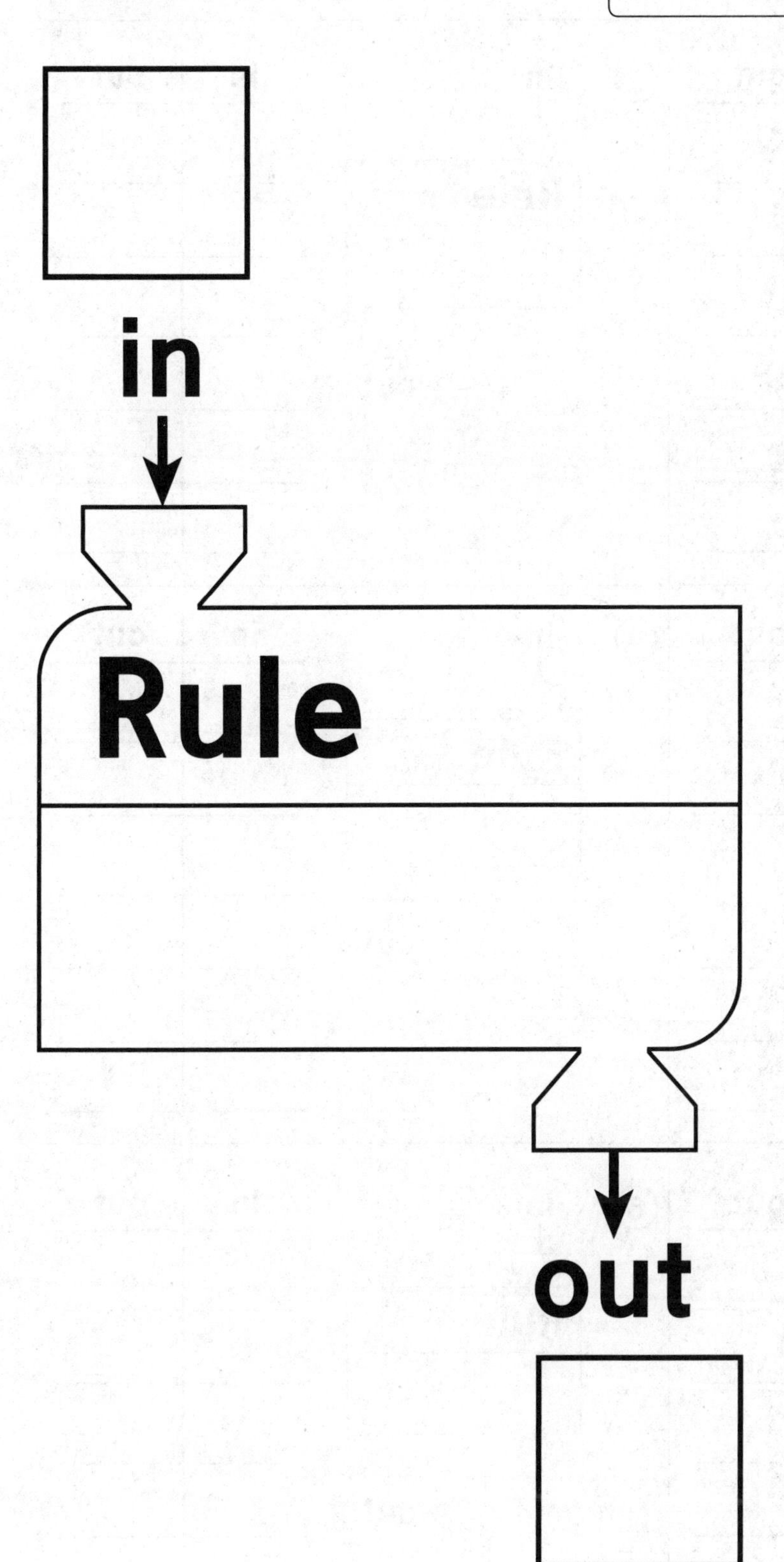

in	out

"What's My Rule?" Tables

NAME DATE TIME

① in → **Rule** → out

in	out

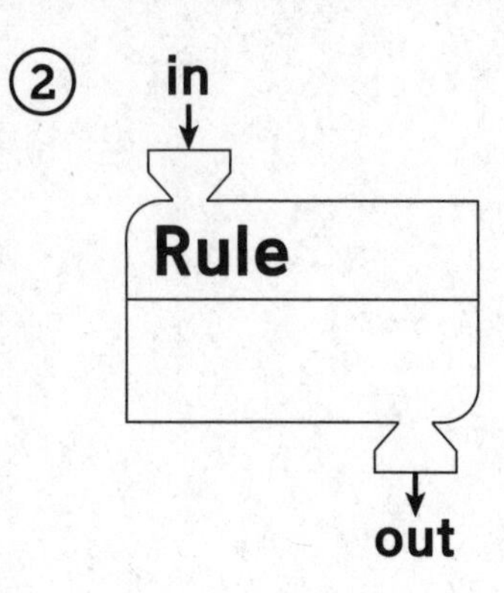

② in → **Rule** → out

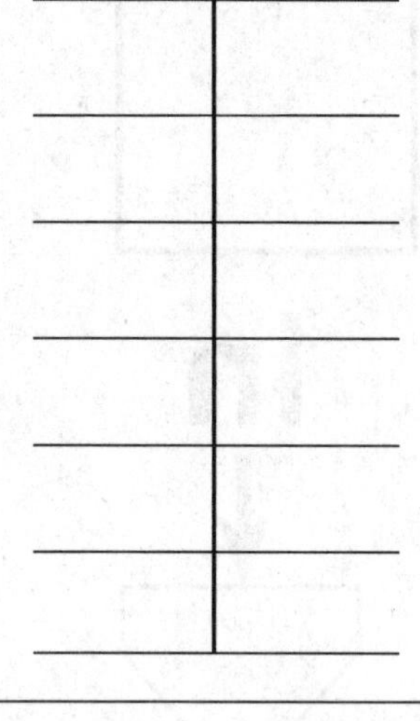

in	out

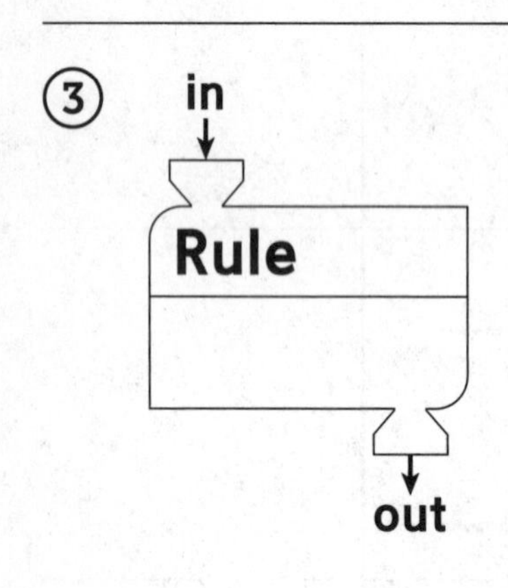

③ in → **Rule** → out

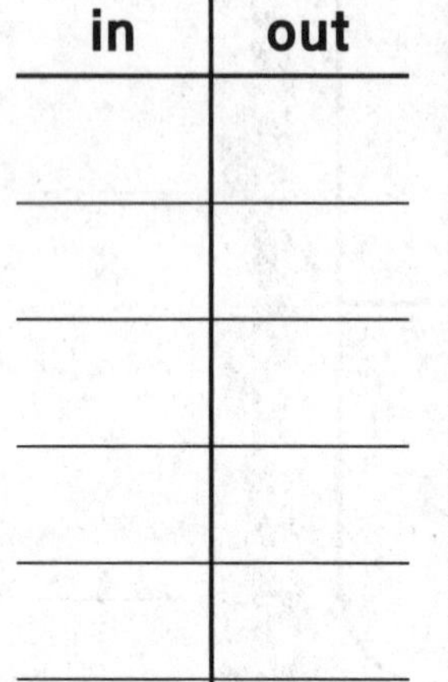

in	out

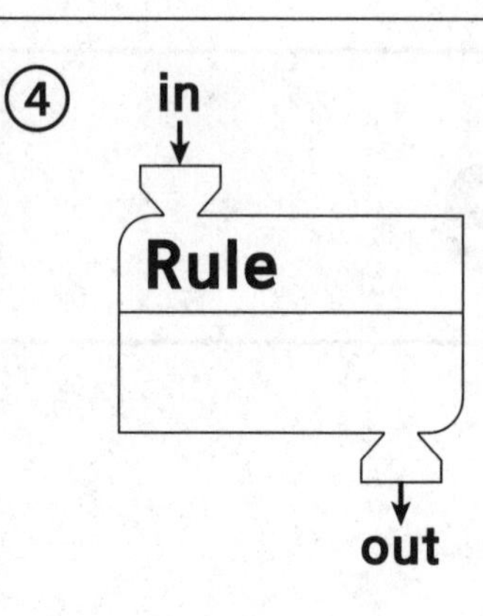

④ in → **Rule** → out

in	out

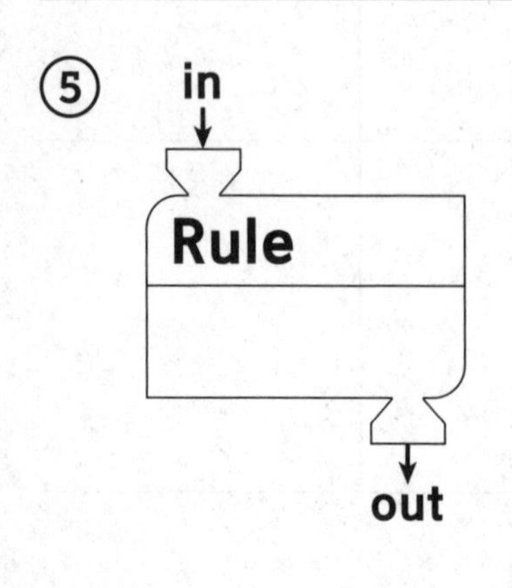

⑤ in → **Rule** → out

in	out

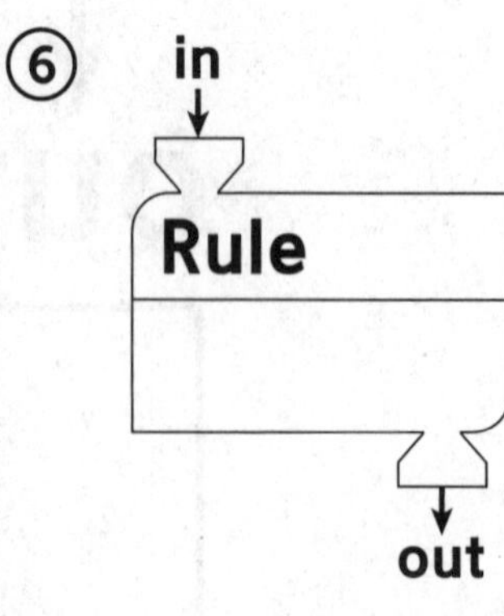

⑥ in → **Rule** → out

in	out

A Picture Graph

NAME DATE TIME

Title: ______________________

Key: Each ________ = ________________

A Picture Graph

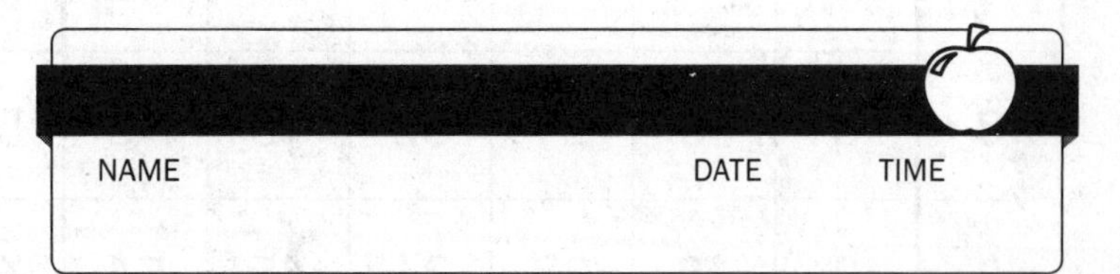

Title: ______________________

Key: Each ________ = ________________

Multiplication/ Division Facts Table

NAME DATE TIME

×,÷	1	2	3	4	5	6	7	8	9	10
1	1	2	3	4	5	6	7	8	9	10
2	2	4	6	8	10	12	14	16	18	20
3	3	6	9	12	15	18	21	24	27	30
4	4	8	12	16	20	24	28	32	36	40
5	5	10	15	20	25	30	35	40	45	50
6	6	12	18	24	30	36	42	48	54	60
7	7	14	21	28	35	42	49	56	63	70
8	8	16	24	32	40	48	56	64	72	80
9	9	18	27	36	45	54	63	72	81	90
10	10	20	30	40	50	60	70	80	90	100

Dominoes

NAME DATE TIME

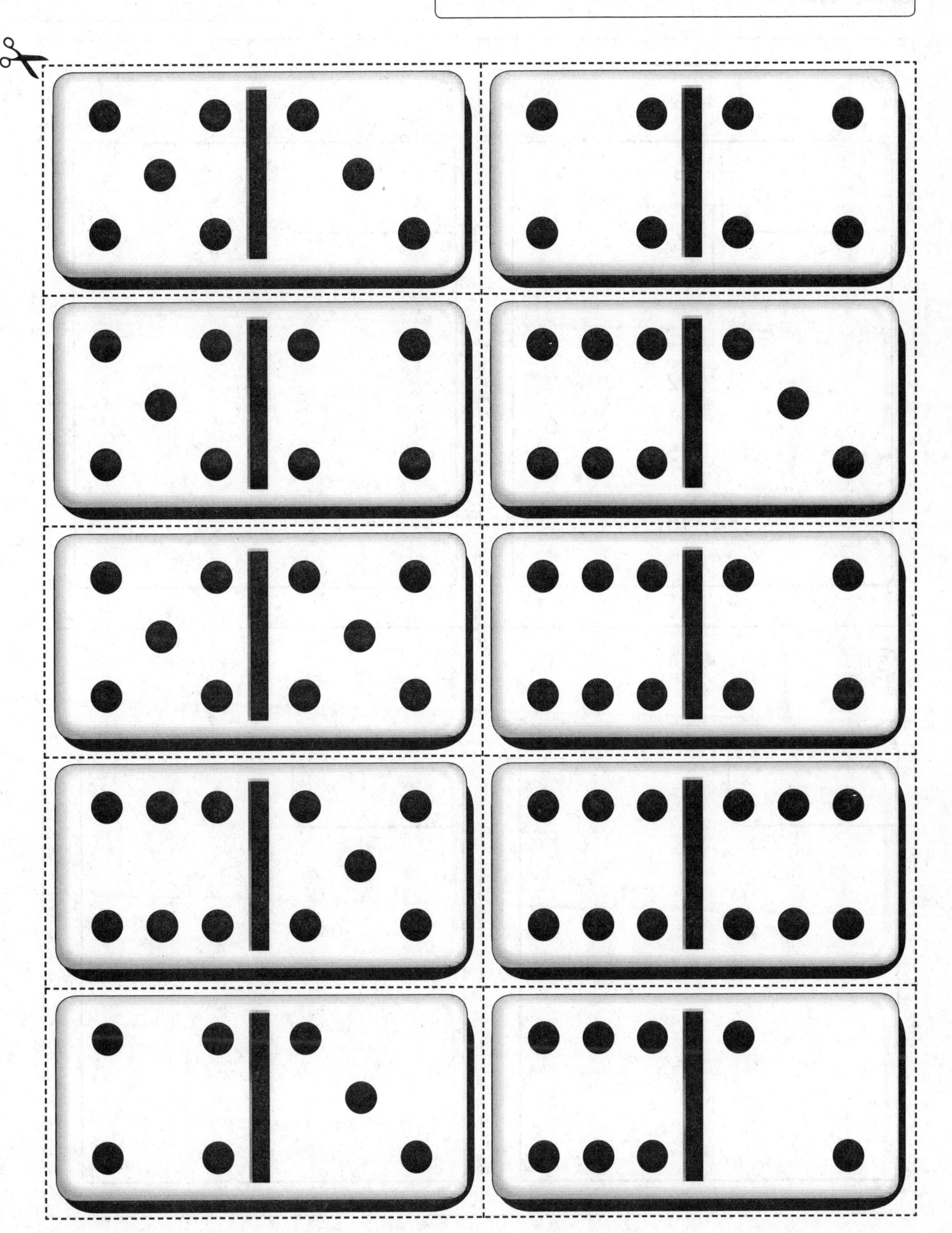

Name-Collection Boxes

NAME DATE TIME

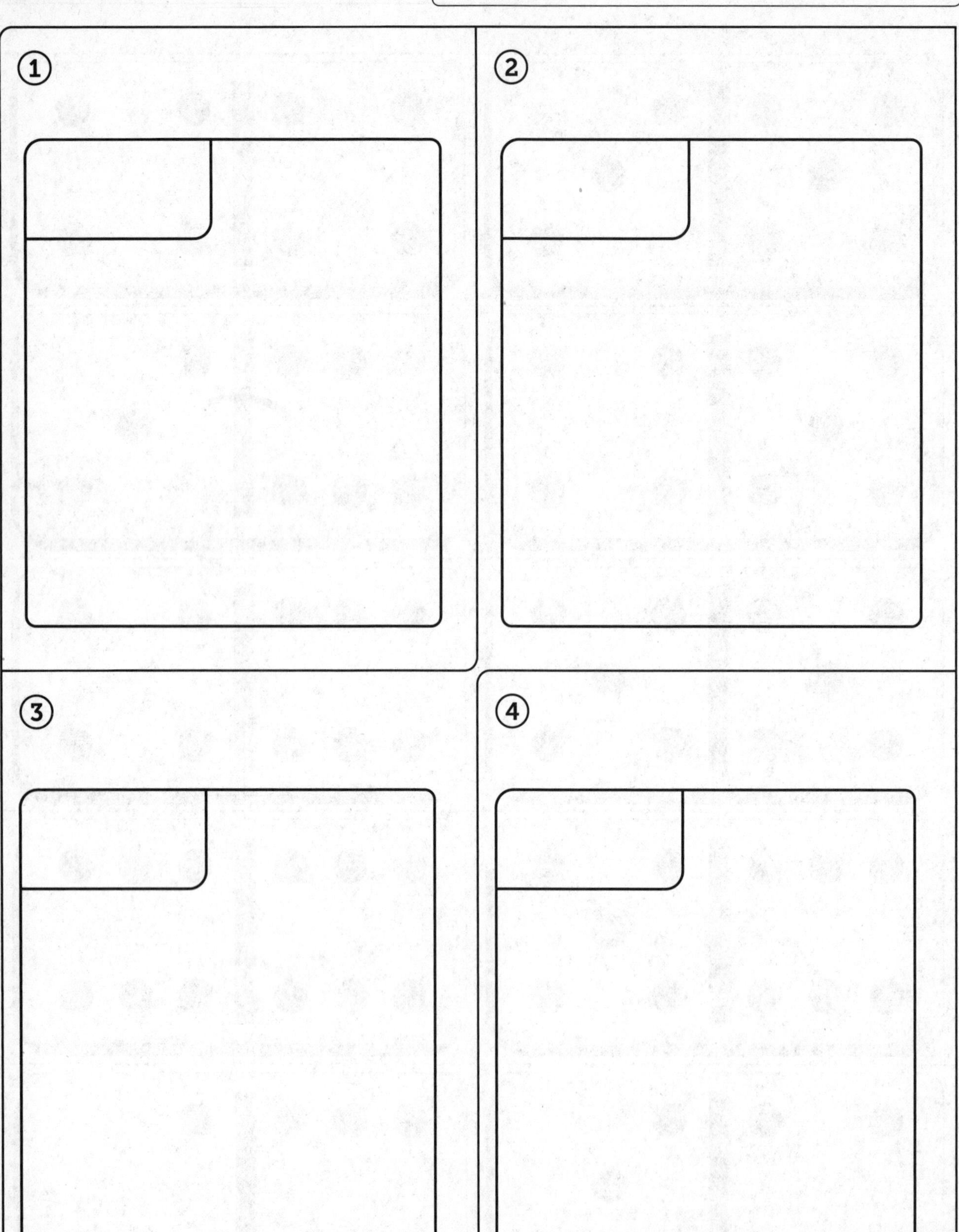

Two-Rule Frames and Arrows

NAME DATE TIME

Solve the Frames-and-Arrows problems.

1
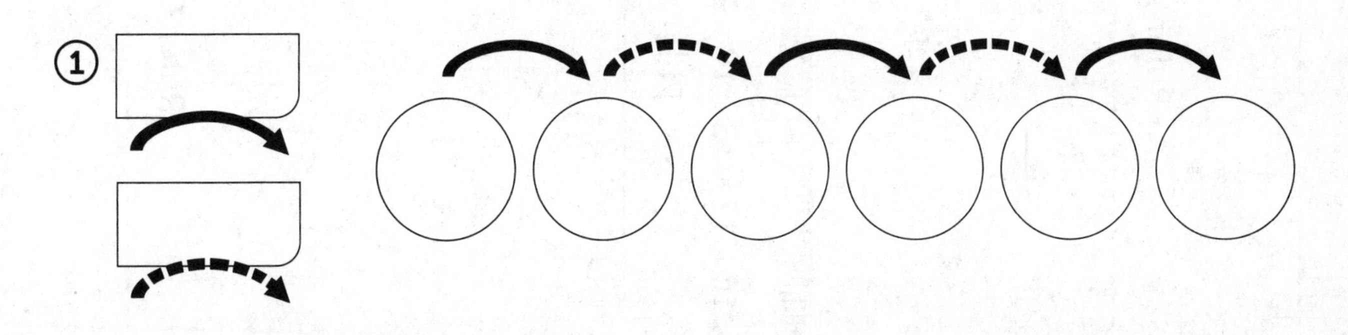

2
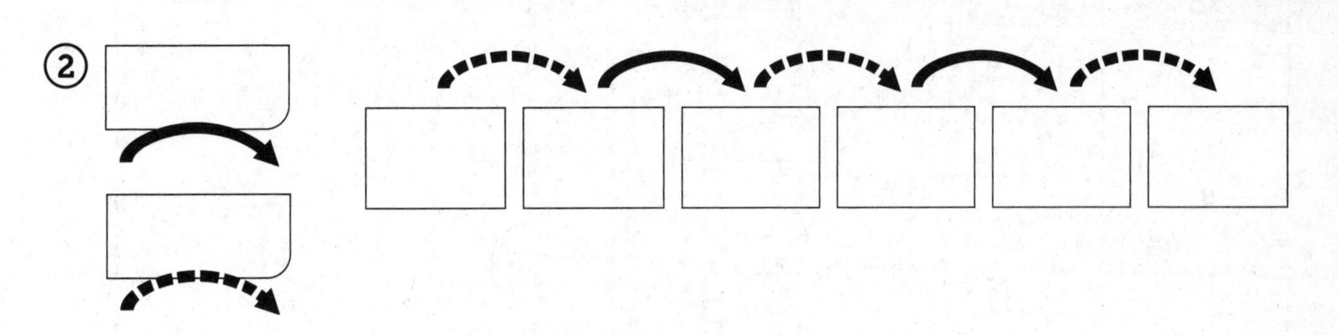

3
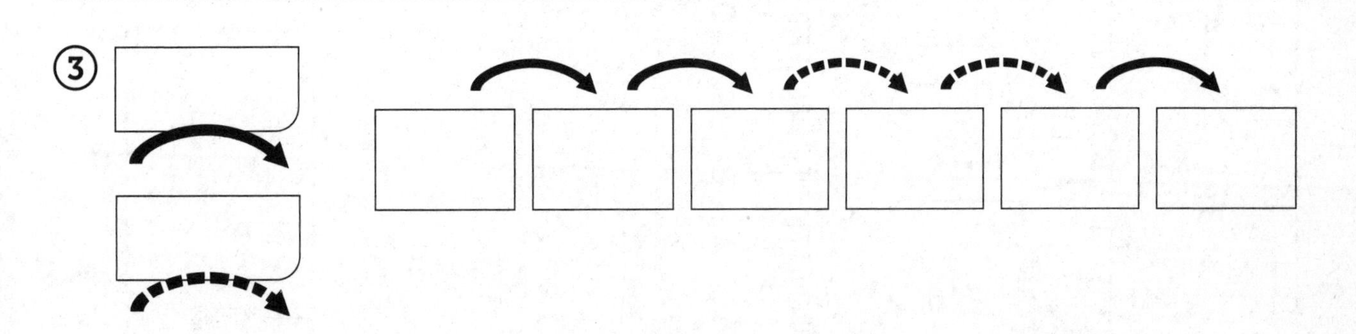

4
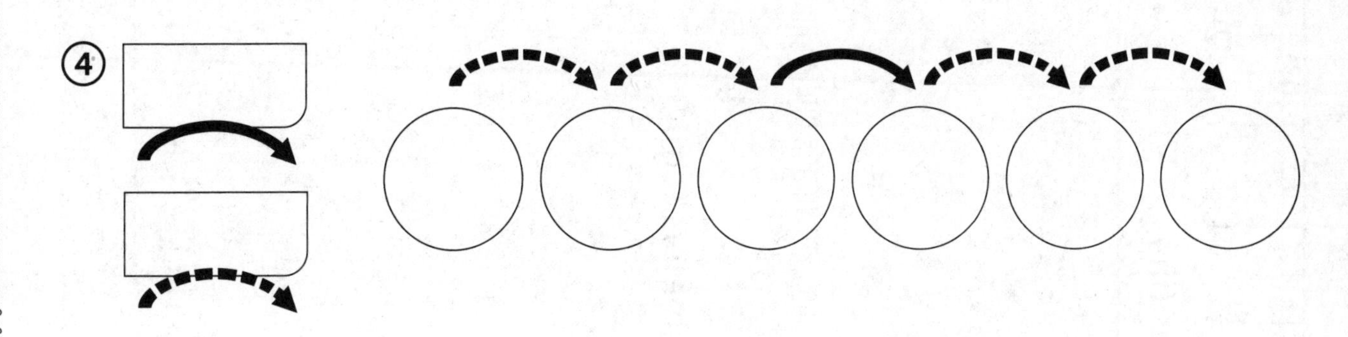

Rulers

NAME DATE TIME

Ruler A

Inches (in.)

0 1/2 1 2 3 4 5 6 7 8

Ruler B

Centimeters (cm)

0 1 2 3 4 5 6 7 8 9 10 11 12 13 14 15 16 17 18 19 20

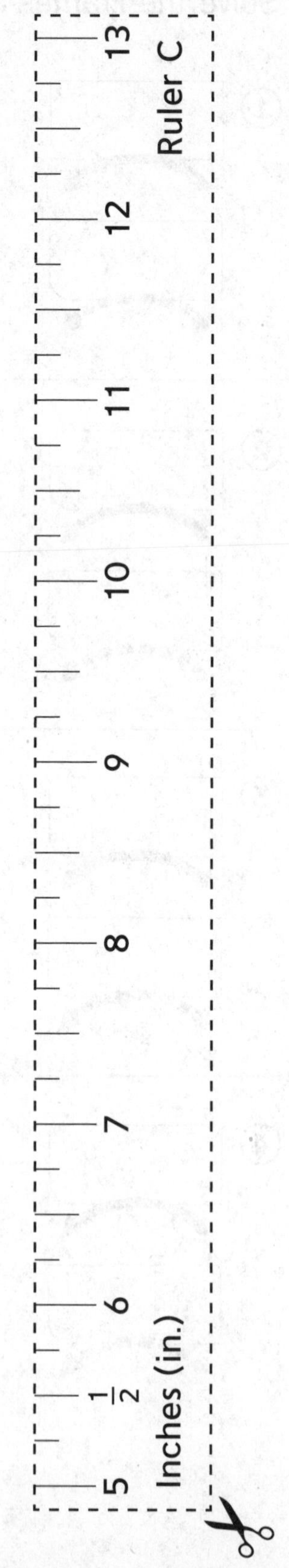

Geoboard Dot Paper

NAME DATE TIME

①

②

③

④

⑤

⑥

Shape Attribute Cards

NAME DATE TIME

exactly 4 equal sides	4 angles	4 sides
1 right angle and 4 sides	2 right angles	4 right angles
some equal length sides	1 pair of parallel sides	2 pairs of parallel sides

Dot Paper

NAME DATE TIME

10 × 10 Grid

NAME DATE TIME

My Exit Slip

NAME DATE TIME

Copyright © McGraw-Hill Education. Permission is granted to reproduce for classroom use.

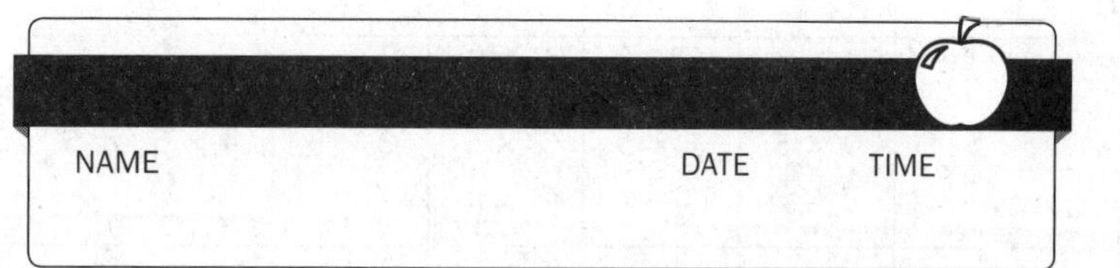

My Exit Slip

NAME DATE TIME

Multiplication Facts Chart

NAME DATE TIME

×	0	1	2	3	4	5	6	7	8	9	10
0	0 × 0	1 × 0	2 × 0	3 × 0	4 × 0	5 × 0	6 × 0	7 × 0	8 × 0	9 × 0	10 × 0
1	0 × 1	1 × 1	2 × 1	3 × 1	4 × 1	5 × 1	6 × 1	7 × 1	8 × 1	9 × 1	10 × 1
2	0 × 2	1 × 2	2 × 2	3 × 2	4 × 2	5 × 2	6 × 2	7 × 2	8 × 2	9 × 2	10 × 2
3	0 × 3	1 × 3	2 × 3	3 × 3	4 × 3	5 × 3	6 × 3	7 × 3	8 × 3	9 × 3	10 × 3
4	0 × 4	1 × 4	2 × 4	3 × 4	4 × 4	5 × 4	6 × 4	7 × 4	8 × 4	9 × 4	10 × 4
5	0 × 5	1 × 5	2 × 5	3 × 5	4 × 5	5 × 5	6 × 5	7 × 5	8 × 5	9 × 5	10 × 5
6	0 × 6	1 × 6	2 × 6	3 × 6	4 × 6	5 × 6	6 × 6	7 × 6	8 × 6	9 × 6	10 × 6
7	0 × 7	1 × 7	2 × 7	3 × 7	4 × 7	5 × 7	6 × 7	7 × 7	8 × 7	9 × 7	10 × 7
8	0 × 8	1 × 8	2 × 8	3 × 8	4 × 8	5 × 8	6 × 8	7 × 8	8 × 8	9 × 8	10 × 8
9	0 × 9	1 × 9	2 × 9	3 × 9	4 × 9	5 × 9	6 × 9	7 × 9	8 × 9	9 × 9	10 × 9
10	0 × 10	1 × 10	2 × 10	3 × 10	4 × 10	5 × 10	6 × 10	7 × 10	8 × 10	9 × 10	10 × 10

Venn Diagram

NAME DATE TIME

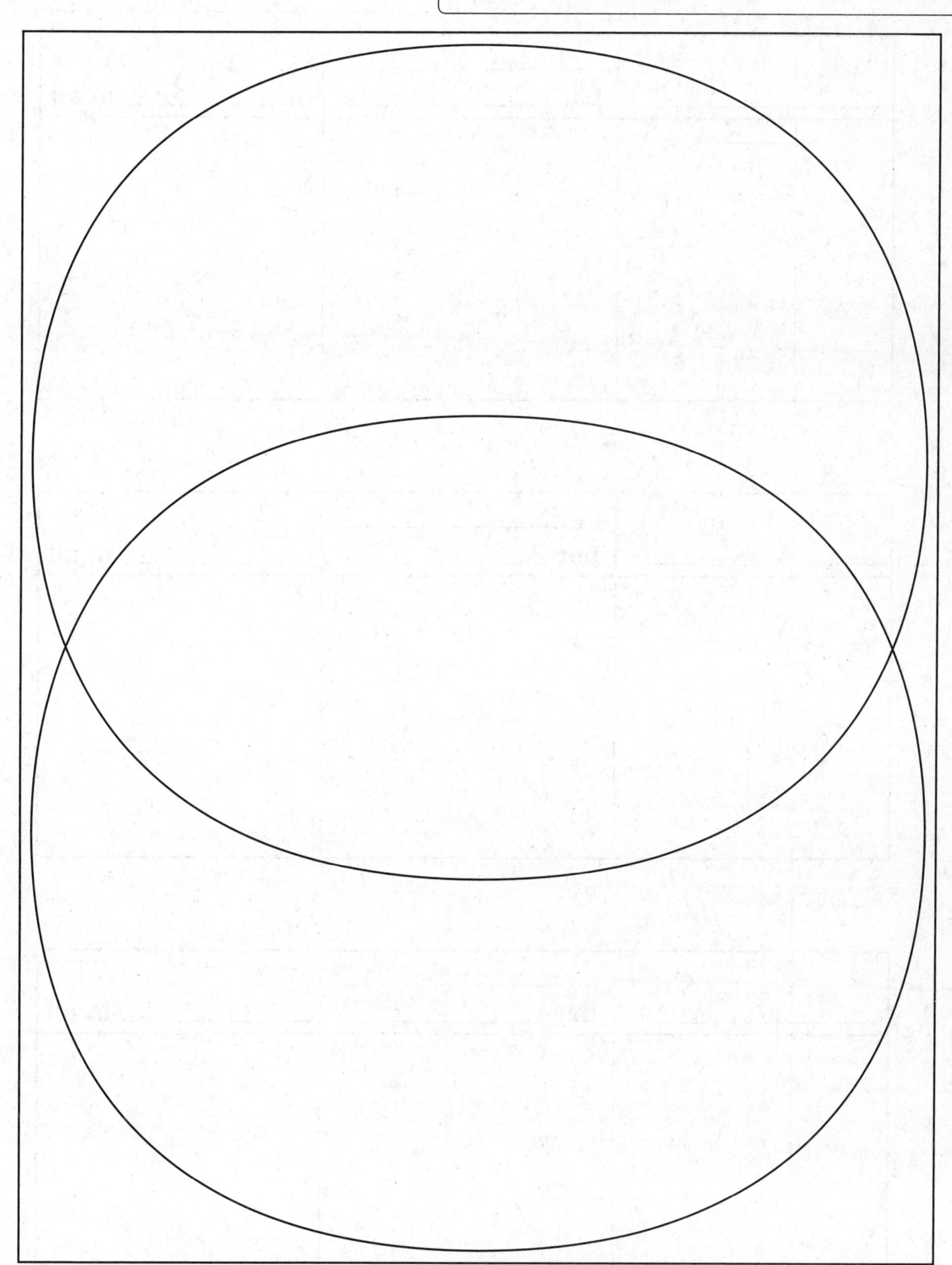

Multiplication/ Division Diagrams

NAME DATE TIME

_______	_______ per _______	_______ in all

_______	_______ per _______	_______ in all

_______	_______ per _______	_______ in all

Fraction Strips

NAME DATE TIME

Fraction Number-Line Poster

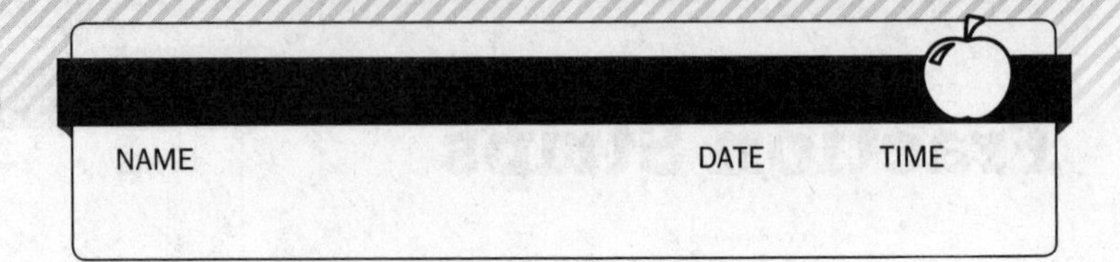

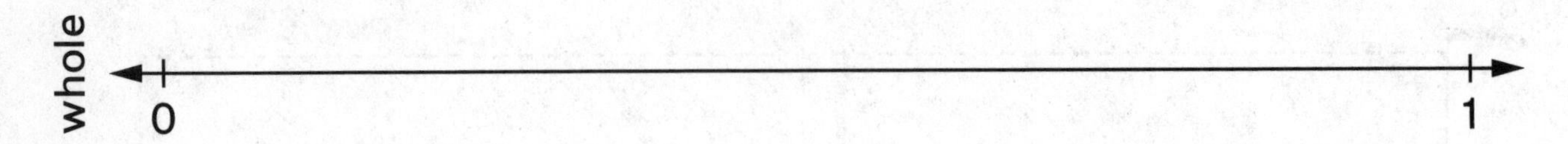

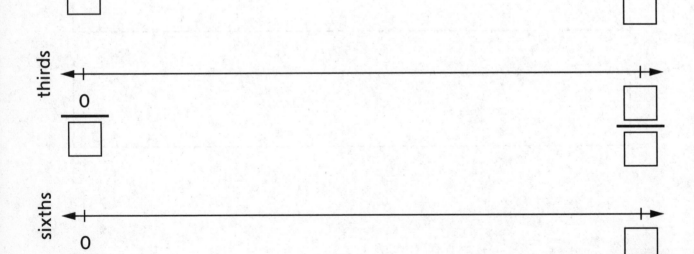

Fraction Strips

NAME DATE TIME

$\frac{1}{2}$	$\frac{1}{2}$

$\frac{1}{4}$	$\frac{1}{4}$	$\frac{1}{4}$	$\frac{1}{4}$

$\frac{1}{8}$	$\frac{1}{8}$	$\frac{1}{8}$	$\frac{1}{8}$	$\frac{1}{8}$	$\frac{1}{8}$	$\frac{1}{8}$	$\frac{1}{8}$

$\frac{1}{3}$	$\frac{1}{3}$	$\frac{1}{3}$

$\frac{1}{6}$	$\frac{1}{6}$	$\frac{1}{6}$	$\frac{1}{6}$	$\frac{1}{6}$	$\frac{1}{6}$

Child-Friendly Rubric

NAME DATE TIME

Goal:				
Meets Expectations	**Child 1**	**Child 2**	**Child 3**	
Exceeds Expectations				

A New Ruler

NAME DATE TIME

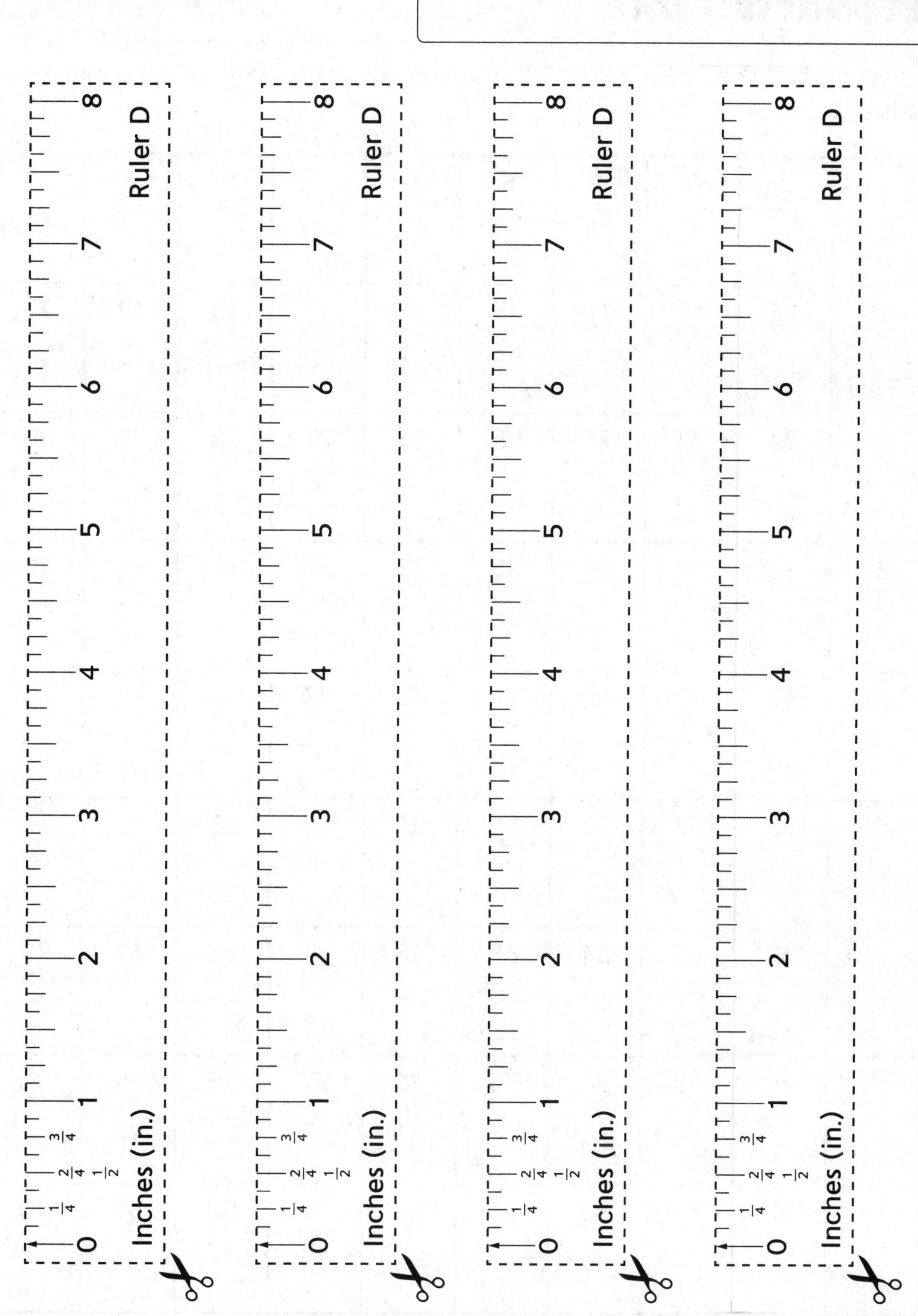

Class Winning-Products Table

NAME DATE TIME

Place a tally mark in each box of the winning products from a *Factor Bingo* game mat.

	2	3	4	5	6	7	8	9	10
11	12	13	14	15	16	17	18	19	20
21	22	23	24	25	26	27	28	29	30
31	32	33	34	35	36	37	38	39	40
41	42	43	44	45	46	47	48	49	50
51	52	53	54	55	56	57	58	59	60
61	62	63	64	65	66	67	68	69	70
71	72	73	74	75	76	77	78	79	80
81	82	83	84	85	86	87	88	89	90

$1 Bills

NAME DATE TIME

$1 Bills

NAME DATE TIME

$10 Bills

NAME DATE TIME

$10 Bills

NAME DATE TIME

Geoboard Dot Paper (10 × 10)

NAME DATE TIME

①

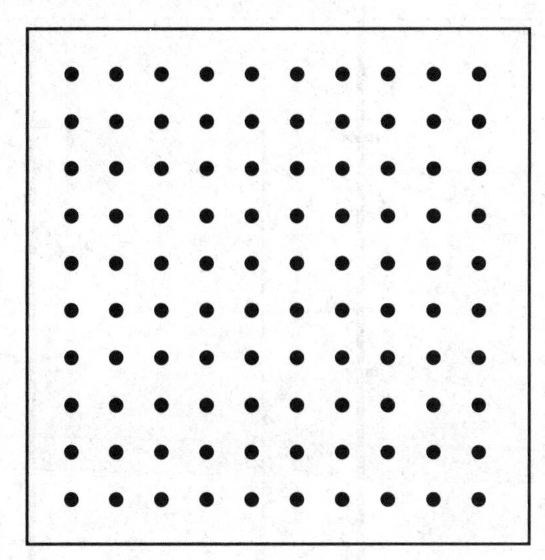

②

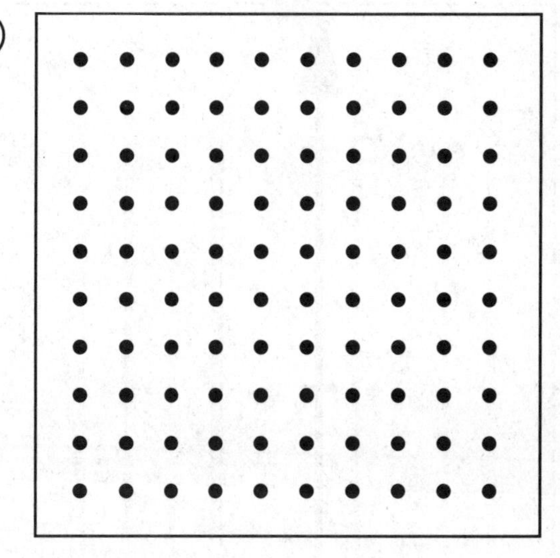

③

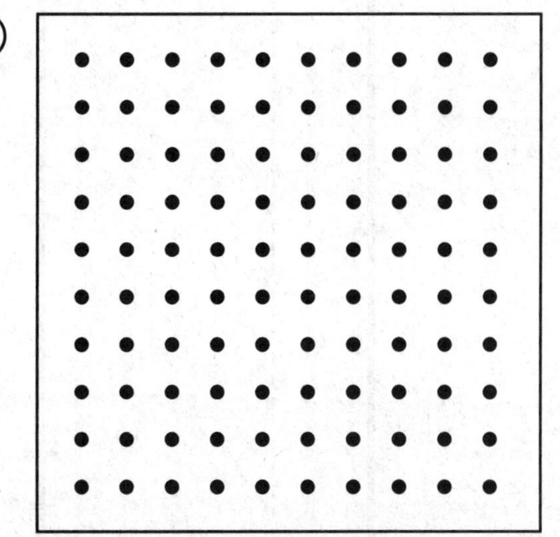

④

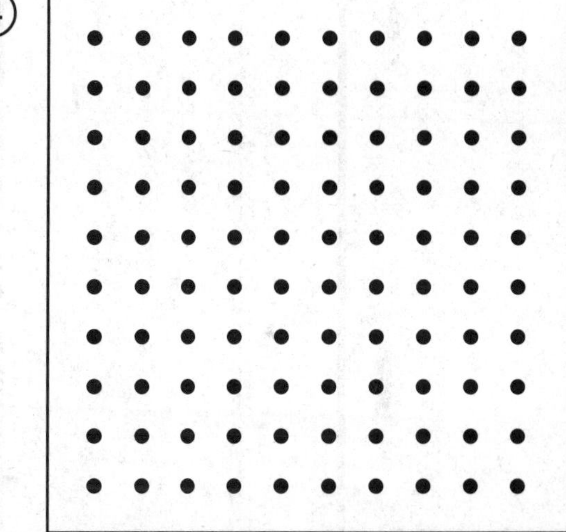

⑤

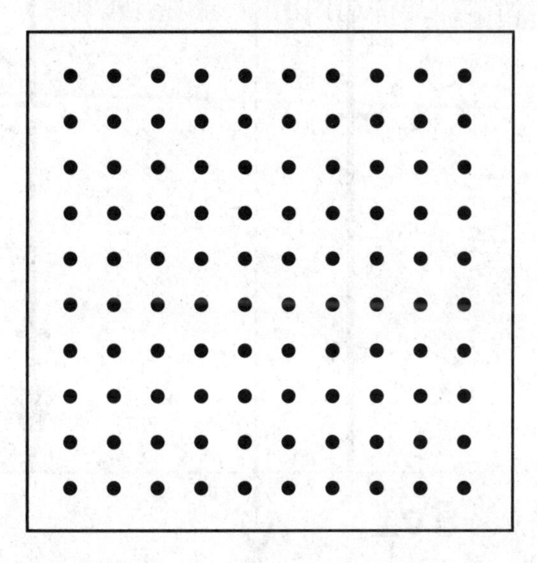

⑥

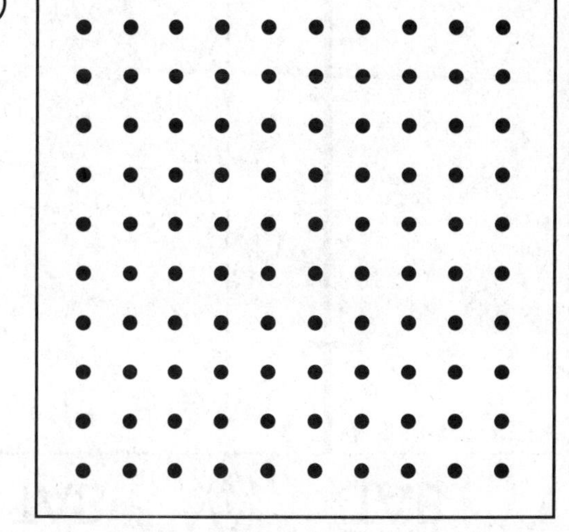

Number-Line Sections

NAME DATE TIME

0

TAB

TAB

TAB

Math Boxes A

NAME DATE TIME

①

②

③

④

Math Boxes B

NAME DATE TIME

①

②

③

④

⑤

⑥

Math Boxes C

NAME DATE TIME

① Fill in the unit.
Solve.

Unit

□ + □ = ___

□ = □ + ___

□ + ___ = □

___ = □ + □

② Write the number that is halfway between □ and □ on the number line.

Find and write □ on the number line.

What is □ rounded to the nearest 10? ___

③ Write the numbers that are 10 less and 10 more.

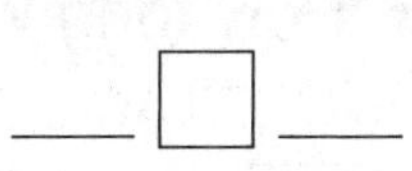

___ □ ___

___ □ ___

④ You may draw open number lines to help.

□ rounded to the nearest 100 is ___.

□ rounded to the nearest 100 is ___.

⑤ Find the difference between each pair of numbers. You may use a number grid.

□ and □ ___

⑥ Fill in the unit.
Solve.

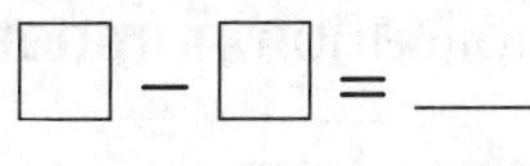

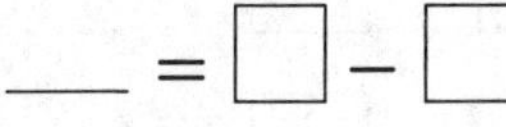

□ − ___ = □

Math Boxes D

NAME DATE TIME

① ☐ children share ☐ ☐ equally. How many ☐ does each child get? You may draw a picture to help.

Answer: ______ (unit)

② Arrange ☐ stars in an array. Write a multiplication number model for the array.

Number model: ______

③ Fill in the unit box. Make an estimate. Solve.

Unit

Estimate: ______

☐ ☐ ☐ = ______

Think: Does my answer make sense?

④ Amy arrived at the ☐ at ☐ : ☐☐.M. She left at ☐ : ☐☐.M. How long was she at the ☐?

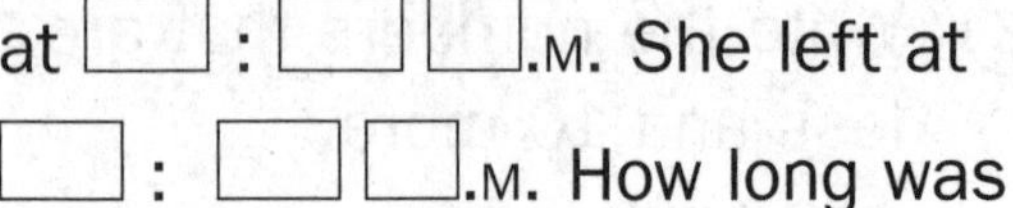

You may use your toolkit clock or draw an open number line to help.

Answer: ______ (unit)

⑤ Write a multiplication number sentence to fit:

- 2 groups of ☐: ______ × ______ = ______
- 5 groups of ☐: ______ × ______ = ______
- 10 groups of ☐: ______ × ______ = ______

Math Boxes E

NAME DATE TIME

① Estimate and Solve:

Unit

Estimate: ______

[ ] [ ] [ ] = ?

Answer: ______

Think: Does my answer make sense?

② Sketch two different quadrilaterals.

Write: On a separate piece of paper write two reasons why your shapes are quadrilaterals.

③ The perimeter of this rectangle is [ ] cm. Figure out the missing side lengths and label them.

______ cm

[ ] cm [ ] cm

______ cm

④ Draw the time on the clock.

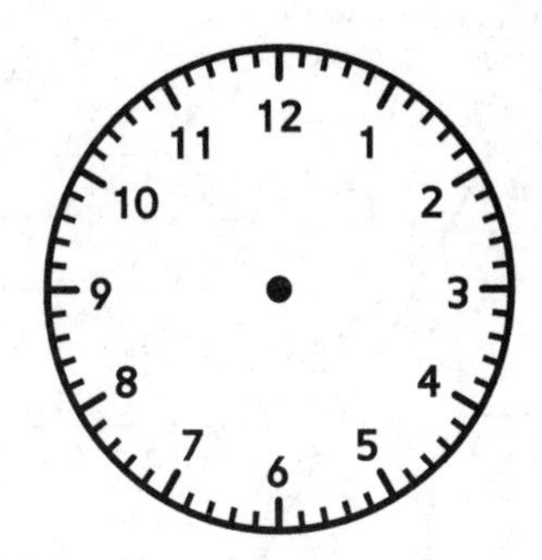

What time will it be in: [ ]

Answer: ______

⑤ Find the area.

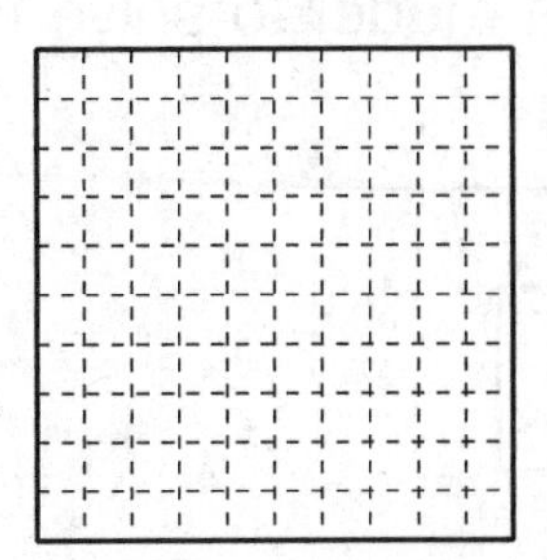

Number sentences: ______

______

______

Area: ______ square ______

⑥ On a separate piece of paper: Draw a path. Use a tool to measure the length of your path to the nearest $\frac{1}{2}$ inch.

Length of my path: ______ inches.

Then draw a path [ ] inches long.

Math Boxes F

NAME DATE TIME

① Partition the number line into equal parts. Write equivalent fractions for 0, 1, and 2.

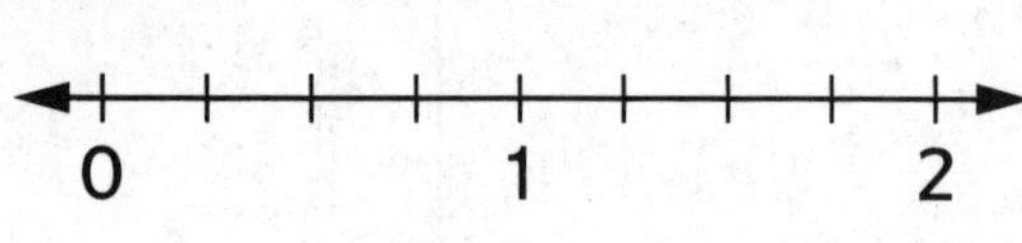

② Partition the rectangle into ☐ equal parts.

Write a unit fraction inside each part.

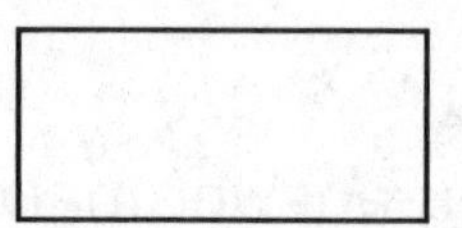

Shade ☐ of the rectangle.

③ Fill in the Fact Triangle. Write the fact family.

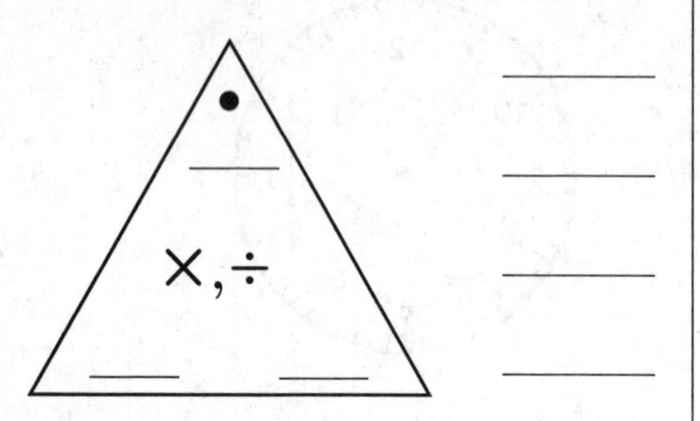

______

______

______

______

④ Philip has ☐ packs of juice boxes. Each pack has ☐ boxes of apple juice and ☐ boxes of berry punch. How many juice boxes does he have in all?

Write an equation to fit the story.

The letter ______ stands for the number of boxes of juice.

Equation: ______

Answer: ______ boxes

⑤ ☐ blocks were used to build ☐ towers.

How many blocks per tower?

towers	blocks per tower	blocks in all

____

⑥ Use an area model to solve ☐ × ☐.

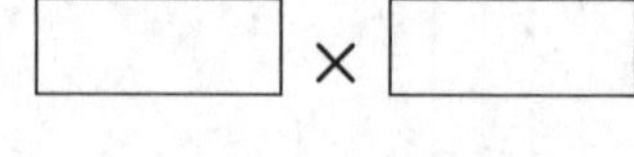

Number Model: ______

Answer: ______

Ten Frame

NAME DATE TIME

Double Ten Frame

NAME DATE TIME

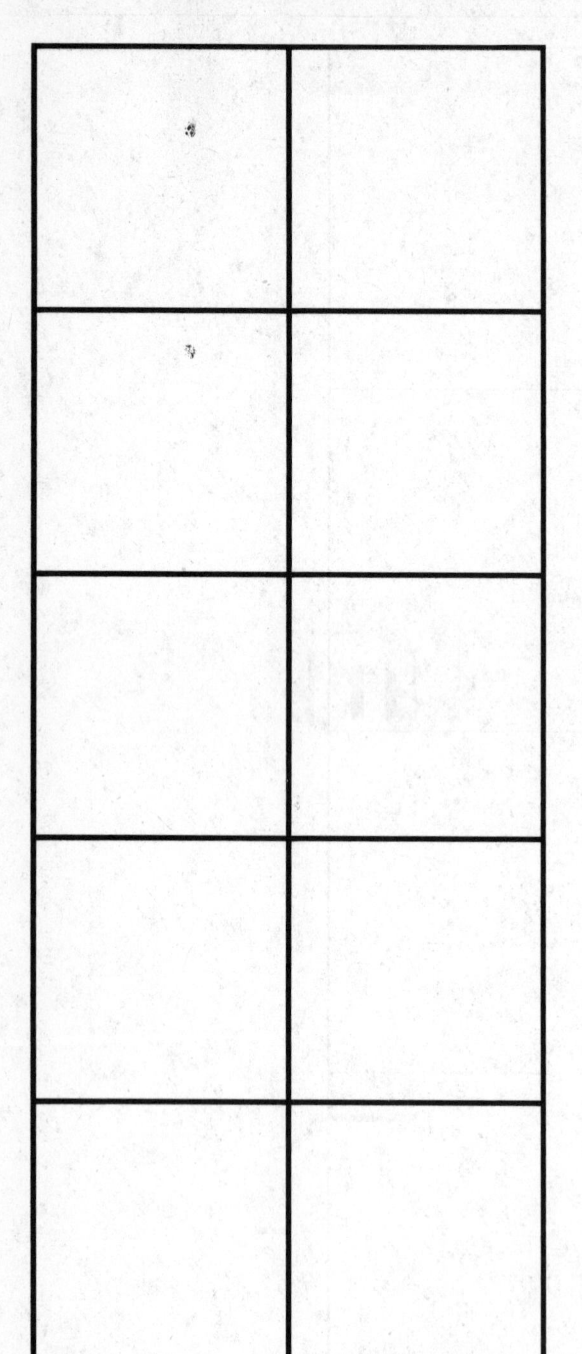

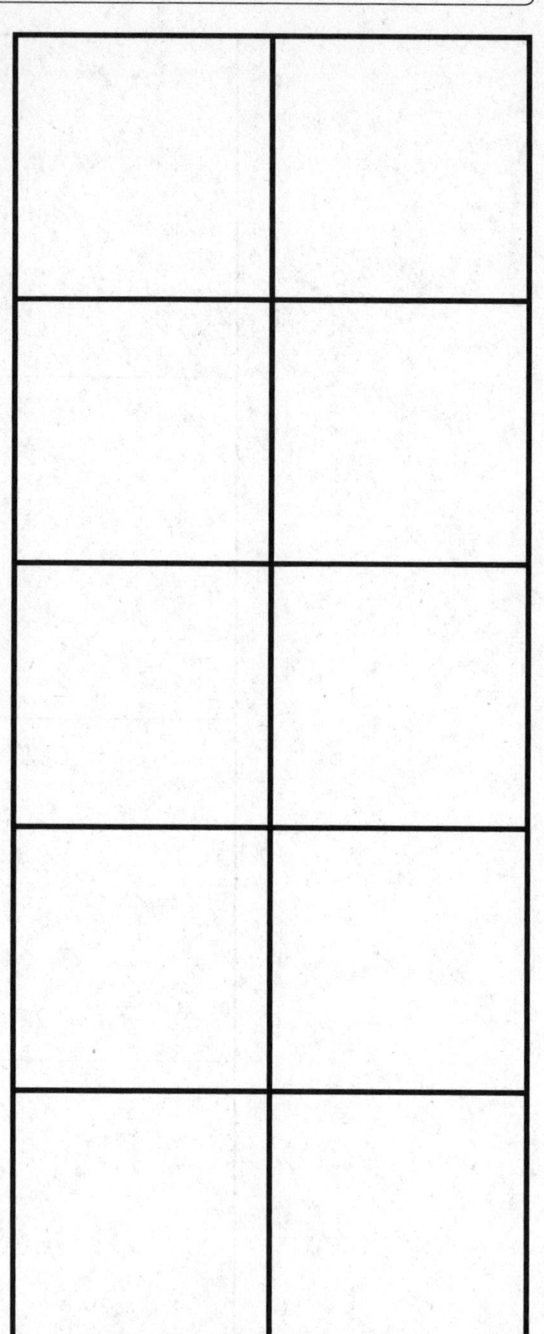

Parts-and-Total Diagram

NAME DATE TIME

Total	
Part	Part

Change Diagrams

NAME DATE TIME

Start

End

Start

End

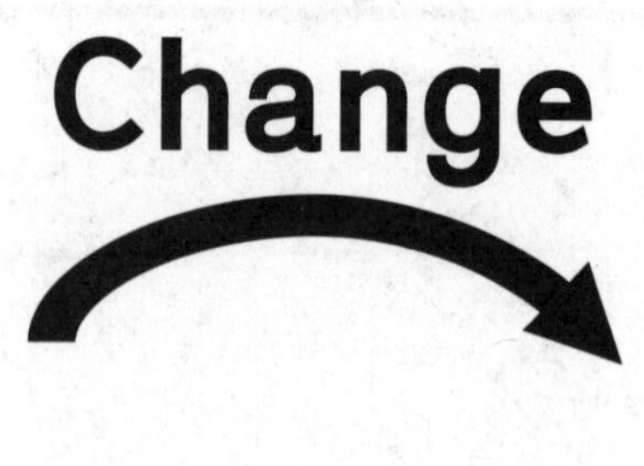

Start

End

Comparison Diagrams

NAME DATE TIME

Quantity

Quantity

Difference

Quantity

Quantity

Difference

Fact Triangles

NAME DATE TIME

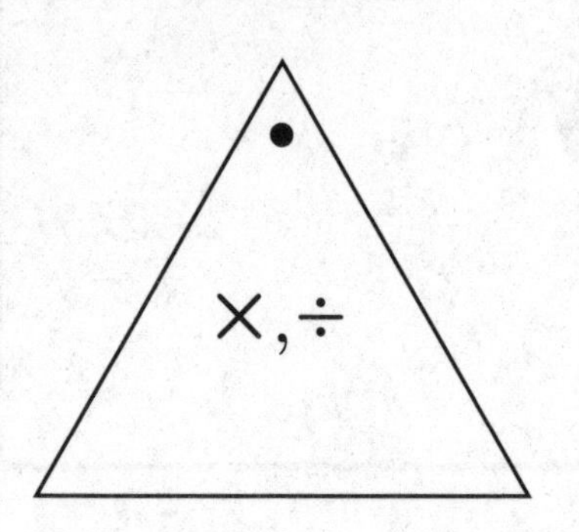

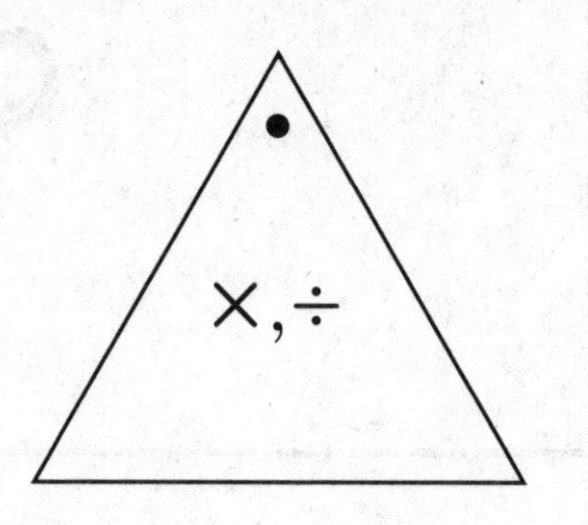

______ × ______ = ______

______ × ______ = ______

______ ÷ ______ = ______

______ ÷ ______ = ______

______ × ______ = ______

______ × ______ = ______

______ ÷ ______ = ______

______ ÷ ______ = ______

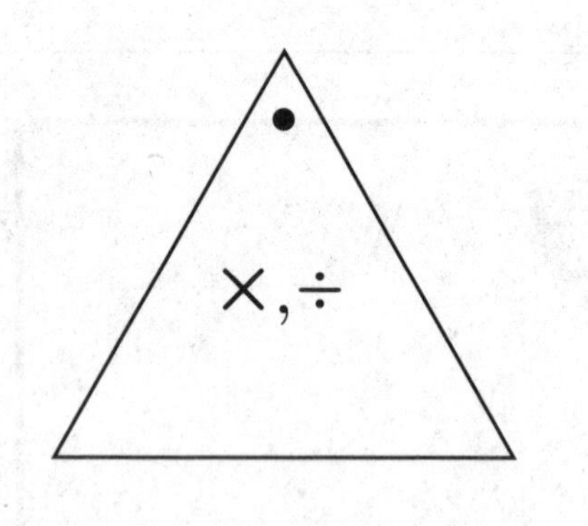

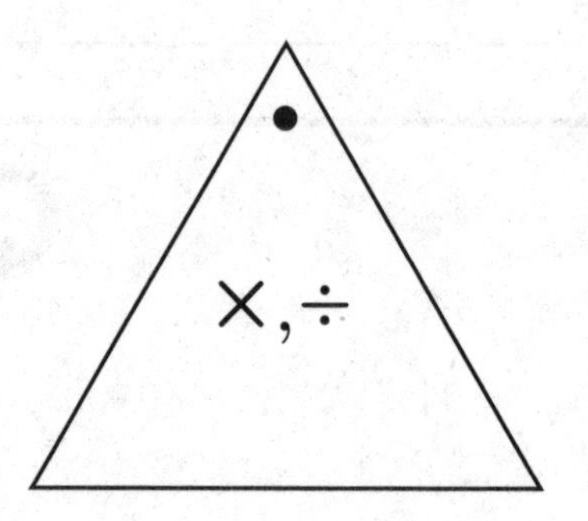

______ × ______ = ______

______ × ______ = ______

______ ÷ ______ = ______

______ ÷ ______ = ______

______ × ______ = ______

______ × ______ = ______

______ ÷ ______ = ______

______ ÷ ______ = ______

Trade-First Template

NAME DATE TIME

100s	10s	1s

Array Dot Paper

NAME DATE TIME

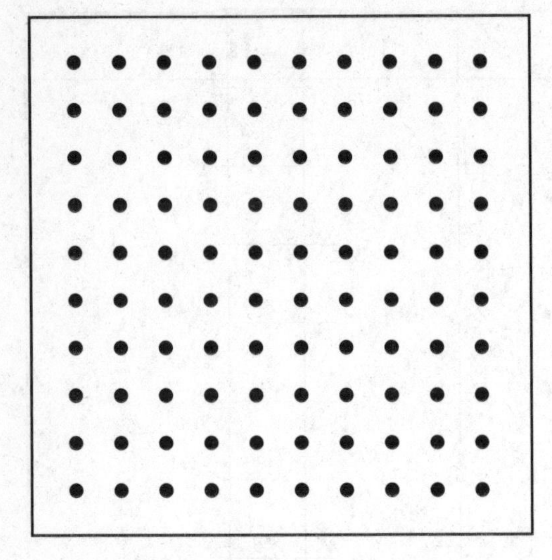

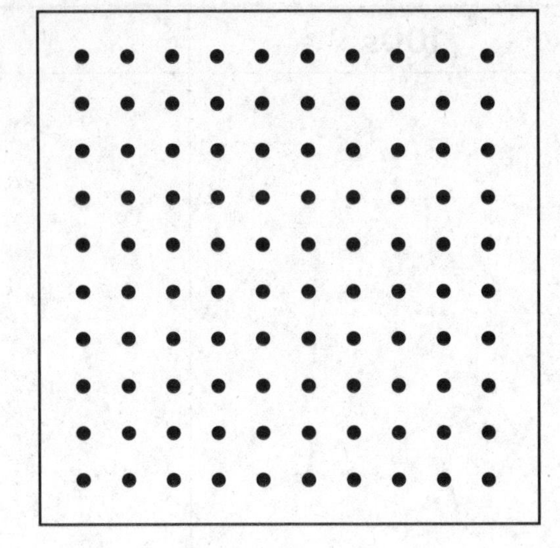

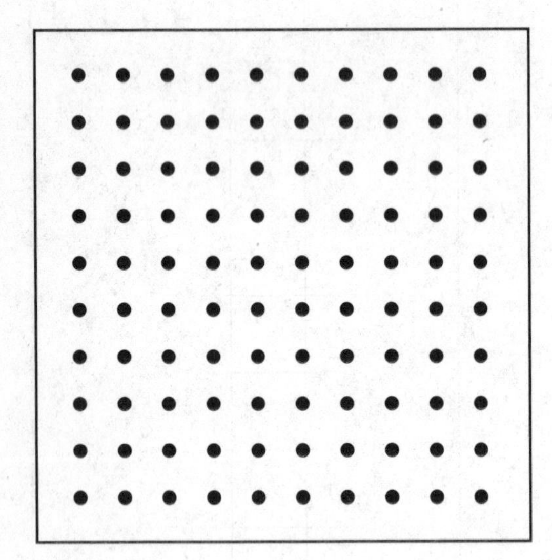

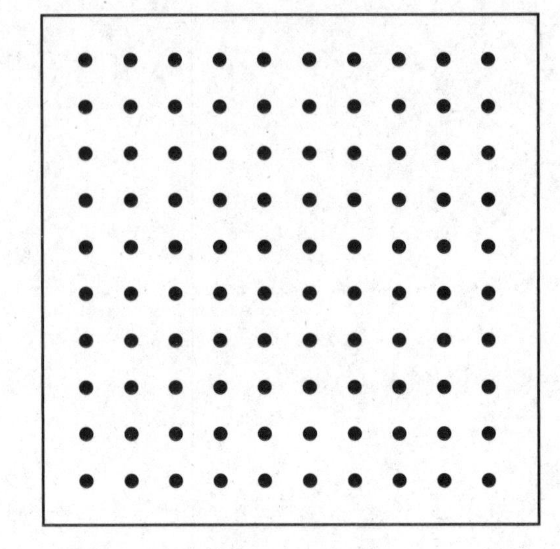

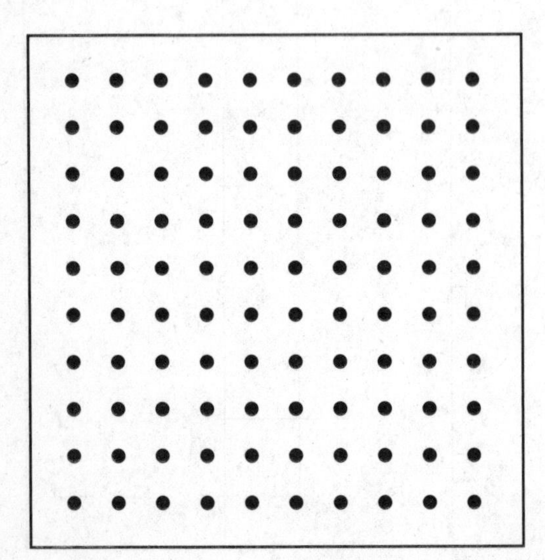

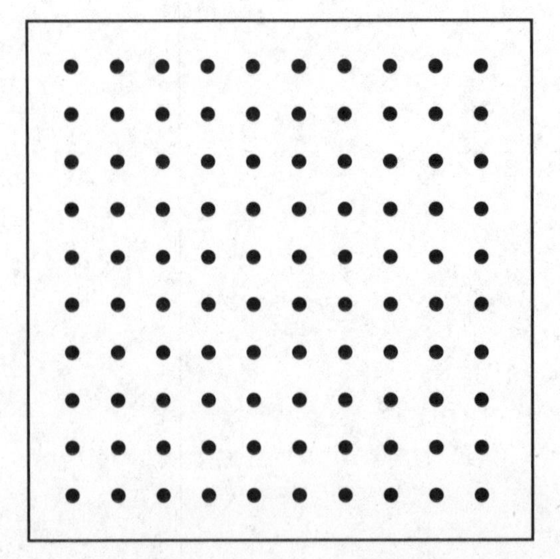

Computation Grid

NAME DATE TIME

Fraction Circle Pieces 1

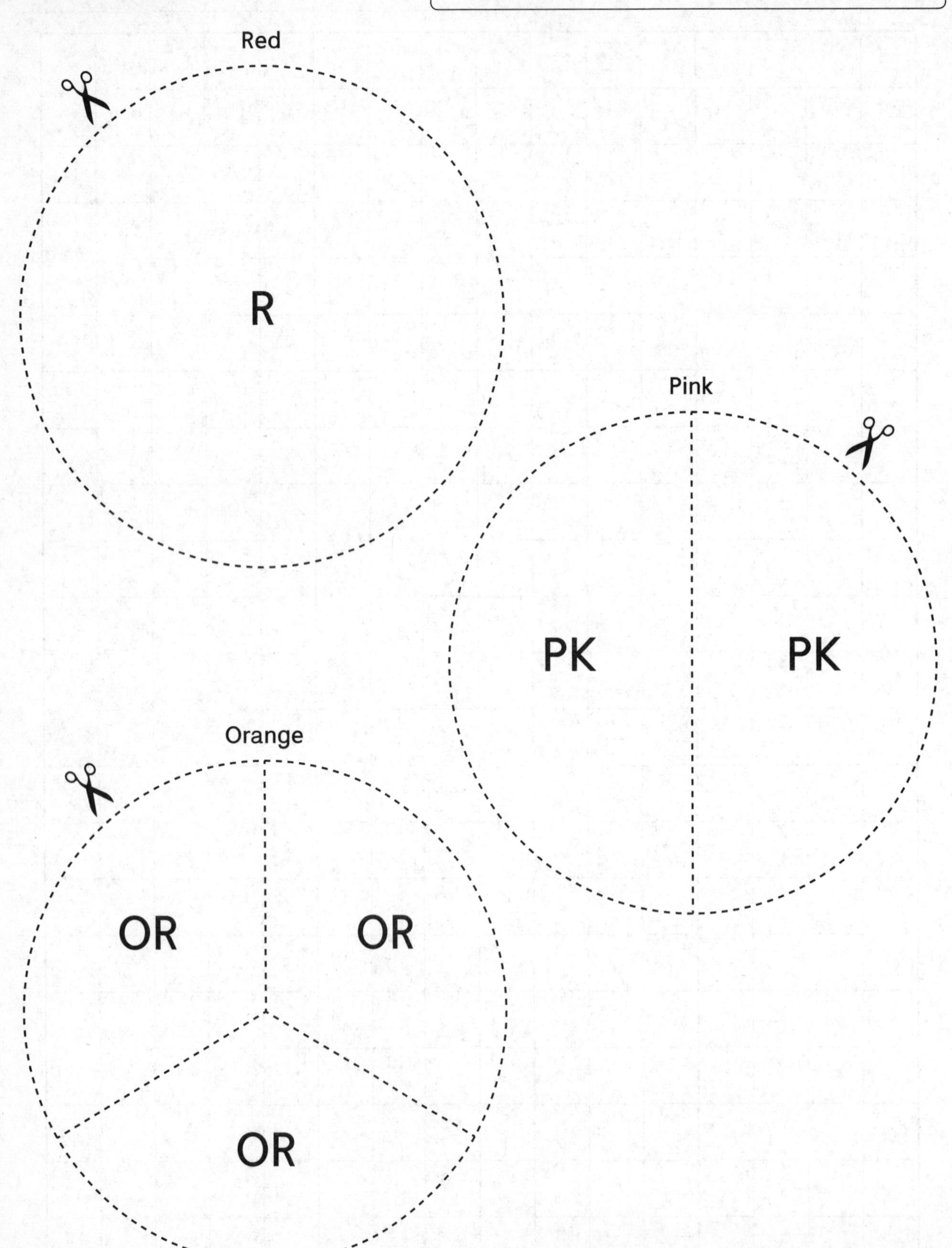

Fraction Circle Pieces 2

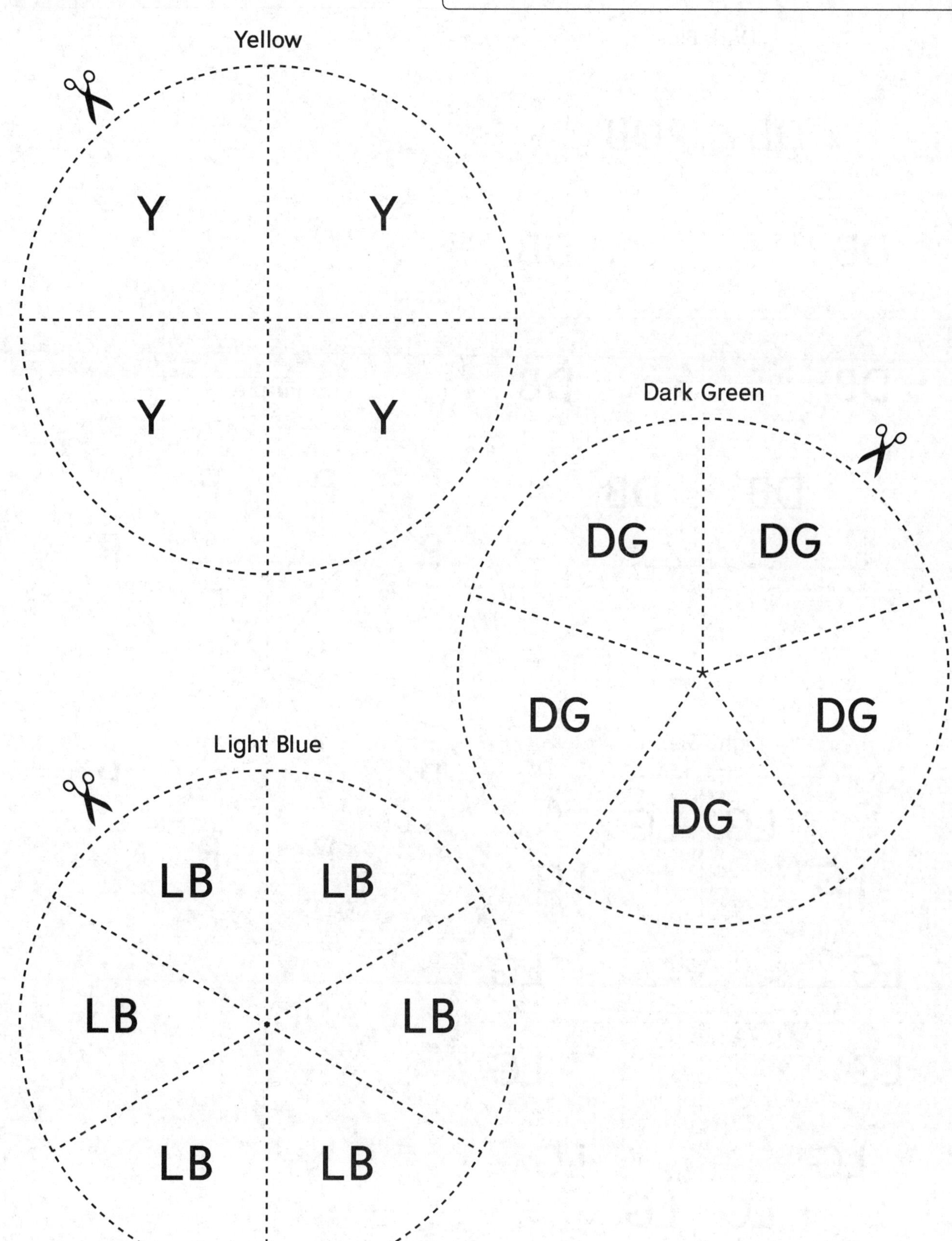

Fraction Circle Pieces 3

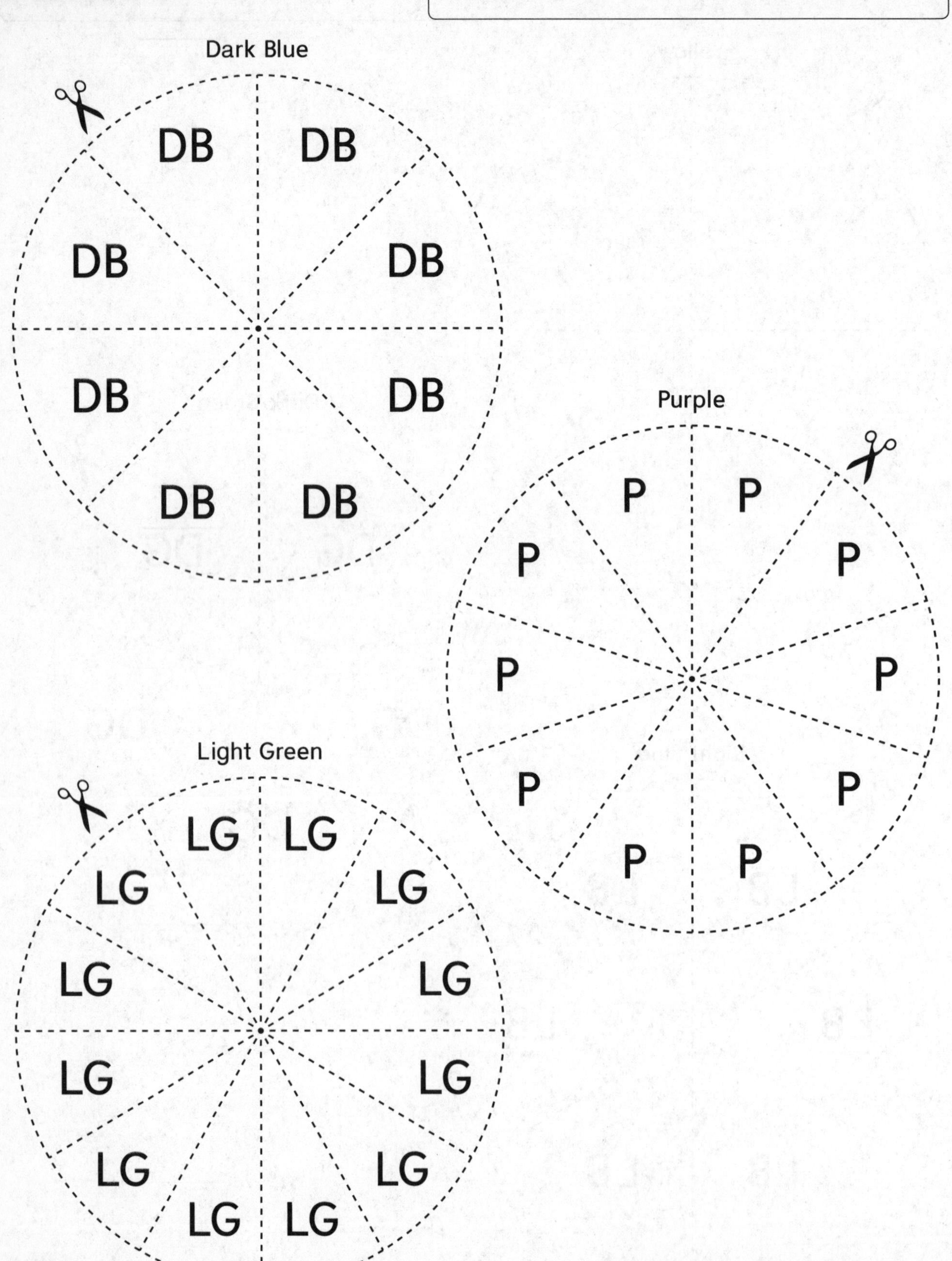

A Number Story

NAME DATE TIME

Unit

Relation Symbols

NAME DATE TIME

$<$ means “is less than.”

$>$ means “is more than.”

$=$ means “is the same as.”

$=$ means “is equal to.”

Game Masters

Number-Grid Difference Record Sheet

NAME DATE TIME

My Record Sheet

Round	My Number	My Partner's Number	Difference (Score)
1	______	______	______
2	______	______	______
3	______	______	______
4	______	______	______
5	______	______	______

Total: ______

Hit the Target Record Sheet

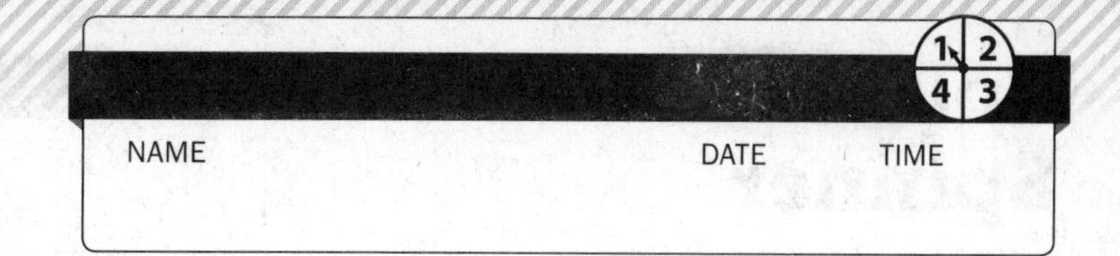

Record three rounds of *Hit the Target.*

Example Round:

Target number: 40

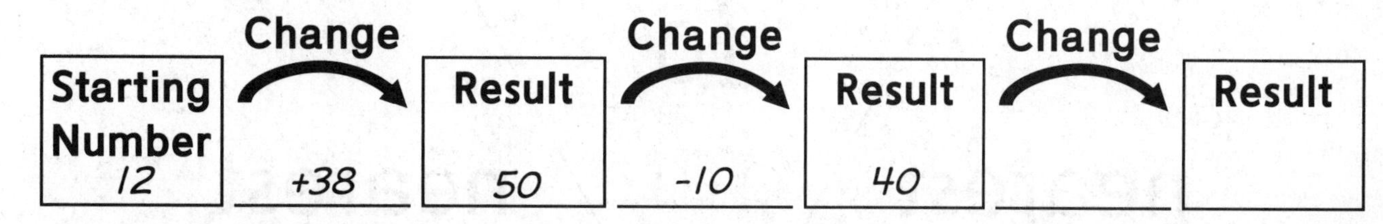

Round 1

Target number: ______

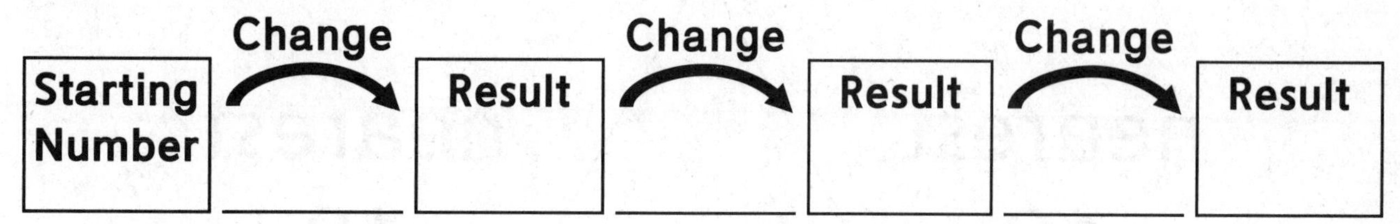

Round 2

Target number: ______

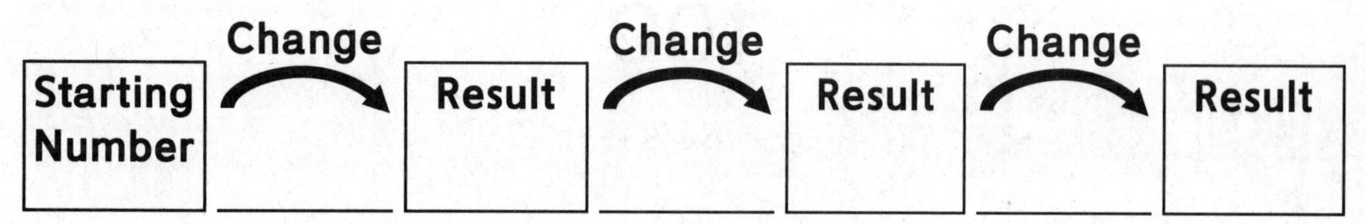

Round 3

Target number: ______

Starting Number	Change	Result	Change	Result	Change	Result

Spin and Round Spinner

NAME DATE TIME

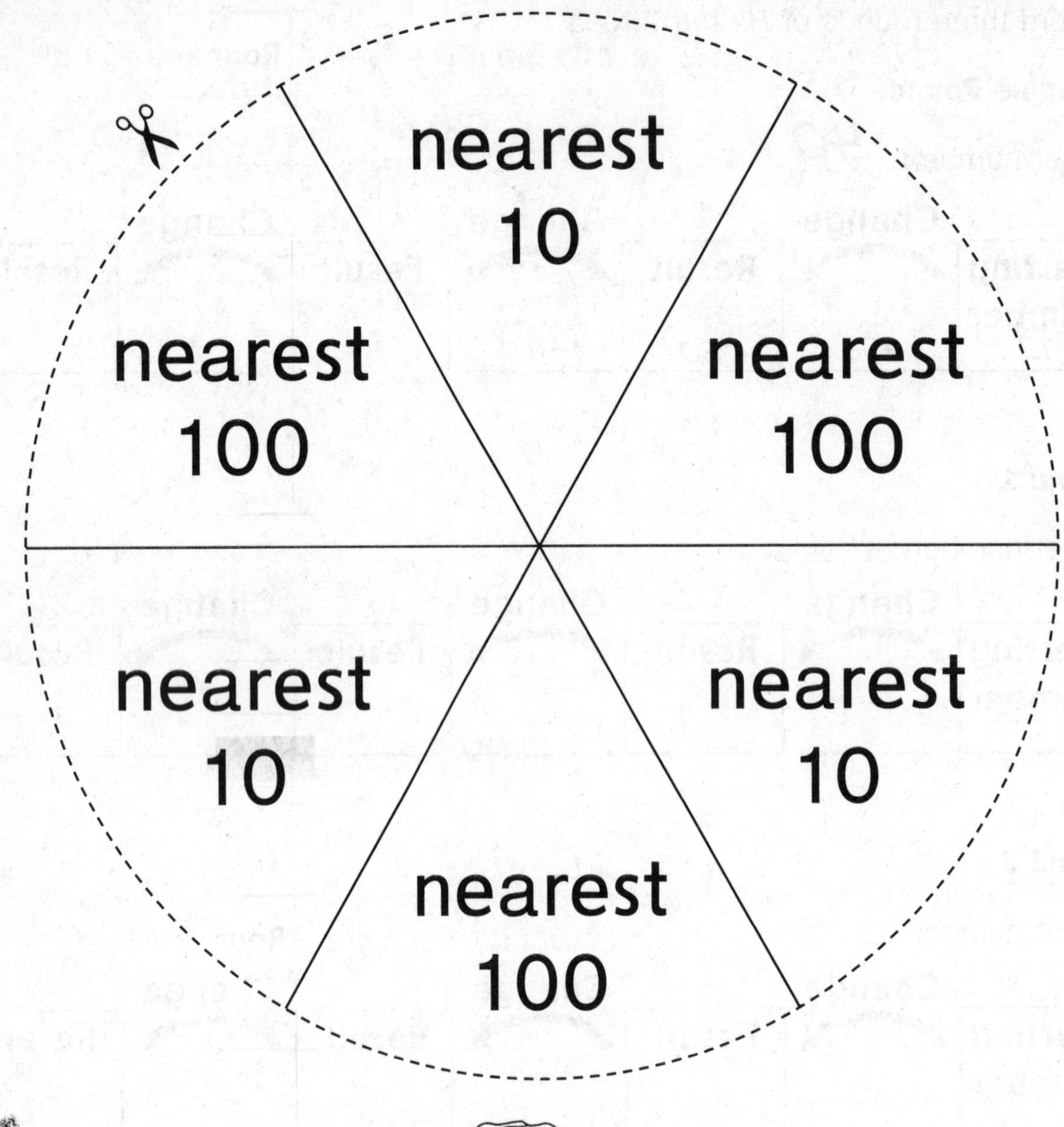

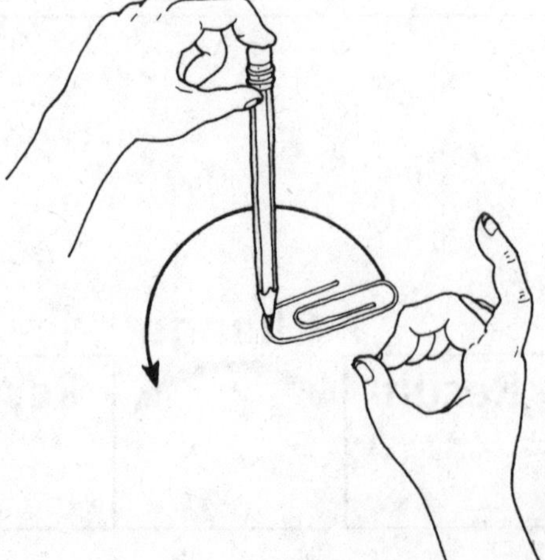

Use a large paper clip and pencil to make a spinner.

Spin and Round Record Sheet

Turn	Nearest 10? Nearest 100?	Original Number	Rounded Number
Example	10	536	540
1			
2			
3			
4			
5			

TOTAL ____________

Spin and Round Record Sheet

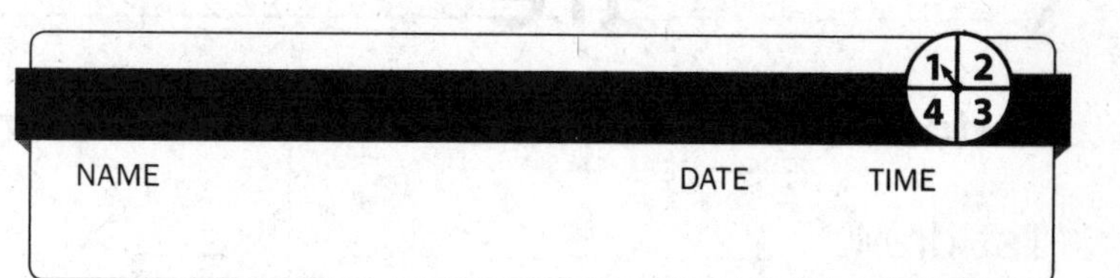

Turn	Nearest 10? Nearest 100?	Original Number	Rounded Number
Example	10	536	540
1			
2			
3			
4			
5			

TOTAL ____________

Multiplication Draw Record Sheet

NAME DATE TIME

Partner 1	Round 1	Round 2	Round 3
1st draw:	___ × ___ = ___	___ × ___ = ___	___ × ___ = ___
2nd draw:	___ × ___ = ___	___ × ___ = ___	___ × ___ = ___
3rd draw:	___ × ___ = ___	___ × ___ = ___	___ × ___ = ___
4th draw:	___ × ___ = ___	___ × ___ = ___	___ × ___ = ___
5th draw:	___ × ___ = ___	___ × ___ = ___	___ × ___ = ___
Sum of products:	______	______	______

Partner 2	Round 1	Round 2	Round 3
1st draw:	___ × ___ = ___	___ × ___ = ___	___ × ___ = ___
2nd draw:	___ × ___ = ___	___ × ___ = ___	___ × ___ = ___
3rd draw:	___ × ___ = ___	___ × ___ = ___	___ × ___ = ___
4th draw:	___ × ___ = ___	___ × ___ = ___	___ × ___ = ___
5th draw:	___ × ___ = ___	___ × ___ = ___	___ × ___ = ___
Sum of products:	______	______	______

Copyright © McGraw-Hill Education. Permission is granted to reproduce for classroom use.

Roll to 1,000 Record Sheet

NAME DATE TIME

Write your score at the end of each turn. The first player to reach or pass 1,000 wins.

Turn	Player 1 ______	Player 2 ______	Player 3 ______	Player 4 ______
1				
2				
3				
4				
5				
6				
7				
8				
9				
10				
11				
12				
13				
14				
15				

Continue recording scores on the back of this page.

Array Bingo Cards

NAME DATE TIME

2 × 2	2 × 3	2 × 4	4 × 4
2 × 5	2 × 6	3 × 5	6 × 3
3 × 3	1 × 5	4 × 3	3 × 6
5 × 3	6 × 1	4 × 5	5 × 4

Division Arrays Record Sheet

NAME DATE TIME

Number card	Die	Array formed	Number model	Leftovers?	Score

Total Score: ______

Shuffle to 100 Record Sheet

NAME DATE TIME

Record your combinations and scores while playing *Shuffle to 100.*

						Score
Round 1:	______	+	______	=	______	______
Round 2:	______	+	______	=	______	______
Round 3:	______	+	______	=	______	______
Round 4:	______	+	______	=	______	______

Total Score: ______

Shuffle to 100 Record Sheet

Record your combinations and scores while playing *Shuffle to 100.*

						Score
Round 1:	______	+	______	=	______	______
Round 2:	______	+	______	=	______	______
Round 3:	______	+	______	=	______	______
Round 4:	______	+	______	=	______	______

Total Score: ______

Shuffle to 1,000 Record Sheet

NAME DATE TIME

Record your combinations and scores while playing *Shuffle to 1,000.*

						Score
Round 1:	____	+	____	=	____	____
Round 2:	____	+	____	=	____	____
Round 3:	____	+	____	=	____	____
Round 4:	____	+	____	=	____	____

Total Score: ____

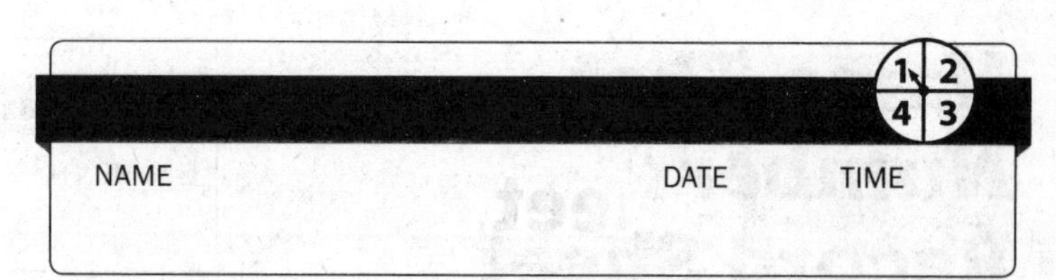

Shuffle to 1,000 Record Sheet

NAME DATE TIME

Record your combinations and scores while playing *Shuffle to 1,000.*

						Score
Round 1:	____	+	____	=	____	____
Round 2:	____	+	____	=	____	____
Round 3:	____	+	____	=	____	____
Round 4:	____	+	____	=	____	____

Total Score: ____

Copyright © McGraw-Hill Education. Permission is granted to reproduce for classroom use.

Name That Number Record Sheet

NAME DATE TIME

Target Number →

Number Sentence Solution

Reminder: Write each step separately.

Name That Number Record Sheet

NAME DATE TIME

Target Number →

Number Sentence Solution

Reminder: Write each step separately.

Copyright © McGraw-Hill Education. Permission is granted to reproduce for classroom use.

Shape Cards 1

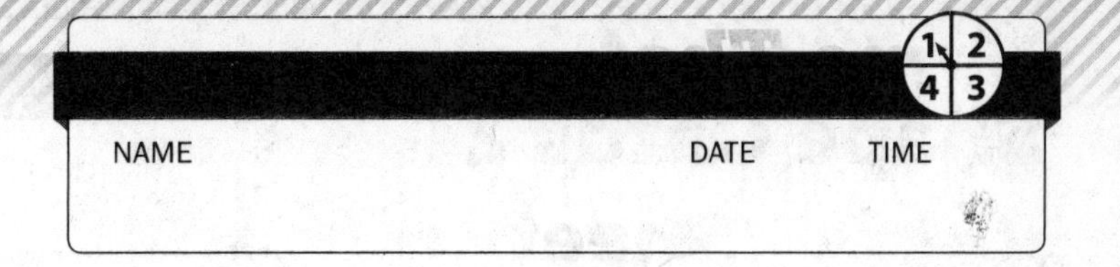

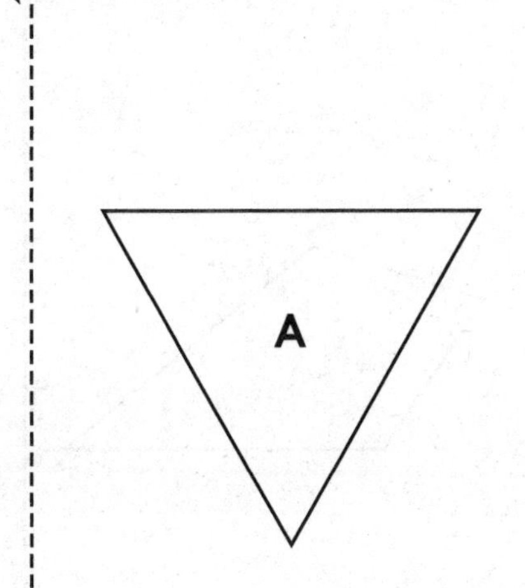

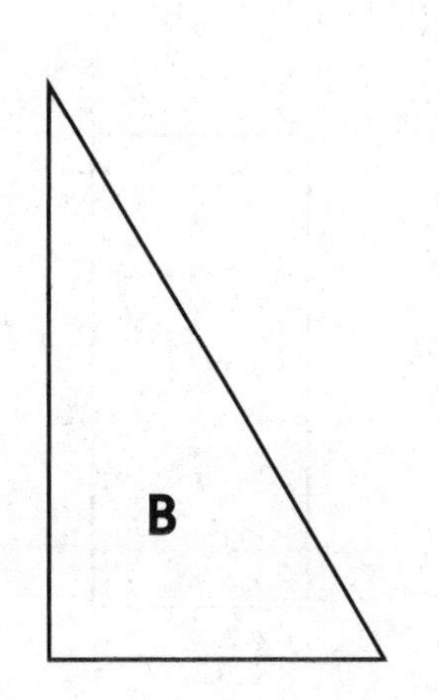

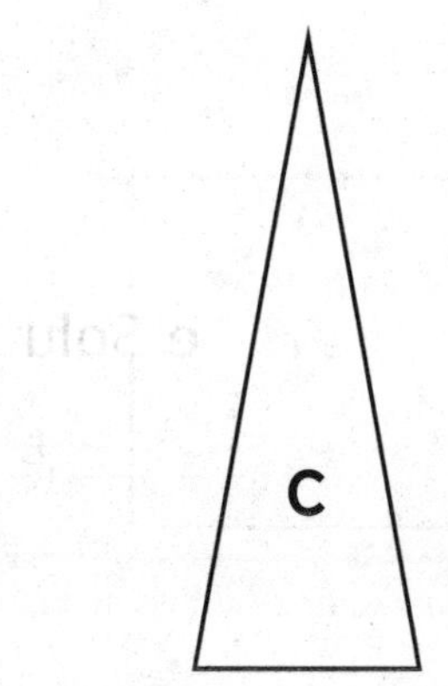

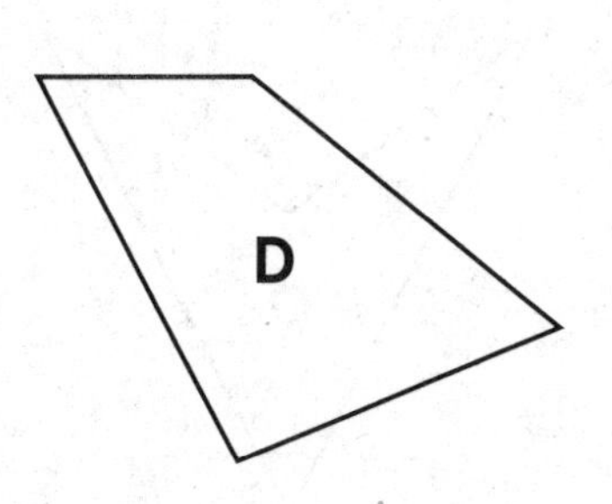

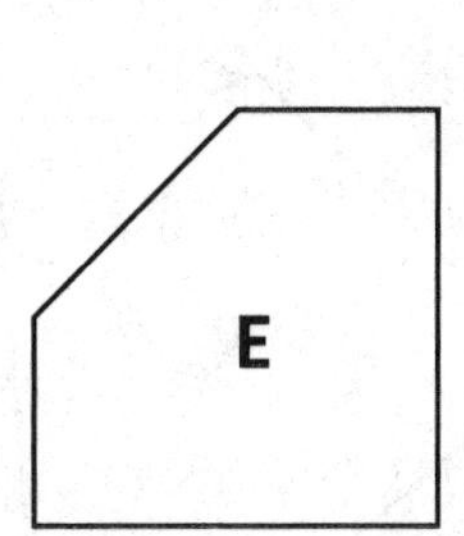

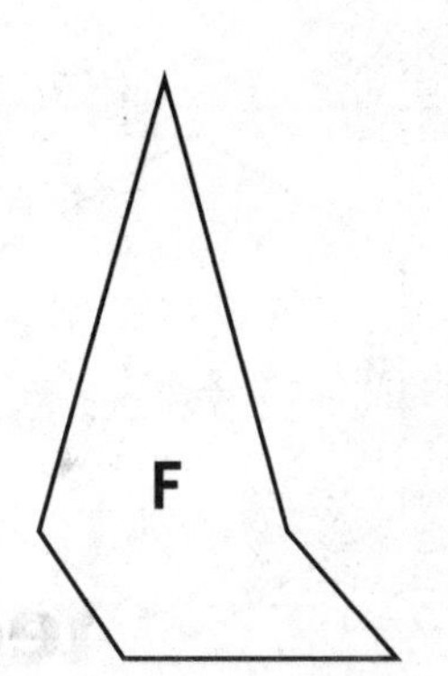

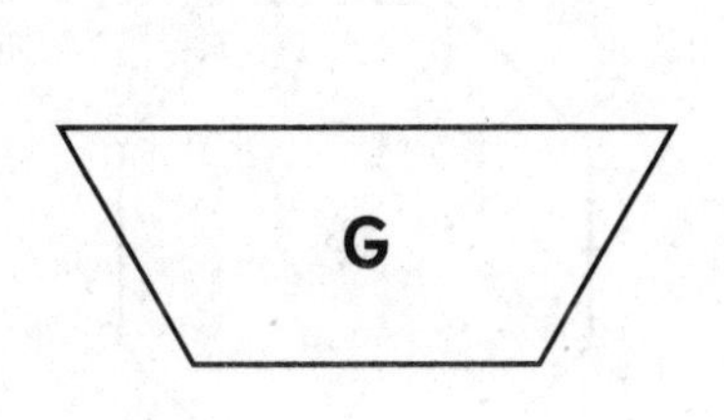

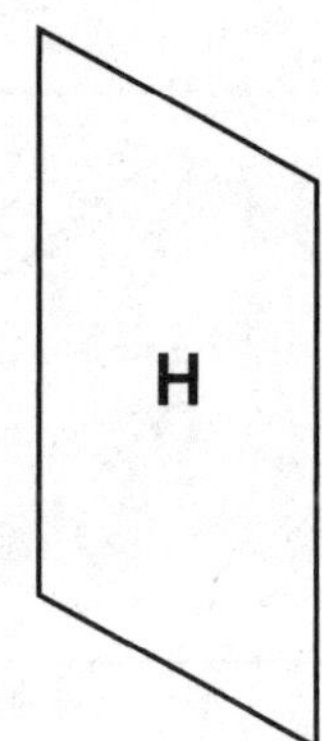

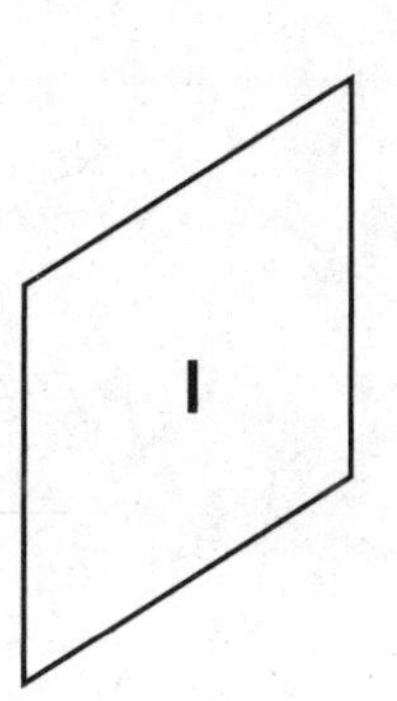

Shape Cards 2

NAME DATE TIME

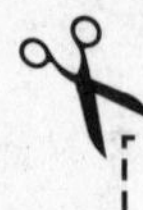

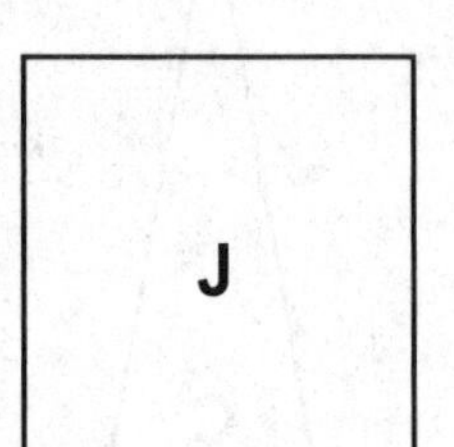

K

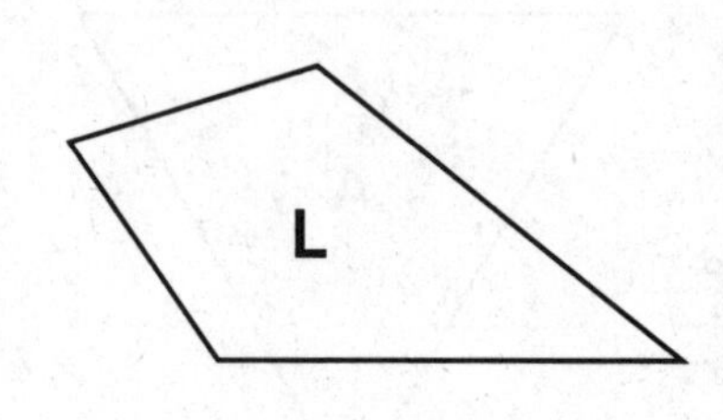

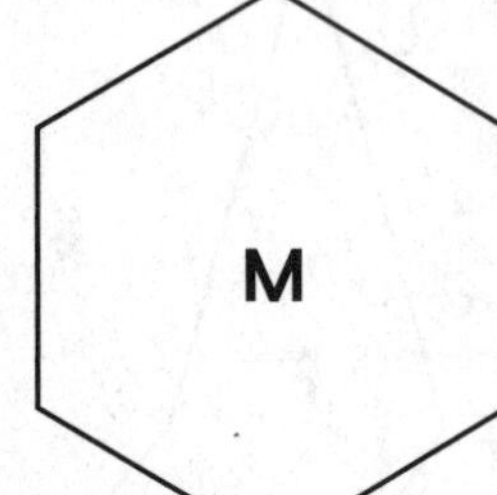

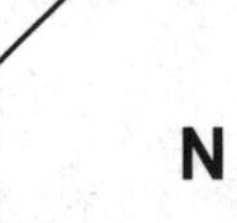

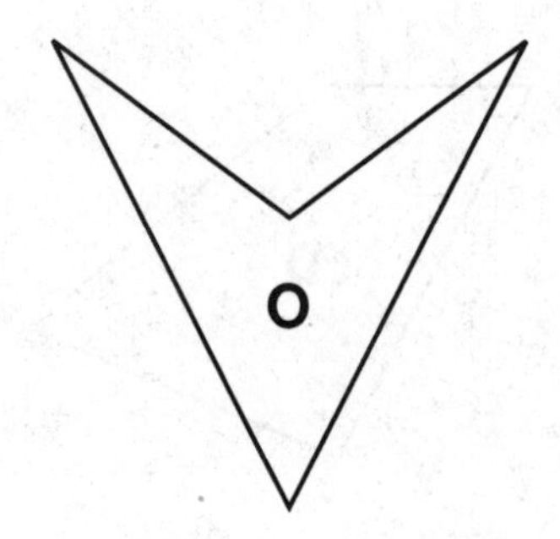

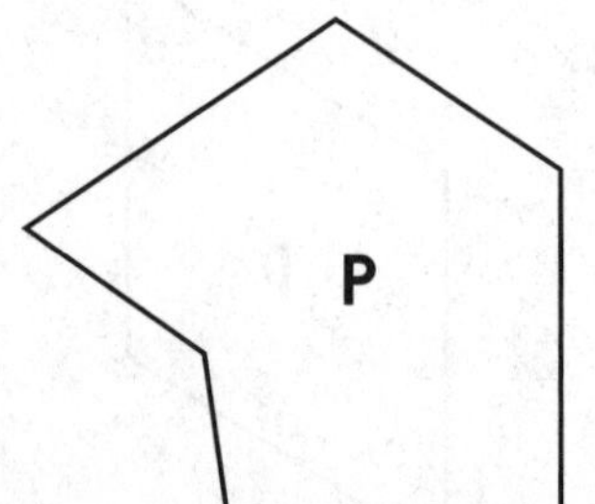

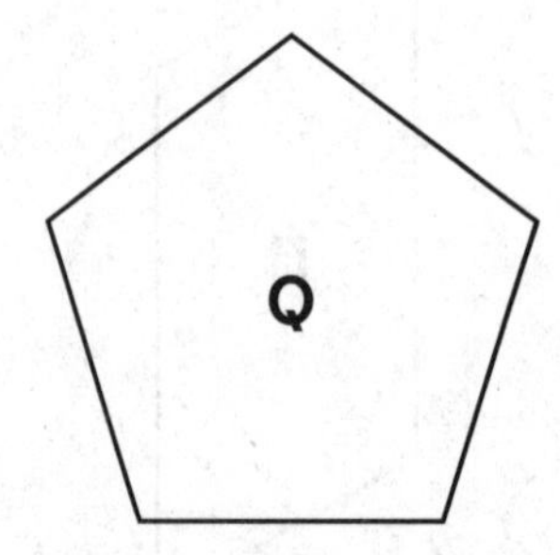

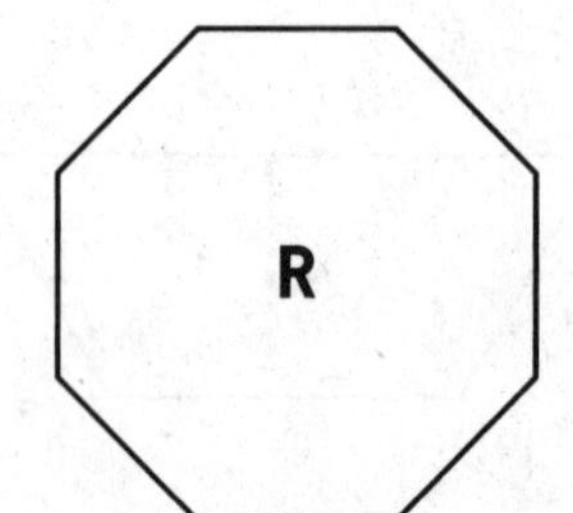

Shading Shapes Gameboard and Reference Page

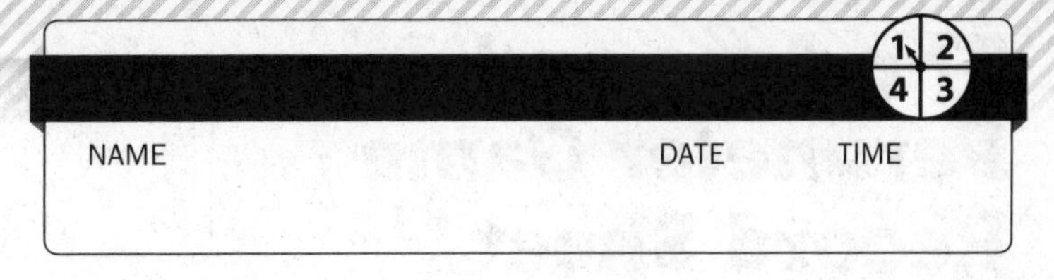

Shading Shapes Gameboard

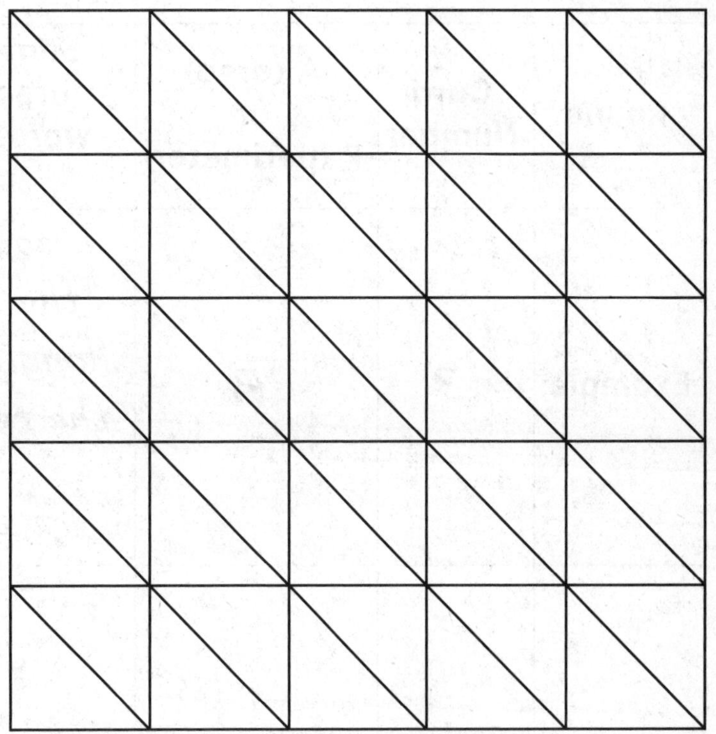

Shading Shapes Reference Page

NAME DATE TIME

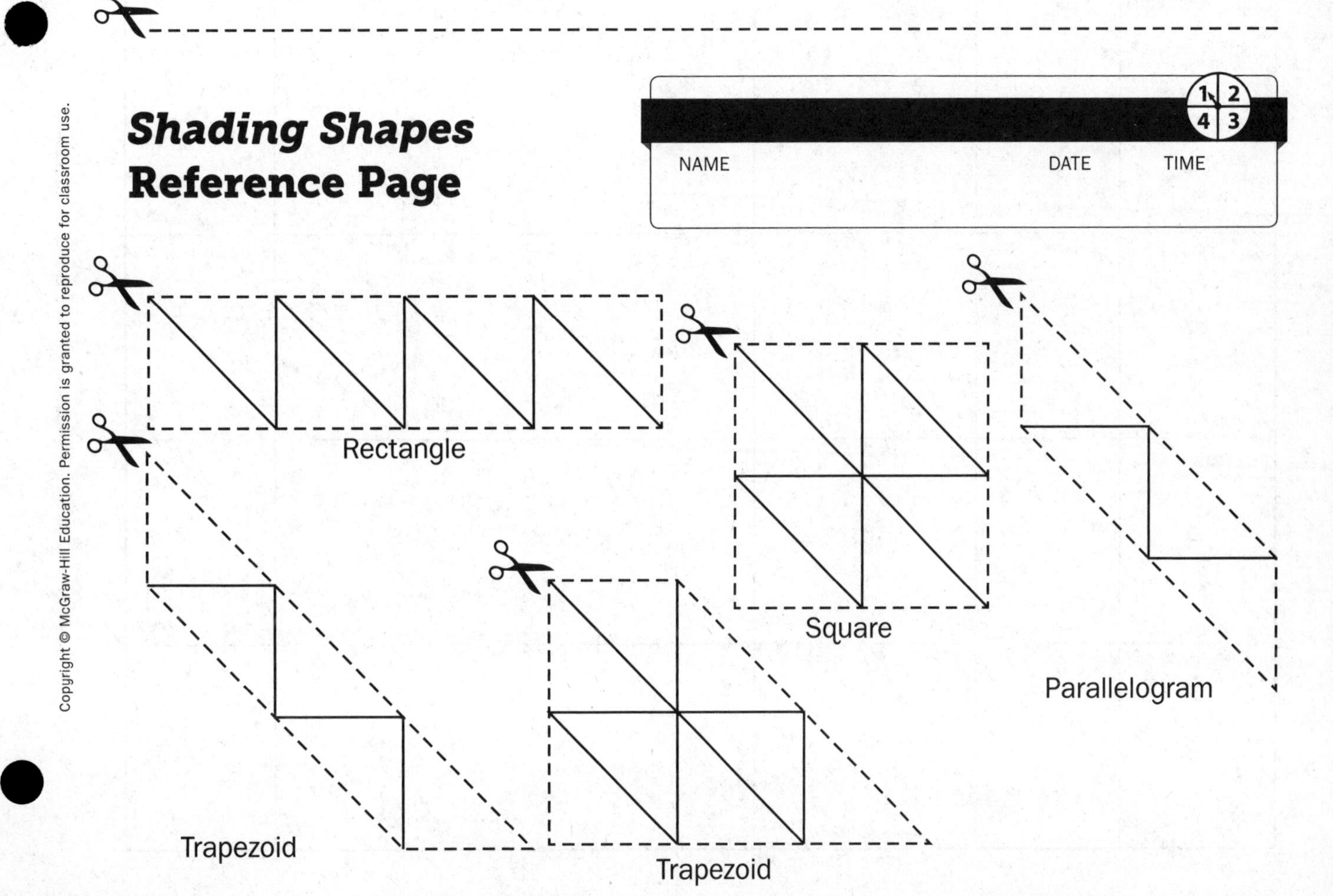

The Area and Perimeter Game Record Sheet

NAME DATE TIME

Round	Card Number	A (area) or P (perimeter)	Show how you found the area or perimeter using words, drawings, and/or a number sentence.	Score
Example:	3	P	I counted the edges of the squares along the long and short sides of the rectangle and added them: $2 + 4 + 2 + 4 = 12$.	12
1				
2				
3				
4				
5				

Baseball Multiplication (with 10-Sided Dice)

NAME DATE TIME

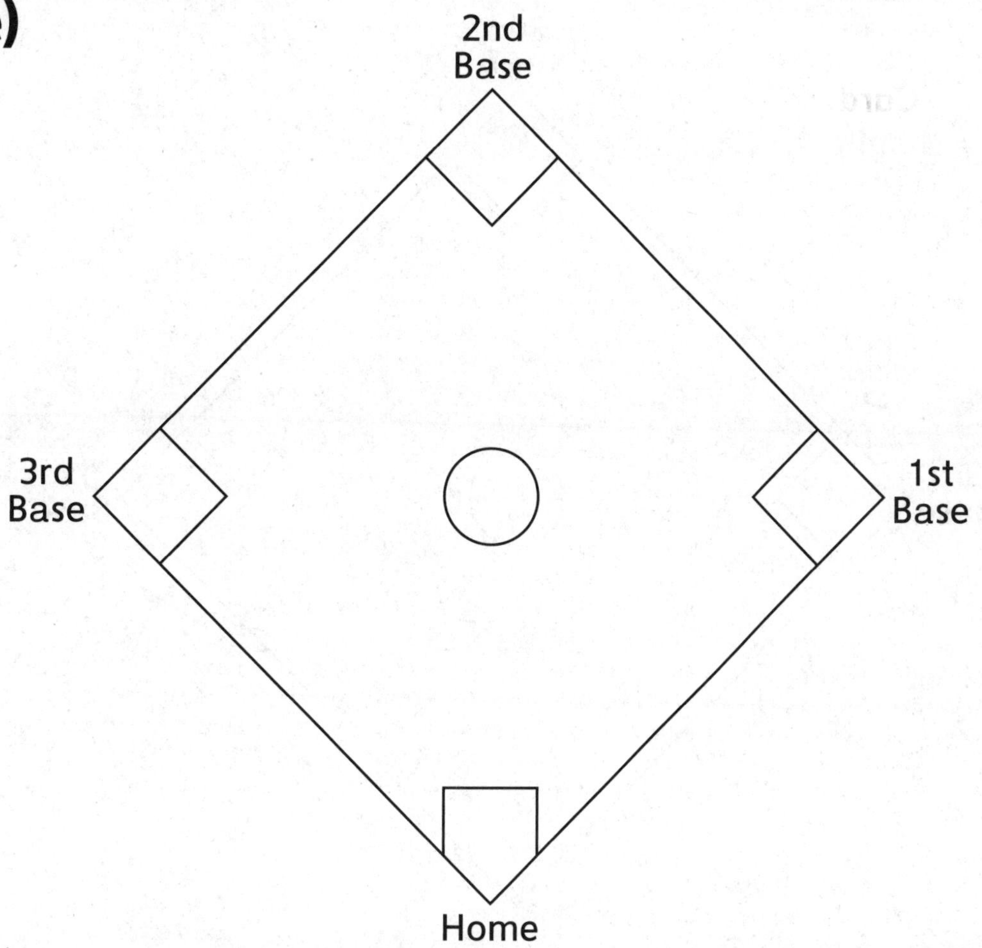

Scoreboard				
Inning	1	2	3	Final
Team 1				
Team 2				

Runs-and-Outs Tally			
Team 1		**Team 2**	
Runs	**Outs**	**Runs**	**Outs**

Scoring Chart (for two 10-sided dice)	
81 to 100 → Home run (score a run)	14 to 36 → Single (go to 1st base)
50 to 80 → Triple (go to 3rd base)	13 or less → Out (record an out)
37 to 49 → Double (go to 2nd base)	

Baseball Multiplication (with 6-Sided Dice)

NAME DATE TIME

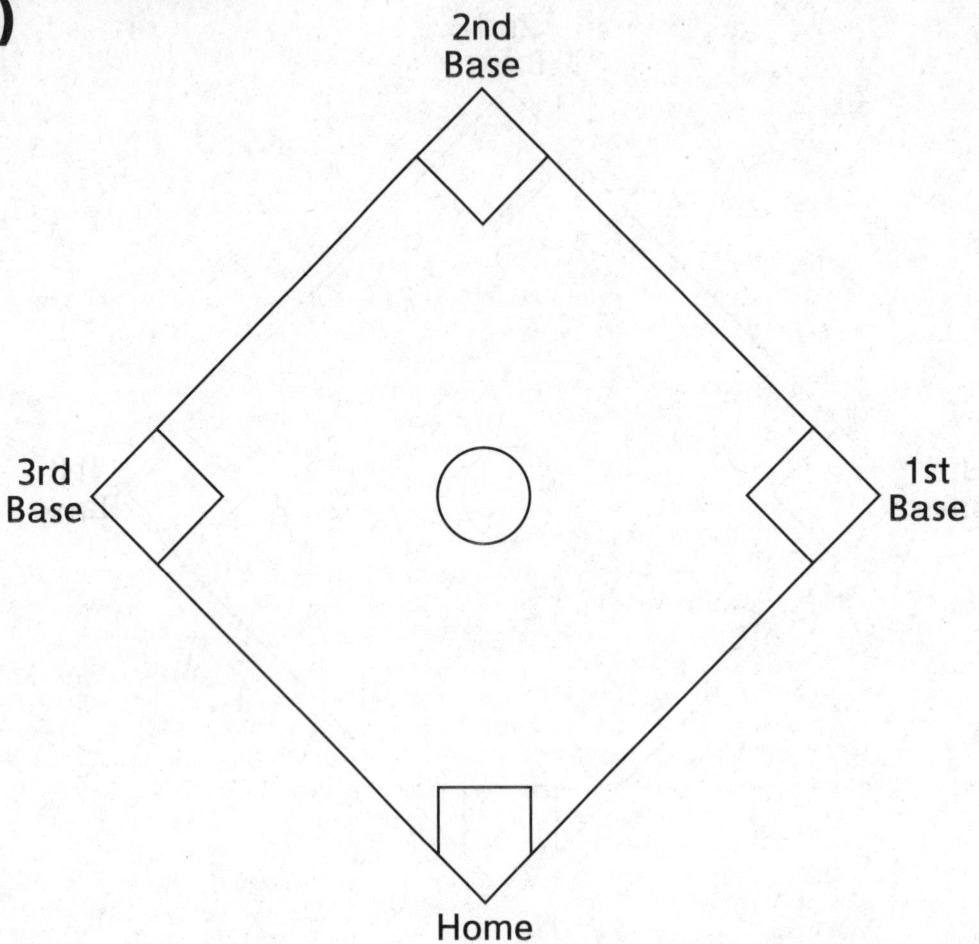

Scoreboard				
Inning	1	2	3	Final
Team 1				
Team 2				

Runs-and-Outs Tally			
Team 1		**Team 2**	
Runs	**Outs**	**Runs**	**Outs**

Scoring Chart (for two 6-sided dice)	
36 → Home run (score a run)	6 to 15 → Single (go to 1st base)
25 to 35 → Triple (go to 3rd base)	5 or less → Out (record an out)
16 to 24 → Double (go to 2nd base)	

Baseball Multiplication (with Tens)

NAME DATE TIME

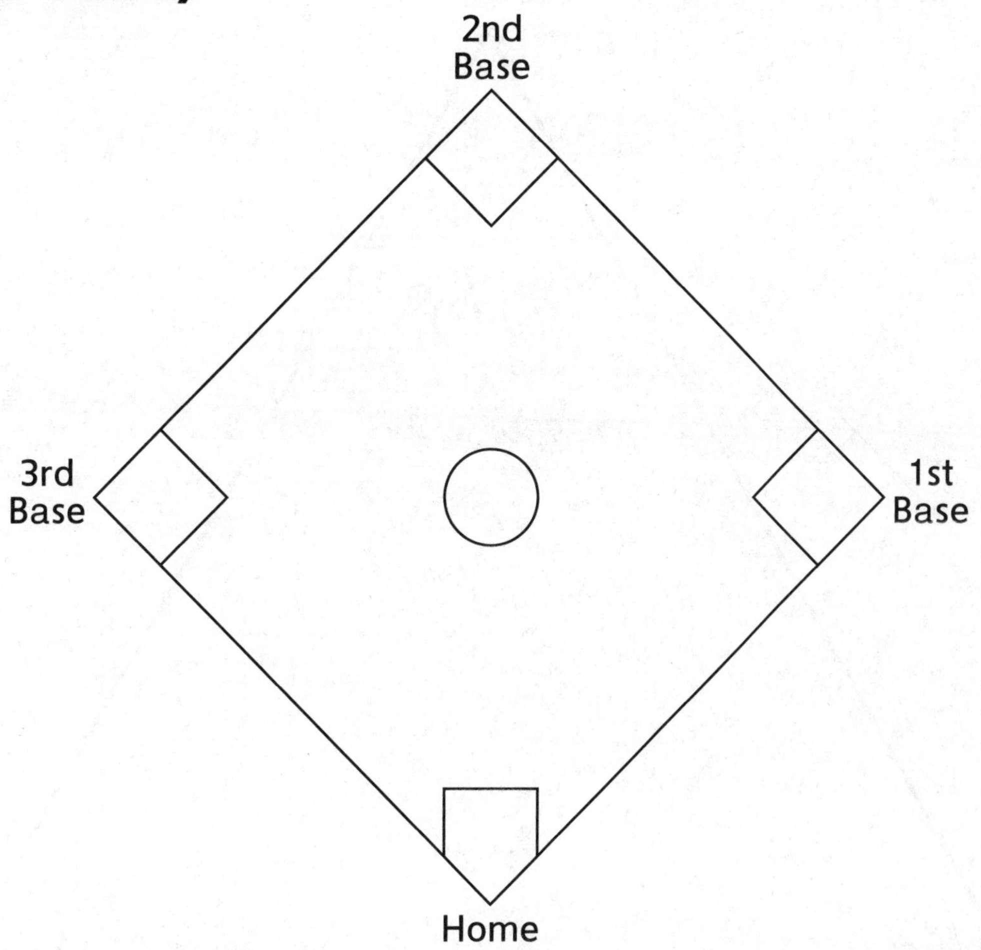

Scoreboard				
Inning	1	2	3	Final
Team 1				
Team 2				

Runs-and-Outs Tally			
Team 1		**Team 2**	
Runs	**Outs**	**Runs**	**Outs**

Scoring Chart (for one 10-sided die and one number card)	
810 to 1,000 → Home run (score a run)	140 to 300 → Single (go to 1st base)
500 to 800 → Triple (go to 3rd base)	130 or less → Out (record an out)
310 to 490 → Double (go to 2nd base)	

Beat the Calculator Triangle

NAME DATE TIME

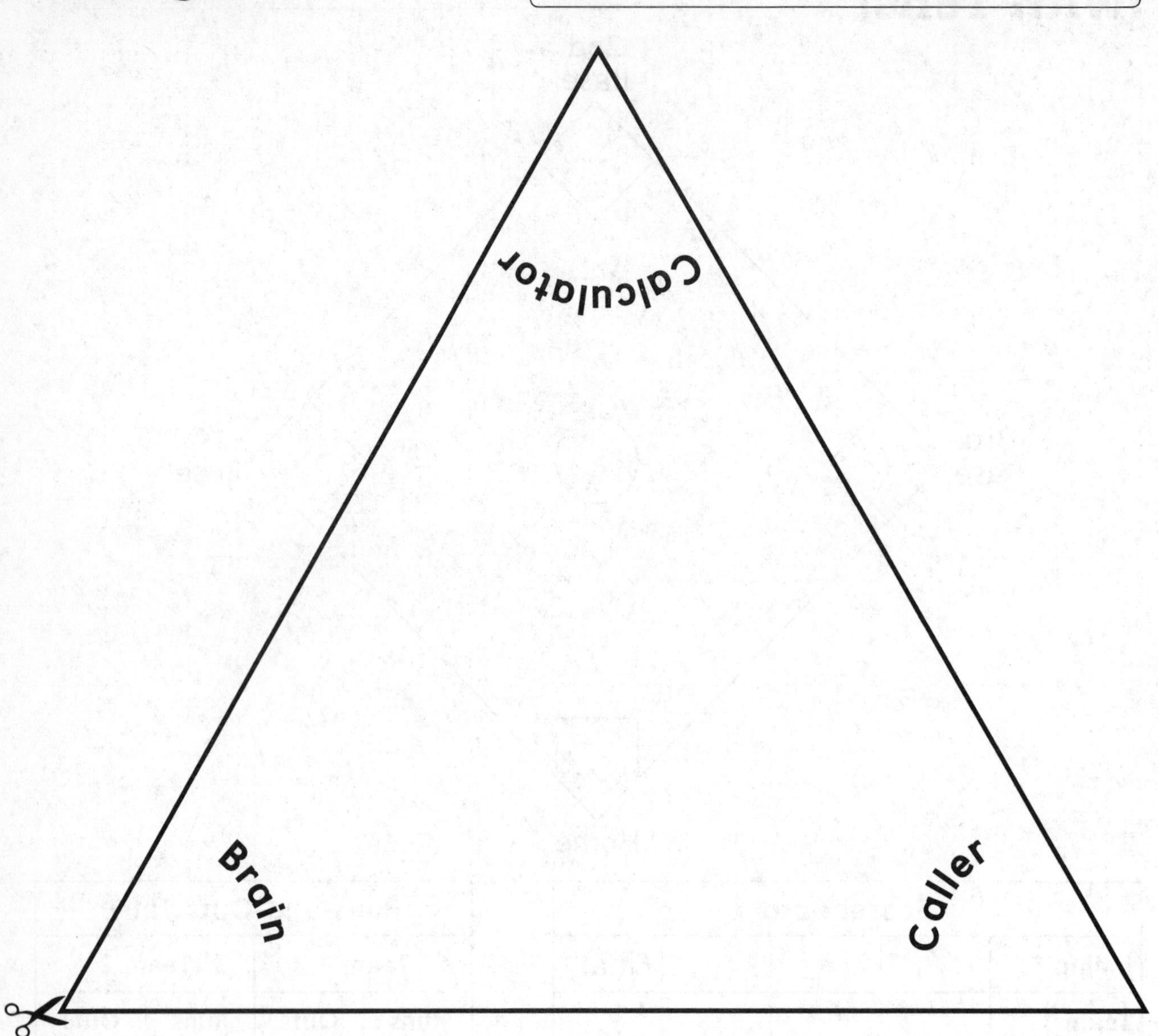

Name That Number Record Sheet 2

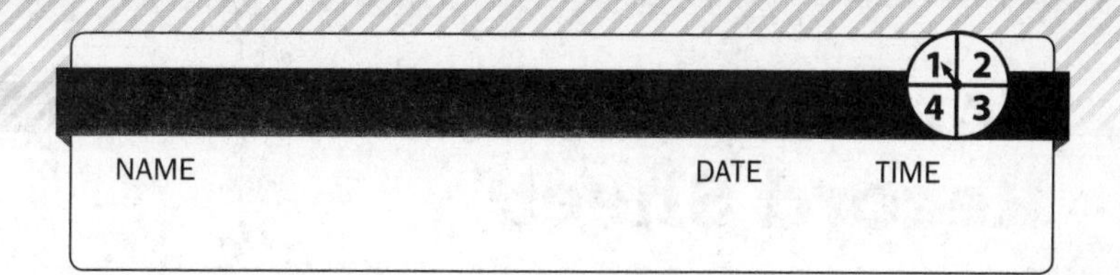

Try to show your steps using one equation with parentheses.

Name That Number Record Sheet 2

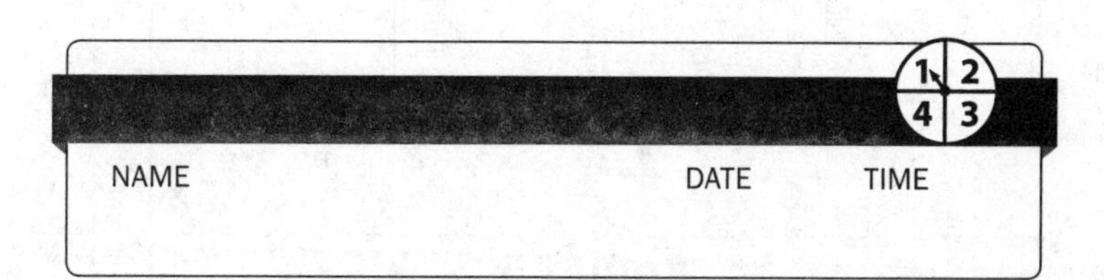

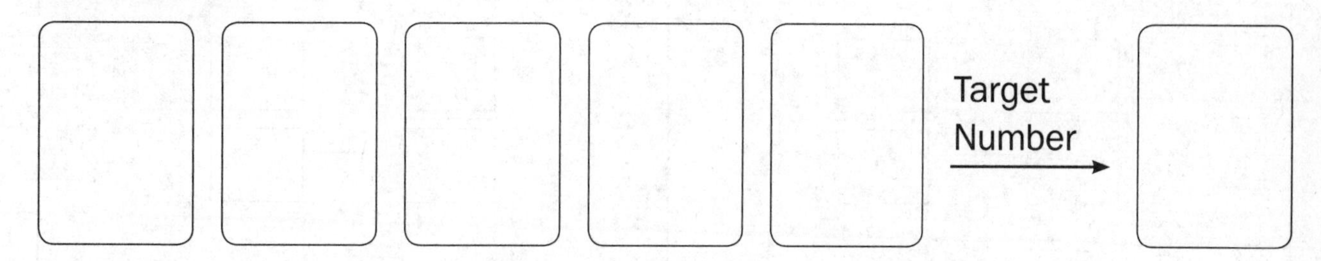

Try to show your steps using one equation with parentheses.

Copyright © McGraw-Hill Education. Permission is granted to reproduce for classroom use.

Fraction Top-It Record Sheet

NAME DATE TIME

Players: ______________ and ______________

Write a number sentence for each round.
Use <, >, and =.

Reminder
= means *is equal to*
< means *is less than*
> means *is greater than*

Round 1	Round 2
$\frac{\square}{\square}$ $\square$ $\frac{\square}{\square}$	$\frac{\square}{\square}$ $\square$ $\frac{\square}{\square}$
Round 3	**Round 4**
$\frac{\square}{\square}$ $\square$ $\frac{\square}{\square}$	$\frac{\square}{\square}$ $\square$ $\frac{\square}{\square}$
Round 5	**Round 6**
$\frac{\square}{\square}$ $\square$ $\frac{\square}{\square}$	$\frac{\square}{\square}$ $\square$ $\frac{\square}{\square}$
Round 7	**Round 8**
$\frac{\square}{\square}$ $\square$ $\frac{\square}{\square}$	$\frac{\square}{\square}$ $\square$ $\frac{\square}{\square}$

Finding Factors Gameboard

NAME DATE TIME

1	2	3	4	5	6
7	8	9	10	12	14
15	16	18	20	21	24
25	27	28	30	32	35
36	40	42	45	48	49
54	56	63	64	72	81

Factor Strip

1	2	3	4	5	6	7	8	9

Factor Bingo Game Mat

NAME DATE TIME

Write any of the numbers 2 through 90 on the grid above.

You may use a number only once.

To help you keep track of the numbers you use, circle them in the list.

	2	3	4	5	6	7	8	9	10
11	12	13	14	15	16	17	18	19	20
21	22	23	24	25	26	27	28	29	30
31	32	33	34	35	36	37	38	39	40
41	42	43	44	45	46	47	48	49	50
51	52	53	54	55	56	57	58	59	60
61	62	63	64	65	66	67	68	69	70
71	72	73	74	75	76	77	78	79	80
81	82	83	84	85	86	87	88	89	90